W9-CJM-506

RAPID PROTOTYPING
OF DIGITAL SYSTEMS
QUARTUS® II EDITION

RAPID PROTOTYPING
OF DIGITAL SYSTEMS
QUARTUS® II EDITION

WITHDRAWN

JAMES O. HAMBLEN
SCHOOL OF ELECTRICAL AND COMPUTER ENGINEERING
GEORGIA INSTITUTE OF TECHNOLOGY

TYSON S. HALL
SCHOOL OF COMPUTING
SOUTHERN ADVENTIST UNIVERSITY

MICHAEL D. FURMAN
DEPARTMENT OF BIOMEDICAL ENGINEERING
UNIVERSITY OF FLORIDA

 Springer

James O. Hamblen
Georgia Institute of Technology
School of Electrical & Computer Engin.
777 Atlantic Drive, N.W.
Atlanta, GA 30332-0250

Tyson S. Hall
Southern Adventist University
School of Computing
481 Taylor Circle
Collegedale, TN 37315-0370

Michael D. Furman
University of Florida
Dept. Biomedical Engineering
141 BME Building
Gainesville, FL 32611-6131

Hamblen, James O., 1954-
    Rapid prototyping of digital systems / James O. Hamblen, Tyson S. Hall, Michael D.
    Furman.-- Quartus II ed.

    Library of Congress Control Number: 2005051723

Printed in the United States of America.

9 8 7 6 5 4 3 2
springer.com

RAPID PROTOTYPING OF DIGITAL SYSTEMS
QUARTUS® II EDITION

Table of Contents

1 Tutorial I: The 15 Minute Design _____ *2*

1.1   Design Entry using the Graphic Editor_____ 7

1.2   Compiling the Design _____ 13

1.3   Simulation of the Design_____ 14

1.4   Downloading Your Design to the UP 3 Board _____ 15

1.5   Downloading Your Design to the UP 2 Board _____ 18

1.6   The 10 Minute VHDL Entry Tutorial_____ 20

1.7   Compiling the VHDL Design _____ 23

1.8   The 10 Minute Verilog Entry Tutorial _____ 24

1.9   Compiling the Verilog Design _____ 26

1.10   Timing Analysis_____ 27

1.11   The Floorplan Editor _____ 28

1.12   Symbols and Hierarchy _____ 30

1.13   Functional Simulation_____ 30

1.14   Laboratory Exercises _____ 31

2 The Altera UP 3 Board _____ *36*

2.1   The UP 3 Cyclone FPGA Features _____ 37

2.2   The UP 3 Board's Memory Features_____ 38

2.3   The UP 3 Board's I/O Features _____ 38

2.4   Obtaining a UP 3 Board and Cables _____ 41

3 Programmable Logic Technology_____ *44*

3.1   CPLDs and FPGAs _____ 47

3.2   Altera MAX 7000S Architecture – A Product Term CPLD Device _____ 48

3.3   Altera Cyclone Architecture – A Look-Up Table FPGA Device _____ 50

3.4   Xilinx 4000 Architecture – A Look-Up Table FPGA Device_____ 53

3.5   Computer Aided Design Tools for Programmable Logic _____ 55

3.6   Next Generation FPGA CAD tools _____ 56

3.7   Applications of FPGAs _____ 57

3.8   Features of New Generation FPGAs _____ 57

3.9   For additional information _____ 58

3.10   Laboratory Exercises _____ 58

4   *Tutorial II: Sequential Design and Hierarchy* _____ *62*

4.1   Install the Tutorial Files and UP3core Library _____ 62

4.2   Open the tutor2 Schematic _____ 63

4.3   Browse the Hierarchy _____ 63

4.4   Using Buses in a Schematic _____ 65

4.5   Testing the Pushbutton Counter and Displays _____ 66

4.6   Testing the Initial Design on the Board _____ 67

4.7   Fixing the Switch Contact Bounce Problem _____ 68

4.8   Testing the Modified Design on the UP 3 Board _____ 69

4.9   Laboratory Exercises _____ 69

5   *UP3core Library Functions* _____ *74*

5.1   UP3core LCD_Display: LCD Panel Character Display _____ 76

5.2   UP3core Debounce: Pushbutton Debounce _____ 77

5.3   UP3core OnePulse: Pushbutton Single Pulse _____ 78

5.4   UP3core Clk_Div: Clock Divider _____ 79

5.5   UP3core VGA_Sync: VGA Video Sync Generation _____ 80

5.6   UP3core Char_ROM: Character Generation ROM _____ 82

5.7   UP3core Keyboard: Read Keyboard Scan Code _____ 83

5.8   UP3core Mouse: Mouse Cursor _____ 84

5.9   For additional information _____ 85

6   *Using VHDL for Synthesis of Digital Hardware* _____ *88*

6.1   VHDL Data Types _____ 88

6.2   VHDL Operators _____ 89

6.3   VHDL Based Synthesis of Digital Hardware _____ 90

6.4   VHDL Synthesis Models of Gate Networks _____ 90

6.5   VHDL Synthesis Model of a Seven-segment LED Decoder _____ 91

6.6   VHDL Synthesis Model of a Multiplexer _____ 93

6.7   VHDL Synthesis Model of Tri-State Output _____ 94

6.8   VHDL Synthesis Models of Flip-flops and Registers _____ 94

6.9   Accidental Synthesis of Inferred Latches _____ 96

6.10 VHDL Synthesis Model of a Counter _____ 96

6.11 VHDL Synthesis Model of a State Machine _____ 97

6.12 VHDL Synthesis Model of an ALU with an Adder/Subtractor and a Shifter _____ 99

6.13 VHDL Synthesis of Multiply and Divide Hardware _____ 100

6.14 VHDL Synthesis Models for Memory _____ 101

6.15 Hierarchy in VHDL Synthesis Models _____ 105

6.16 Using a Testbench for Verification _____ 107

6.17 For additional information _____ 108

6.18 Laboratory Exercises _____ 108

7   Using Verilog for Synthesis of Digital Hardware _____ *112*

7.1   Verilog Data Types _____ 112

7.2   Verilog Based Synthesis of Digital Hardware _____ 112

7.3   Verilog Operators _____ 113

7.4   Verilog Synthesis Models of Gate Networks _____ 114

7.5   Verilog Synthesis Model of a Seven-segment LED Decoder _____ 114

7.6   Verilog Synthesis Model of a Multiplexer _____ 115

7.7   Verilog Synthesis Model of Tri-State Output _____ 116

7.8   Verilog Synthesis Models of Flip-flops and Registers _____ 117

7.9   Accidental Synthesis of Inferred Latches _____ 118

7.10 Verilog Synthesis Model of a Counter _____ 118

7.11 Verilog Synthesis Model of a State Machine _____ 119

7.12 Verilog Synthesis Model of an ALU with an Adder/Subtractor and a Shifter _____ 120

7.13 Verilog Synthesis of Multiply and Divide Hardware _____ 121

7.14 Verilog Synthesis Models for Memory _____ 122

7.15 Hierarchy in Verilog Synthesis Models _____ 125

7.16 For additional information _____ 126

7.17 Laboratory Exercises _____ 126

8   State Machine Design: The Electric Train Controller _____ *130*

8.1   The Train Control Problem _____ 130

8.2   Track Power (T1, T2, T3, and T4) _____ 132

8.3   Track Direction (DA1-DA0, and DB1-DB0) _____ 132

8.4   Switch Direction (SW1, SW2, and SW3) _____ 133

8.5   Train Sensor Input Signals (S1, S2, S3, S4, and S5) _____ 133

8.6   An Example Controller Design _____ 134

8.7   VHDL Based Example Controller Design _____ 138

8.8   Simulation Vector file for State Machine Simulation _____ 140

8.9   Running the Train Control Simulation _____ 142

8.10   Running the Video Train System (After Successful Simulation)_____ 142

8.11   Laboratory Exercises _____ 144

9   *A Simple Computer Design: The μP 3* _____ *148*

9.1   Computer Programs and Instructions _____ 149

9.2   The Processor Fetch, Decode and Execute Cycle _____ 150

9.3   VHDL Model of the μP 3_____ 157

9.4   Simulation of the μP3 Computer _____ 161

9.5   Laboratory Exercises _____ 162

10 *VGA Video Display Generation* _____ *168*

10.1   Video Display Technology_____ 168

10.2   Video Refresh _____ 168

10.3   Using an FPGA for VGA Video Signal Generation _____ 171

10.4   A VHDL Sync Generation Example: UP3core VGA_SYNC _____ 172

10.5   Final Output Register for Video Signals _____ 174

10.6   Required Pin Assignments for Video Output _____ 174

10.7   Video Examples _____ 175

10.8   A Character Based Video Design _____ 176

10.9   Character Selection and Fonts _____ 176

10.10 VHDL Character Display Design Examples _____ 179

10.11 A Graphics Memory Design Example _____ 181

10.12 Video Data Compression_____ 182

10.13 Video Color Mixing using Dithering _____ 183

10.14 VHDL Graphics Display Design Example_____ 183

10.15 Higher Video Resolution and Faster Refresh Rates _____ 185

10.16 Laboratory Exercises _____ 185

11 *Interfacing to the PS/2 Keyboard and Mouse*_____ *188*

11.1   PS/2 Port Connections _____ 188

11.2   Keyboard Scan Codes _____ 189

11.3   Make and Break Codes _____ 189

11.4   The PS/2 Serial Data Transmission Protocol_____ 190

11.5   Scan Code Set 2 for the PS/2 Keyboard _____ 192

11.6   The Keyboard UP3core _____ 194

11.7   A Design Example Using the Keyboard UP3core_____ 197

11.8   Interfacing to the PS/2 Mouse _____ 198

11.9   The Mouse UP3core _____ 200

11.10 Mouse Initialization _____ 200

11.11 Mouse Data Packet Processing _____ 201

11.12 An Example Design Using the Mouse UP3core _____ 202

11.13 For Additional Information _____ 202

11.14 Laboratory Exercises _____ 203

12 Legacy Digital I/O Interfacing Standards _____ 206

12.1   Parallel I/O Interface _____ 206

12.2   RS-232C Serial I/O Interface _____ 207

12.3   SPI Bus Interface _____ 209

12.4 I^2C Bus Interface_____ 211

12.5   For Additional Information _____ 213

12.6   Laboratory Exercises _____ 213

13 UP 3 Robotics Projects _____ 216

13.1   The UP3-bot Design _____ 216

13.2   UP3-bot Servo Drive Motors _____ 216

13.3   Modifying the Servos to make Drive Motors _____ 217

13.4   VHDL Servo Driver Code for the UP3-bot_____ 218

13.5   Low-cost Sensors for a UP 3 Robot Project _____ 220

13.6   Assembly of the UP3-bot Body_____ 233

13.7   I/O Connections to the UP 3's Expansion Headers _____ 240

13.8   Robot Projects Based on R/C Toys, Models, and Robot Kits _ 242

13.9   For Additional Information _____ 248

13.10 Laboratory Exercises _____ 250

14 A RISC Design: Synthesis of the MIPS Processor Core ___ 256

14.1   The MIPS Instruction Set and Processor _____ 256

14.2   Using VHDL to Synthesize the MIPS Processor Core_____ 259

14.3   The Top-Level Module _____ 260

14.4   The Control Unit_____ 263

14.5   The Instruction Fetch Stage _____ 265

14.6   The Decode Stage _____ 268

14.7   The Execute Stage _____ 270

14.8   The Data Memory Stage _____ 272

14.9   Simulation of the MIPS Design _____ 273

14.10 MIPS Hardware Implementation on the UP 3 Board_____ 274

14.11 For Additional Information _____ 275

14.12 Laboratory Exercises_____ 276

15 Introducing System-on-a-Programmable-Chip _____ 282

15.1   Processor Cores _____ 282

15.2   SOPC Design Flow_____ 283

15.3   Initializing Memory _____ 285

15.4   SOPC Design versus Traditional Design Modalities_____ 287

15.5   An Example SOPC Design _____ 288

15.6   Hardware/Software Design Alternatives _____ 289

15.7   For additional information _____ 289

15.8   Laboratory Exercises_____ 290

16 Tutorial III: Nios II Processor Software Development _____ 294

16.1   Install the UP 3 board files _____ 294

16.2   Starting a Nios II Software Project _____ 294

16.3   The Nios II IDE Software _____ 296

16.4   Generating the Nios II System Library _____ 297

16.5   Software Design with Nios II Peripherals _____ 298

16.6   Starting Software Design – main() _____ 301

16.7   Downloading the Nios II Hardware and Software Projects_____ 302

16.8   Executing the Software _____ 303

16.9   Starting Software Design for a Peripheral Test Program _____ 303

16.10 Handling Interrupts_____ 306

16.11 Accessing Parallel I/O Peripherals_____ 307

16.12 Communicating with the LCD Display _____ 308

16.13 Testing SRAM _____ 311

16.14 Testing Flash Memory_____ 312

16.15 Testing SDRAM _____ 313

16.16 Downloading the Nios II Hardware and Software Projects_____ 318

16.17 Executing the Software _____ 319

16.18 For additional information _____ 320

16.19 Laboratory Exercises _____ 320

17 Tutorial IV: Nios II Processor Hardware Design _____ 324

17.1   Install the UP 3 board files _____ 324

17.2   Creating a New Project _____ 324

17.3   Starting SOPC Builder _____ 325

17.4   Adding a Nios II Processor _____ 327

17.5   Adding UART Peripherals _____ 329

17.6   Adding an Interval Timer Peripheral _____ 330

17.7   Adding Parallel I/O Components _____ 331

17.8   Adding a SDRAM Memory Controller _____ 332

17.9   Adding an External Bus _____ 333

17.10 Adding Components to the External Bus _____ 334

17.11 Global Processor Settings _____ 335

17.12 Finalizing the Nios II Processor _____ 337

17.13 Add the Processor Symbol to the Top-Level Schematic _____ 337

17.14 Create a Phase-Locked Loop Component _____ 338

17.15 Add the UP 3 External Bus Multiplexer Component _____ 339

17.16 Complete the Top-Level Schematic _____ 339

17.17 Design Compilation _____ 339

17.18 Testing the Nios II Project _____ 341

17.19 For additional information _____ 341

17.20 Laboratory Exercises _____ 341

Appendix A: Generation of Pseudo Random Binary Sequences _____ 345

Appendix B: Quartus II Design and Data File Extensions _____ 347

Appendix C: UP 3 Pin Assignments _____ 349

Appendix D: ASCII Character Code _____ 355

Appendix E: Programming the UP 3's Flash Memory _____ 357

Glossary _____ 359

Index _____ 367

About the Accompanying CD-ROM _____ 371

PREFACE

Changes to the Quartus Edition

Rapid Prototyping of Digital Systems provides an exciting and challenging laboratory component for undergraduate digital logic and computer design courses using FPGAs and CAD tools for simulation and hardware implementation. The more advanced topics and exercises also make this text useful for upper level courses in digital logic, programmable logic, and embedded systems. The third edition now uses Altera's new Quartus II CAD tool and includes laboratory projects for Altera's UP 2 and the new UP 3 FPGA board. Student laboratory projects provided on the book's CD-ROM include video graphics and text, mouse and keyboard input, and three computer designs.

Rapid Prototyping of Digital Systems includes four tutorials on the Altera Quartus II and Nios II tool environment, an overview of programmable logic, and IP cores with several easy-to-use input and output functions. These features were developed to help students get started quickly. Early design examples use schematic capture and IP cores developed for the Altera UP FPGA boards. VHDL is used for more complex designs after a short introduction to VHDL-based synthesis. Verilog is also now supported more as an option for the student projects.

New chapters in this edition provide an overview of System-on-a-Programmable Chip (SOPC) technology and SOPC design examples for the UP 3 using Altera's new Nios II Processor hardware and C software development tools. A full set of Altera's FPGA CAD tools is included on the book's CD-ROM.

Intended Audience

This text is intended to provide an exciting and challenging laboratory component for an undergraduate digital logic design class. The more advanced topics and exercises are also appropriate for consideration at schools that have an upper level course in digital logic or programmable logic. There are a number of excellent texts on digital logic design. For the most part, these texts do not include or fully integrate modern CAD tools, logic simulation, logic synthesis using hardware description languages, design hierarchy, and current generation field programmable gate array (FPGA) technology and SOPC design. The goal of this text is to introduce these topics in the laboratory portion of the course. Even student laboratory projects can now implement entire digital and computer systems with hundreds of thousands of gates.

Over the past eight years, we have developed a number of interesting and challenging laboratory projects involving serial communications, state machines with video output, video games and graphics, simple computers, keyboard and mouse interfaces, robotics, and pipelined RISC processor cores.

Source files and additional example files are available on the CD-ROM for all designs presented in the text. The student version of the PC based CAD tool on the CD-ROM can be freely distributed to students. Students can purchase their own UP 3 board for little more than the price of a contemporary textbook. As an alternative, a few of the low-cost UP 3 boards can be shared among students in a laboratory. Course instructors should contact the Altera University Program for detailed information on obtaining full versions of the CAD tools for laboratory PCs and UP 3 boards for student laboratories.

Topic Selection and Organization

Chapter 1 is a short CAD tool tutorial that covers design entry, simulation, and hardware implementation using an FPGA. The majority of students can enter the design, simulate, and have the design successfully running on the UP 3 board in less than thirty minutes. After working through the tutorial and becoming familiar with the process, similar designs can be accomplished in less than 10 minutes.

Chapter 2 provides an overview of the UP 3 FPGA development boards. The features of the board are briefly described. Several tables listing pin connections of various I/O devices serve as an essential reference whenever a hardware design is implemented on the UP 3 board.

Chapter 3 is an introduction to programmable logic technology. The capabilities and internal architectures of the most popular CPLDs and FPGAs are described. These include the Cyclone FPGA used on the UP 3 board, and the Xilinx 4000 family FPGAs.

Chapter 4 is a short CAD tool tutorial that serves as both a hierarchical and sequential design example. A counter is clocked by a pushbutton and the output is displayed in the seven-segment LED's. The design is downloaded to the UP 3 board and some real world timing issues arising with switch contact bounce are resolved. It uses several functions from the UP3core library which greatly simplify use of the UP 3's input and output capabilities.

Chapter 5 describes the available UP3core library I/O functions. The I/O devices include switches, the LCD, a multiple output clock divider, VGA output, keyboard input, and mouse input.

Chapter 6 is an introduction to the use of VHDL for the synthesis of digital hardware. Rather than a lengthy description of syntax details, models of the commonly used digital hardware devices are developed and presented. Most VHDL textbooks use models developed for simulation only and they frequently use language features not supported in synthesis tools. Our easy to understand synthesis examples were developed and tested on FPGAs using the Altera CAD tools.

Chapter 7 is an introduction to the use of Verilog for the synthesis of digital hardware. The same hardware designs as Chapter 6 as modeled in Verilog. It is optional, but is included for those who would like an introduction to Verilog.

Chapter 8 is a state machine design example. The state machine controls a virtual electric train system simulation with video output generated directly by the FPGA. Using track sensor input, students must control two trains and three track switches to avoid collisions.

Chapter 9 develops a model of a simple computer. The fetch, decode, and execute cycle is introduced and a brief model of the computer is developed using VHDL. A short assembly language program can be entered in the FPGA's internal memory and executed in the simulator.

Chapter 10 describes how to design an FPGA-based digital system to output VGA video. Numerous design examples are presented containing video with both text and graphics. Fundamental design issues in writing simple video games and graphics using the UP 3 board are examined.

Chapter 11 describes the PS/2 keyboard and mouse operation and presents interface examples for integration in designs on the UP 3 board. Keyboard scan code tables, mouse data packets, commands, status codes, and the serial communications protocol are included. VHDL code for a keyboard and mouse interface is also presented.

Chapter 12 describes several of the common I/O standards that are likely to be encountered in FPGA systems. Parallel, RS232 serial, SPI, and I^{2}C standards and interfacing are discussed.

Chapter 13 develops a design for an adaptable mobile robot using the UP 3 board as the controller. Servo motors and several sensor technologies for a low cost mobile robot are described. A sample servo driver design is presented. Commercially available parts to construct the robot described can be obtained for as little as $60. Several robots can be built for use in the laboratory. Students with their own UP 3 board may choose to build their own robot following the detailed instructions found in section 13.6.

Chapter 14 describes a single clock cycle model of the MIPS RISC processor based on the hardware implementation presented in the widely used Patterson and Hennessy textbook, *Computer Organization and Design The Hardware/Software Interface*. Laboratory exercises that add new instructions, features, and pipelining are included at the end of the chapter.

Chapters 15, 16, and 17 introduce students to SOPC design using the Nios II RISC processor core. Chapter 15 is an overview of the SOPC design approach. Chapter 16 contains a tutorial for the Nios II IDE software development tool and examples using the Nios II C/C++ compiler. Chapter 17 contains a tutorial on the processor core hardware configuration tool, SOPC builder. A UP 3 board is required for this new material since it is not supported on the UP 2's FPGA.

We anticipate that many schools will still choose to begin with TTL designs on a small protoboard for the first few labs. The first chapter can also be started at this time since only OR and NOT logic functions are used to introduce the CAD tool environment. The CAD tool can also be used for simulation of TTL labs, since a TTL parts library is included.

Even though VHDL and Verilog are complex languages, we have found after several years of experimentation that students can write HDL models to synthesize hardware designs after a short overview with a few basic hardware design examples. The use of HDL templates and online help files in the CAD tool makes this process easier. After the initial experience with HDL synthesis, students dislike the use of schematic capture on larger designs since it can be very time consuming. Experience in industry has been much the same since huge productivity gains have been achieved using HDL based synthesis tools for application specific integrated circuits (ASICs) and FPGAs.

Most digital logic classes include a simple computer design such as the one presented in Chapter 9 or a RISC processor such as the one presented in Chapter 14. If this is not covered in the first digital logic course, it could be used as a lab component for a subsequent computer architecture class.

A typical quarter or semester length course could not cover all of the topics presented. The material presented in Chapters 7 through 17 can be used on a selective basis. The keyboard and mouse are supported by UP3core library functions, and the material presented in Chapter 11 is not required to use these library functions for keyboard or mouse input. A UP 3 board is required for the SOPC Nios designs in Chapters 16 and 17.

A video game based on the material in Chapter 10 can serve as the basis for a team design project. For a final team design project, we use robots with sensors from chapter 13 that are controlled by the simple computer in chapter 9. Our students really enjoyed working with the robot described in Chapter 13, and it presents almost infinite possibilities for an exciting design competition. A more advanced class could develop projects based on the Nios II processor reference designs in Chpater 16 and 17 using C/C++ code.

Software and Hardware Packages

The new 5.0 SP1 web version of Quartus II FPGA CAD tool is included with this book. Software was tested using this version and it is recommended. UP 3 boards are available from Altera at special student pricing. A board can be shared among several students in a lab, or some students may wish to purchase their own board. Details and suggestions for additional cables that may be required for a laboratory setup can be found in Section 2.4. Source files for all designs presented in the text are available on the CD-ROM.

Additional Web Material and Resources

There is a web site for the text with additional course materials, slides, text errata, and software updates at:

http://www.ece.gatech.edu/users/hamblen/book/bookte.htm

Acknowledgments

Over three thousand students and several hundred teaching assistants have contributed to this work during the past eight years. In particular, we would like to acknowledge Doug McAlister, Michael Sugg, Jurgen Vogel, Greg Ruhl, Eric Van Heest, Mitch Kispet, and Evan Anderson for their help in testing and developing several of the laboratory assignments and tools. Mike Phipps, Joe Hanson, Tawfiq Mossadak, and Eric Shiflet at Altera provided software, hardware, helpful advice, and encouragement.

CHAPTER 1

Tutorial I:
The 15-Minute Design

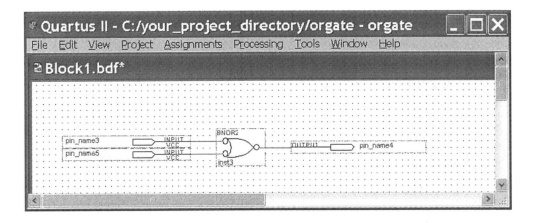

1 Tutorial I: The 15 Minute Design

The purpose of this tutorial is to introduce the user to the Altera CAD tools and the University Program (UP 3 or UP 2) Development Board in the shortest possible time. The format is an aggressive introduction to schematic, VHDL, and Verilog entry for those who want to get started quickly. The approach is tutorial and utilizes a path that is similar to most digital design processes.

Once completed, you will understand and be able to:

- Navigate the Altera schematic entry environment,
- Compile a VHDL or Verilog design file,
- Simulate, debug, and test your designs,
- Generate and verify timing characteristics, and
- Download and run your design on a UP 3 or UP 2 board.

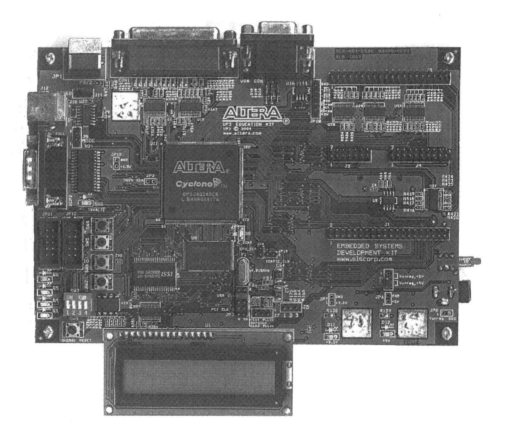

Figure 1.1 The Altera UP 3 FPGA Development board.

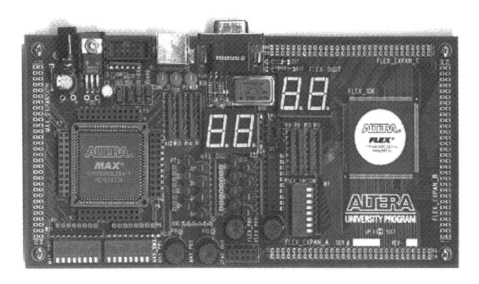

Figure 1.2 The Altera UP 2 FPGA development board.

In this tutorial, an OR function will be demonstrated to provide an introduction to the Altera Quartus II CAD tools. After simulation, the design will then be used to program a field programmable gate array (FPGA) on a UP 3 or UP 2 development board.

ALTERA'S NEWEST BOARD IS THE UP 3. IF YOU HAVE A UP 2 BOARD AS SEEN IN FIGURE 1.2, YOU SHOULD ALWAYS FOLLOW THE INSTRUCTIONS AND PROCEDURES OUTLINED IN THE CD-ROM AND TEXT FOR A UP 2 BOARD. THE CD-ROM CONTAINS ADDITIONAL INFORMATION AND FILES FOR UP 2 BOARD USERS.

The inputs to the OR logic will be two pushbuttons and the output will be displayed using a light emitting diode (LED). Both the pushbuttons and the LED are part of the development board, and no external wiring is required.

Of course, any actual design will be more complex, but the objective here is to quickly understand the capabilities and flow of the design tools with minimal effort and time.

More complex designs including computers will be introduced later in this text after you have become familiar with the development tools and hardware description languages (HDLs) used in digital designs.

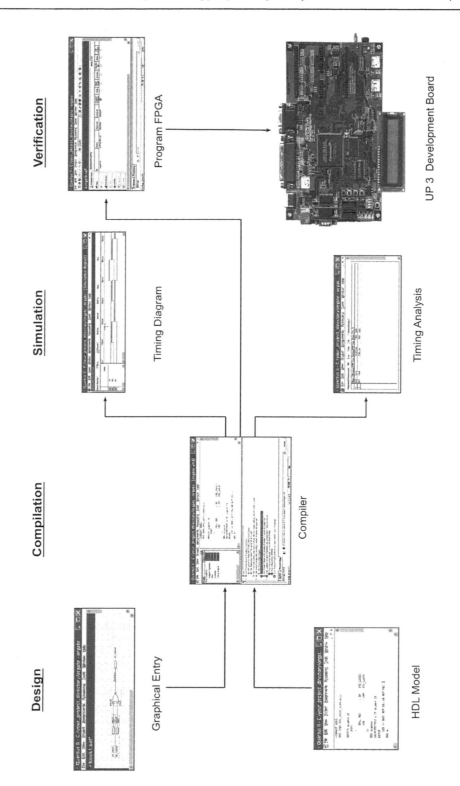

Design

Graphical Entry

HDL Model

Compilation

Compiler

Simulation

Timing Diagram

Timing Analysis

Verification

Program FPGA

UP 3 Development Board

Figure 1.3 Design process for schematic or HDL entry.

Granted, all this may not be accomplished in just 15 minutes; however, the skills acquired from this demonstration tutorial will enable the first-time user to duplicate similar designs in less time than that!

INSTALL THE QUARTUS II WEB VERSION SOFTWARE USING THE CD-ROM AND OBTAIN A WEB LICENSE FILE FROM ALTERA. CHECK FOR ALTERA QUARTUS II WEB VERSION SOFTWARE UPDATES AT WWW.ALTERA.COM.

Designs can be entered via schematic capture or by using a HDL such as VHDL or Verilog. It is also possible to combine blocks with different entry methods into a single design. As seen in Figure 1.3, tools can then be used to simulate, calculate timing delays, synthesize logic, and program a hardware implementation of the design on an FPGA.

The Board

The board that will be used is the Altera UP 3. Although the following tutorial can be done with either the UP 3 or a UP 2, some modifications (mainly device and pin number assignments) will be needed for the UP 2.

The Pushbuttons

The UP 3's two pushbutton switch inputs, PB1 and PB2, are connected to pins 62 (labeled SW7 on board) and 48 (labeled SW4 on board). On the UP 2, PB1 and PB2 are connected directly to the FLEX FPGA chip at pins 28 and 29 respectively. Each pushbutton input is tied High with a pull-up resistor and pulled Low when the respective pushbutton is pressed. One needs to remember that when using the on-board pushbuttons, this "active low" condition ties zero volts to the input when the button is pressed and the V_{cc} high supply to the input when not pressed. See Figure 1.4. V_{cc} is 3.3V on the UP 3 and 5V on the UP 2. As seen in Figure 1.4, on the UP 2 board a logic "0" turns on the LED.

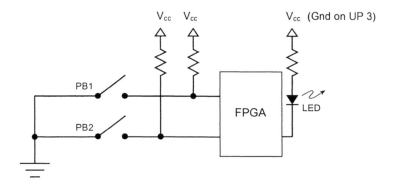

Figure 1.4 Connections between the pushbuttons, the LEDs, and the Altera FPGA.

The LED Outputs

The UP 3 has four discrete LEDs located on the lower left side of the board. On the UP 3 board, the LED in Figure 1.4 is reversed and connected to ground so that a logic "1" on the FPGA's output pin turns on the LED and a logic "0" turns off the LED.

The Problem Definition

To illustrate the capabilities of the software in the simplest terms, we will be building a circuit that turns off the LED when one OR the other pushbutton is pushed. In a simple logic equation, one could write:

$$LED\_OFF = PB1\_HIT + PB2\_HIT$$

At first, this may seem too simple; however, the active low inputs and outputs add just enough complication to illustrate some of the more common errors, and it provides an opportunity to compare some of the different syntax features of VHDL and Verilog. (Students needing an exercise in DeMorgan's Law will also find these exercises particularly enlightening.)

We will first build this circuit with the graphical editor and then implement it in VHDL and Verilog. As you work through the tutorial, note how the design entry method is relatively independent of the compile, timing, and simulation steps.

Resolving the Active Low Signals

Since the pushbuttons generate inverted signals and the LED will require an inverted or low level logic signal to turn off (UP 3), we could build an OR logic circuit using the layout in Figure 1.5a. Recalling that a bubble on a gate input or output indicates inversion, careful examination shows that the two circuits in Figure 1.5 are functionally equivalent; however, the circuit in Figure 1.5a uses more gates and would take a bit longer to enter in the schematic editor. We will therefore use the single gate circuit illustrated in Figure 1.5b.

(a)   (b)

Figure 1.5a and 1.5b. Equivalent circuits for ORing active low inputs and outputs.

This form of the OR function is known as a "negative-logic OR." If you are confused, try writing a truth table to show this Boolean equality. (In Exercise 1 at the end of the chapter, this circuit will be compared with its DeMorgan's equivalent, the "positive-logic AND."). On the UP 2 board, the LED's output state will appear inverted since its LED output circuit is inverted, so pushing one of the UP 2 pushbuttons will turn on the UP 2's LED.

1.1 Design Entry using the Graphic Editor

Examine the CAD tool overview diagram in Figure 1.3. The initial path in this section will be from schematic capture (Graphical Entry) to downloading the design to the UP 3 board. On the way, we will pass through some of the nuances of the Compiler along with setting up and controlling a simulation. Later, after having actually tested the design, we will examine the Timing Analysis information of the design. Although relatively short, each step is carefully illustrated and explained. Install the Altera Quartus II software on your PC using the book's CD-ROM, if it is not already installed.

New Project Creation

Start the Quartus II program. In Quartus II, the New Project wizard is used to create a new project. Choose **File ⇨ New Project Wizard.** Click next in the Introduction window, if it appears to continue. A second dialog box will appear asking for the working directory for your new project. Enter an appropriate directory. For the project name and top-level design entity boxes, enter **orgate**. Click **Next**. If you need to create a new project directory with that name, click **Yes**. An Add Files dialog box then appears. This page is used to enter all of the design files (other than the top-level file). Since this simple project will only use a single top-level design file, click **Next**.

Figure 1.6 Creating a new Quartus II Project.

Select the Device to be Used

The next dialog box is used to select the FPGA type. If you are using the UP 3 board, select **Cyclone** family and for the UP 1 or 2 select FLEX10K. You will then need to select the specific FPGA on your board. The UP 3 is available with two different sizes of Cyclone FPGAs: an **EP1C6Q240C8** or the larger **EP1C12Q240C8**. On the UP 2, it will be a EPF10K70RC240-*X* (-*X* is the speed grade of the chip). Check the large square chip in the middle of the board to verify the FPGA part number. The last digit in the FPGA part number is the speed grade. The correct speed grade is needed for accurate delays in timing simulations. You may need to change the setting of the Speed Grade dialog box to **Any** to display your specific device. Always choose the correct speed grade to match your board's FPGA.

If you choose the wrong device type, you will have errors when you attempt to download your design to the FPGA. (In any existing project, it is a good idea to always verify the correct FPGA setting for your UP board by selecting **Assignments** ⇨ **Device** in any new design before compiling it for the first time.)

Figure 1.7 Setting the FPGA Device Type.

After selecting the correct FPGA part number, click **Next** on the third-party EDA tools settings box since we will not be using any third-party EDA tools – only Quartus II. Double check the information summary page that appears and click **Finish**. In case of problems, use the back option to make changes.

Establishing Graphics (Schematic) as the Input Format

After creating your new project, choose **File ⇨ New,** and a popup menu will appear. Select **Block Diagram/Schematic File,** then click **OK**. This will create a blank schematic worksheet – a graphics display file (*.gdf file). Note that the toolbar options in Quartus II are context sensitive and change as different tools are selected. An empty schematic window with grids will appear named Block1.bdf.

Enter and Place the OR Symbol in Your Schematic

Click on the **AND gate icon** on the left-side toolbar. This selects the symbol tool. In the symbol library window, click the **library path** to expand the options. Find the library named **primitives** and click on it to expand it. Then click to expand the **logic library**. Scroll down the list of logic symbols and select **BNOR2**. An OR gate with inverted inputs and outputs should appear in the symbol window. (The naming convention is **B**-bubbled **NOR** with **2** inputs. Although considered to be a NOR with active low inputs, it is fundamentally an OR gate with active low inputs and output.) Click **OK** at the bottom of the Symbol window.

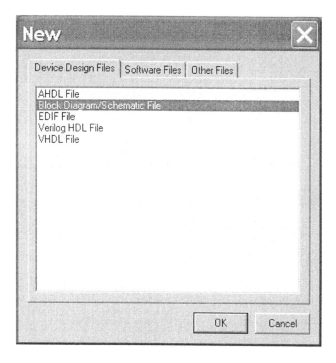

Figure 1.8 Creating the top-level project schematic design file.

Select the Block1.bdf window and the BNOR2 symbol will appear
in the schematic. Drag the symbol to the middle of the window and
left click to place it. Click on the arrow icon on the left side
toolbar or hit escape to stop inserting that symbol.

TO USE THE ONLINE HELP SYSTEM, CLICK **HELP** ON THE TOP MENU, SELECT SEARCH AND
THEN ENTER **BNOR**. AT ANY POINT IN THE TUTORIAL, EXTENSIVE ONLINE HELP IS ALWAYS
AVAILABLE. TO SEARCH BY TOPIC OR KEYWORD SELECT THE **HELP** MENU AND FOLLOW THE
INSTRUCTIONS THERE.

Assigning the Output Pin for Your Schematic

Select the **AND gate symbol** again on the left side toolbar, expand the **pin
library**, select **output**, and click **OK.** Using the mouse and the left mouse
button, drag the output symbol to the right of the **BNOR2** symbol leaving
space between them – they will be connected later.

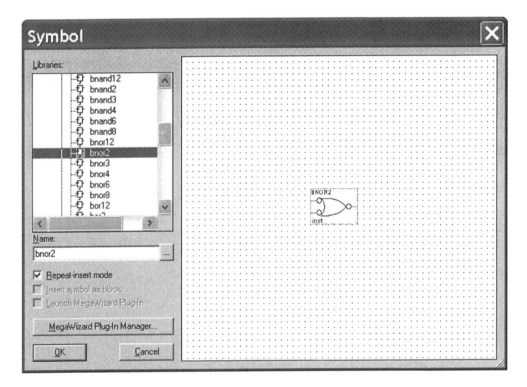

Figure 1.9 Selecting a new symbol with the Symbol Tool.

Assigning the Input Pins for Your Schematic

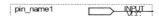

Find and place two pin **input** symbols to the left of the **BNOR2** symbol in the same way that you just selected and placed the **output** symbol. (Another hint: Once selected, a symbol can be copied with **Right Click⇨Copy** and pasted multiple times using the **Right Click⇨Paste** function.). Hit the arrow symbol on the left tool bar and deselect the new symbol by moving the cursor away and clicking the left mouse button a second time.

Connecting the Signal Lines

Using the mouse, move to the end of one of the wires. A cross-symbol mouse cursor should appear when the mouse is near a wire. Move to one end of a wire you need to add and push and hold down the left mouse button. Hold down the left mouse button and drag the end of the wire to the other point that you want to connect. Release the left button to connect the wire. If you need to delete a wire, click on it – the wire should turn blue when selected. Hit the delete key to remove it. You can also use the Right Click⇨ Delete function. Connect the other wires using the same process so that the diagram looks something like Figure 1.10

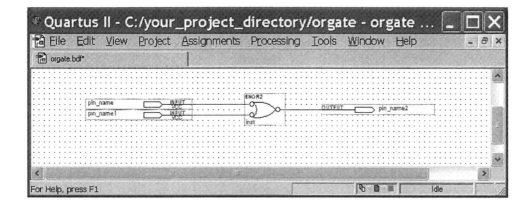

Figure 1.10 Active low OR-gate schematic example with I/O pins connected.

Enter the PIN Names

Right click on the first ⊏⊐ **INPUT** symbol. It will be outlined in blue and a menu will appear. Select **Properties**. Type **PB1** for the pin name and click **OK**. Name the other input pin **PB2** and the output pin for the **LED** in a similar fashion.

Assign the PIN Numbers to Connect the Pushbuttons and the LED

Since the FPGA chip on the UP 3 or UP 2 board is prewired to the pushbuttons and the LED, you need to look up the pin numbers and designate them in your design. The information in Table 1.1 is from the documentation on the pinouts for the UP 3 and UP 2 board user's manuals. (See Table 2.4.)

Table 1.1 Hardwired connections on the FPGA chips for the design.

I/O Device	UP 3 Pin Number Connections	UP 1 & UP 2 Pin Number Connections
PB1	62 (SW7)	28 (FLEX PB1)
PB2	48 (SW4)	29 (FLEX PB2)
LED	56 (D3)	14 (7Seg LED DEC. PT.)

In the main menu, select **Assignments** ➪ **Pin**. (If the option to select the pin is unavailable, you need to go back and select **Assignments** ➪ **Device,** and make sure that your device is selected correctly.) In the To column, type the name of the new pin, **PB1**. In the Location column, just enter **62** in the space provided (NOTE: pin numbers will be different on the UP 2). The software adds PIN_ to the pin number. Repeat this process assigning **PB2** to **PIN_48** and **LED** to **PIN_56**. After assigning all three pins and verifying your entries, **close** the assignment editor and click **Yes** to save. Device and pin information is stored in the project's *.qsf file. Pin names are case sensitive.

CAUTION: BE SURE TO USE UNIQUE NAMES FOR DIFFERENT PINS. PINS AND WIRES WITH THE SAME NAME ARE AUTOMATICALLY CONNECTED EVEN WITHOUT A VISIBLE WIRE SHOWING UP ON THE SCHEMATIC DIAGRAM.

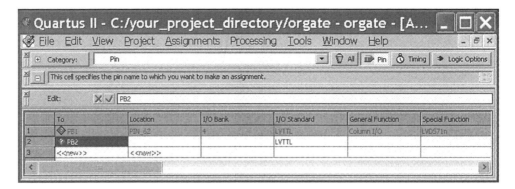

Figure 1.11 Assigning Pins with the Assignment Editor.

Saving Your Schematic

Select **File** ➪ **Save As** and your project directory. Name the file **ORGATE**. Throughout the remainder of this tutorial you should always refer to the same project directory path when saving or opening files. A number of other files are automatically created by the Quartus II tools and maintained in your project directory.

Set Unused Pins as Inputs

The memory chips on the UP 3 board could all be turned on at the same time by unused pins on the FPGA, causing their tri-state output drivers to try to force output data bus bits to different states. This causes high currents, which can overheat and damage devices after several minutes. To eliminate the possibility of any damage to the board, the following option should always be set in a new project. On the menu bar, select **Assignments ⇨Device ⇨ Device and Pin Options**. Click on the **Unused Pins** tab and check the **As inputs, tri-stated** option. Click **OK** and then **OK** in the first window. This setting is saved in the projects *.qsf file. Any time you create a new project repeat this step.

1.2 Compiling the Design

Compiling your design checks for syntax errors, synthesizes the logic design, produces timing information for simulation, fits the design on the selected FPGA, and generates the file required to download the program. After any changes are made to the design files or pin assignments, the project should always be re-compiled prior to simulation or downloading.

Compiling your Project

Compile by selecting **Processing ⇨ Start Compilation**. The compilation report window will appear in the Quartus II screen and can be used to monitor the compilation process, view warnings, and errors.

Checking for Compile Warnings and Errors

The project should compile with **0 Errors**. If a popup window appears that states, "Full Compilation was Successful," then you have not made an error. Info messages will appear in green in the message window. Warnings appear in blue in the message window and Errors will be red. Errors must be corrected. If you forget to assign pins, the compiler will select pins based on the best performance for internal timing and routing. Since the pins for the pushbuttons and the LED are pre-wired on the UP 3 or UP 2 board, their assignment cannot be left up to the compiler.

Examining the Report File

After compilation, the compiler window shows a summary of the compiled design including the FPGA logic and memory resources used by the design.

Select the orgate.bdf schematic window. Use **View ⇨ Show Pin and Location Assignments** and check the pins to verify the correct pin numbers have been assigned. If a pin is not assigned you may have a typo somewhere in one of the pin names or you did not save your pin assignments earlier. You will need to recompile whenever you change pin assignments.

You can also check all of the FPGA's pins by going to the compiler report window with **Processing** ⇨ **Compilation Report**, expanding the **Fitter entry**, and clicking on the **Pin-out file**.

1.3 Simulation of the Design

For complex designs, the project is normally simulated prior to downloading to a FPGA. Although the OR example is straightforward, we will take you through the steps to illustrate the simulation of the circuit.

Set Up the Simulation Traces

Choose **File** ⇨ **New**, select the **Other Files** tab, and then from the popup window select **Vector Waveform File** and click **OK**. A blank waveform window should be displayed. Right click on the **Name** column on the left side. Select **Insert Nodes or Bus**. Click on the **Node Finder** and then the **LIST** button. PB1, PB2 and LED should appear as trace values in the window. Then click on the center >> button and click **OK** and **OK** again. The signals should appear in the waveform window.

Generate Test Vectors for Simulation

A simulation requires external input data or "stimulus" data to test the circuit. Since the PB1 and PB2 input signals have not been set to a value, the simulator sets them to a default of zero. The 'X' on the LED trace indicates that the simulator has not yet been run. (If the simulator *has* been run and you still get an 'X,' then the simulator was unable to determine the output condition.)

Right click on **PB1**. The PB1 trace will be highlighted. Select **Value** ⇨ **Count Value...**, click on the **Timing** tab and change the entry for **Multiplied By** from **1** to **5** and click **OK**. An alternating pattern of Highs and Lows should appear in the PB1 trace. Right click on **PB2**. Select **Value** ⇨ **Count Value...**, click on the **Timing** tab and change the entry for **Multiplied By** from **1** to **10,** and click **OK**. **PB2** should now be an alternating pattern of ones and zeros but at twice the frequency of PB1. (Other useful options in the **Value** menu will generate a clock and set a signal High or Low. It is also possible to highlight a portion of a signal trace with the mouse and set it High or Low manually.)

When you need a longer simulation time in a waveform, you can change the default simulation end time using **Edit** ⇨ **End Time**.

Performing the Simulation with Your Timing Diagram

Select **File** ⇨ **Save** and click the **Save** button to save your project's vector waveform file. Select **Processing** ⇨ **Start Simulation** and click **OK** on the window that appears. The simulation should run and the output waveform for LED should now appear in the Simulation Report window. You may want to right click on the timing display and use the Zoom options to adjust the time scale as seen in Figure 1.12. Note that the simulation includes the actual timing delays through the device and that it takes almost 10 ns (ns = 10^{-9} sec.)

for the delayed output to reflect the new inputs. Taking this LED output delay into account, examine the Simulation Waveform to verify that the LED output is Low only when either PB1 OR PB2 inputs are Low.

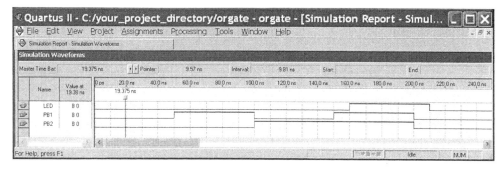

Figure 1.12 Active low OR-gate timing simulation with time delays.

1.4 Downloading Your Design to the UP 3 Board

Hooking Up the UP 3 Board to the Computer

If you have a UP 2 board skip to Section 1.5. To try your design on a UP 3 board, plug the Byteblaster™ II cable into the UP 3 board's JTAG connector (innermost of the two connectors on the left side of the board) and attach the other end to the parallel port on the PC (USB port if you are using a USB Blaster). If you have not done so already, make sure that the PC's BIOS settting for the printer port is ECP or EPP mode. Using the 6V AC to DC wall transformer attach power to the DC power connector (DC_IN) located on the lower right side of the UP 3 board. Press in the power switch located on the right edge of the board above the power connector. When properly powered, two LEDs on the bottom of the UP 3 board near the power connector should light up.

Preparing for Downloading

After checking to make sure that the cables and jumpers are hooked up properly, you are ready to download the compiled circuit to the UP 3 board. Select **Tools** ⇨ **Programmer**. Click on **Hardware Setup**, select the proper hardware, a **ByteBlasterII** on **LPT1**. (If a window comes up that displays, "**No Hardware**" to the right of the Hardware Setup button, use the Hardware Setup button to change currently selected hardware from "**No Hardware**" to "**ByteblasterII**". If a red JTAG error message appears or the start button is not working, close down the Programmer window and reopen it. If this still doesn't correct the problem, then there is something else wrong with the setup or cable connection. Go back to the beginning of this section and check each step and connection carefully.)

Final Steps to Download

The filename orgate.sof should be displayed in the programmer window. The *.sof file contains the FPGA's configuration (programming) data for your design. To the right of the filename in the Program/Configure column, check the **Program/Configure** box. To start downloading your design to the board, click on the **Start** button. Just a few seconds are required to download. If download is successful, a green info message displays in the system window notifying you the programming was successful.

Testing Your Design

The locations of PB1, PB2, and the decimal LED are indicated in Figure 1.13. After downloading your program to the UP 3 board, the LED in the lower right corner should turn off whenever a pushbutton is hit. Since the output of the OR gate is driving the LED signal, it should be on when no pushbuttons are hit. Since the buttons are active low, and the **BNOR2** gate also has active low inputs and output, hitting either button should turn off the LED.

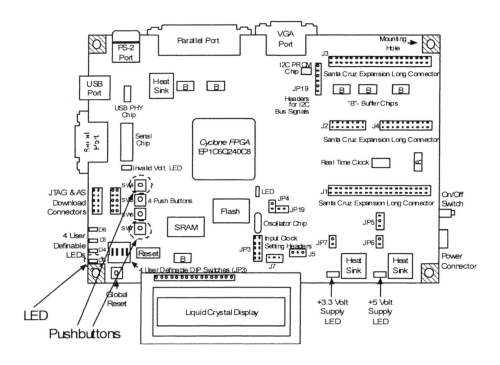

Figure 1.13 ALTERA UP 3 board showing Pushbutton and LED locations used in design.

Congratulations! You have just entered, compiled, simulated, downloaded a design to a FPGA device, and verified its operation. Since you are using a UP 3 board, you can skip the next section on the UP 2 board and go directly to Section 1.6.

COMPLETED TUTORIAL FILES ARE AVAILABLE ON THE TEXT'S CD-ROM.

THE SUBDIRECTORY \1C12 CONTAINS VERSIONS FOR THE LARGER

1C12 UP 3 BOARD AND THE SUBDIRECTORY \UP2 CONTAINS UP 2 FILES.

1.5 Downloading Your Design to the UP 2 Board

Hooking Up the UP 1 or UP 2 Board to the Computer

To try your design on a UP 1 or UP 2 board, plug the ByteBlaster cable into the UP board and attach the other end to the parallel port on a PC. If you have not done so already, make sure that the PC's BIOS settting for the printer port is ECP or EPP mode. Using a 9V AC to DC wall transformer or another 7 to 9V DC power source, attach power to the DC power connector (DC_IN) located on the upper left-hand corner of the UP 3 board. When properly powered, one of the green LEDs on the board should light up.

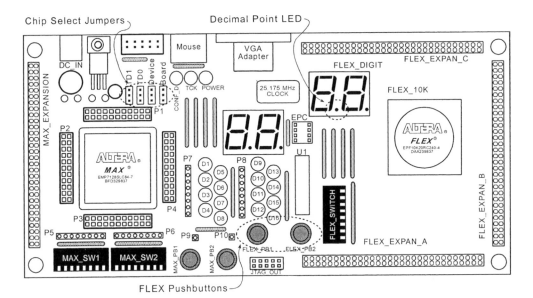

Figure 1.14 ALTERA UP 2 board with jumper settings and PB1, PB2, and LED locations.

Verify that the device jumpers are set for the FLEX chip as shown in Table 1.2. The locations of the pushbuttons, PB1 and PB2, and the LED decimal point are also highlighted in Figure 1.14. (Note that for the MAX EPM7128 chip, the jumper pins are all set to the top position as indicated in Table 1.2.)

Table 1.2 Jumper settings for downloading to the MAX and FLEX devices.

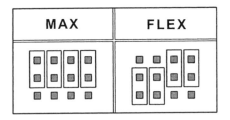

Preparing for Downloading

After checking to make sure that the cables and jumpers are hooked up properly, you are ready to download the compiled circuit to the UP 2 board. Select **Tools** ⇨ **Programmer**. Click on **Hardware Setup**, select the proper hardware, a **ByteBlasterII** on **LPT1**. (If a window comes up that displays, "**No Hardware**" to the right of the Hardware Setup button, use the Hardware Setup button to change currently selected hardware from "**No Hardware**" to "**ByteBlasterII**". If a red JTAG error message appears or the start button is not working, close down the Programmer window and reopen it. If this still doesn't correct the problem, then there is something else wrong with the setup or cable connection. Go back to the beginning of this section and check each step and connection carefully.)

Final Steps to Download

Make sure that the **Device Name** has changed to **EP10K20** or **EPF10K20RC240** for the UP 1 or **EPF10K70RC240** for the UP 2 (depending on the UP board and Quartus II version that you are running). Make sure you have also assigned the pin numbers for a UP 2 board and not the UP 3 (see Table 1.1). If it does not display the correct device, then return to your schematic, assign the correct device first and then the pin numbers (See section 1.1.), recompile, and try again. Next, check the **Program/Configure** box.

The **Start** button in the programming window should now be highlighted. Click on the **Start** button to download to the UP 2 board. Just a few seconds are required to download. If download is successful, a green info successful programmer operation message displays in the system window. (If the **Start** button is not highlighted, click **Hardware Setup** from the programmer window. Confirm the port settings and click **OK**. If you still have problems confirm that the printer port BIOS settings use ECP or EPP mode.)

Testing Your Design

The locations of PB1, PB2, and the decimal LED are indicated in Figure 1.14. On the UP 2, one of the seven-segment LED's decimal points is used for monitoring the output.

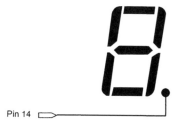

Pin 14

Figure 1.15 UP 2's FLEX FPGA pin connection to seven-segment display decimal point.

All of these LEDs are pre-wired to the FPGA chip with a pull-up resistor as illustrated earlier in Figure 1.4. This configuration allows the external resistor to control the amount of current through the LED; however, it also requires the FPGA chip to output a Low signal to turn on the LED. (Students regularly forget this point and have a fully working project with an inverted pattern on the LEDs.). V_{cc} is 5V on the UP2.

Figure 1.15 shows the UP 2's Flex FPGA pin number 14 hard wired to the seven-segment LED's decimal point. On the UP 2, in this tutorial, only the decimal point will be used for output.

After downloading your program to the UP 2 board, locate the two rightmost seven-segment displays. Since the output of the BNOR2 gate is driving the decimal LED signal on the left digit of the two seven-segment displays, it should be off (LED state is inverted on UP3). Since the buttons are active low, and the **BNOR2** gate also has active low inputs and output, hitting either button should turn on the LED.

Congratulations! You have just entered, compiled, simulated, and downloaded a design to a FPGA device, and verified its operation.

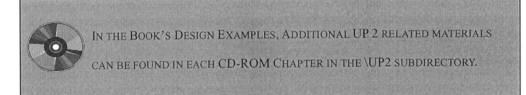

IN THE BOOK'S DESIGN EXAMPLES, ADDITIONAL UP 2 RELATED MATERIALS CAN BE FOUND IN EACH CD-ROM CHAPTER IN THE \UP2 SUBDIRECTORY.

1.6 The 10 Minute VHDL Entry Tutorial

As an alternative to schematic capture, a hardware description language such as VHDL or Verilog can be used. In large designs, these languages greatly increase productivity and reduce design cycle time. Logic minimization and synthesis to a netlist are automatically performed by the compiler and synthesis tools. (A netlist is a textual representation of a schematic.) As an example, to perform addition, the VHDL statement:

$$A <= B + C;$$

will automatically generate an addition logic circuit with the correct number of bits to generate the new value of A. Using the OR-gate design from the Schematic Entry Tutorial, we will now create the same circuit using VHDL.

PRIOR KNOWLEDGE OF VHDL IS NOT NEEDED TO COMPLETE THIS TUTORIAL.

Using a Template to Begin the Entry Process

Choose **File ⇨ New**, select **VHDL File** and **OK**. Place the cursor within the text area, right click the mouse, and select **Insert Template.** Make sure **VHDL** is selected. (Note the different prewritten templates. These are provided to expedite the entry of VHDL.) Select **Entity Declaration**– this template is the one you will generally start with since it also sets up the input and output declarations. The template for the ENTITY declaration appears in the Insert Template preview window. Click **OK** to paste the template in your VHDL window. Since the editor knows that it is a VHDL source file, the text will appear in different context-sensitive colors. VHDL keywords appear in blue and strings in green. The coloring information should be used to detect syntax errors while still in the text editor.

Saving the VHDL Source File

Select **File ⇨ Save As** and save the file as **orgate.vhd** – click **Save**.

Replacing Comments in the VHDL Code

The entire string indicating the position of the entity name, **__entity_name**, should be set to the name used for the filename – in this case, **orgate**. There are two occurrences of **__entity_name** in the text. Find and change both accordingly.

Declaring the I/O Pins

The input and output pins, PB1, PB2 and LED need to be specified in the PORT declaration. Since there are no input vectors, bi-directional I/O pins, or GENERIC declarations in this design, remove all of these lines. The source file should look like Figure 1.16.

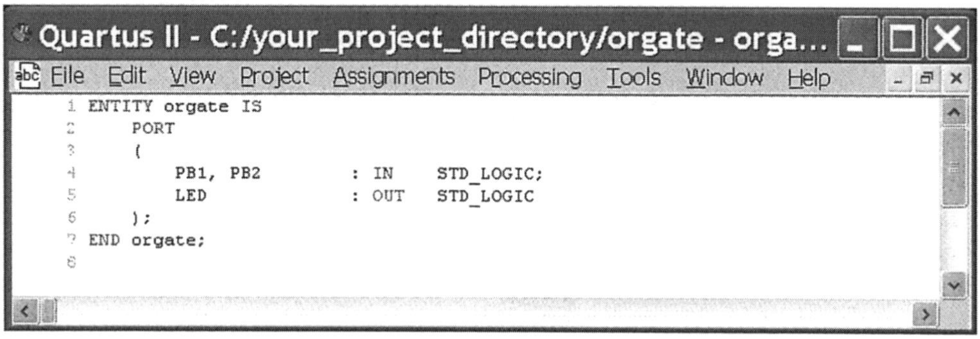

```
Quartus II - C:/your_project_directory/orgate - orga...
 File   Edit   View   Project   Assignments   Processing   Tools   Window   Help
1  ENTITY orgate IS
2      PORT
3      (
4          PB1, PB2        : IN     STD_LOGIC;
5          LED             : OUT    STD_LOGIC
6      );
7  END orgate;
8
```

Figure 1.16 VHDL Entity declaration text.

Setting up the Architecture Body

Click the mouse at the bottom of the text field. (We will be inserting another template here.) Following the earlier procedure for selecting a VHDL template (start with a right click), select an **Architecture Body**. (The **Architecture Body** specifies the internal logic of the design.) The syntax for the **Architecture Body** appears in the text window after the other text. (You can now see why the template is left highlighted – had you not placed your cursor first, text would have appeared at your last cursor position. If you do misplace the template, hitting the Edit Undo key removes the new text.)

Editing the Architecture Body

Change the entity name in the **ARCHITECTURE** statement to **orgate**. Template lines with a "--" preceding a comment, need to be edited for each particular design. Delete the two signal declaration lines since this simple design does not require internal signals. Delete the remaining comment lines that start with "--", and insert **LED <= NOT (NOT PB1 OR NOT PB2)** as a single line. (This line contains a deliberate syntax error that will be detected and fixed later.) Insert the following two lines at the beginning of the text file to define the libraries for the STD_LOGIC data type.

LIBRARY IEEE;
USE IEEE.STD_LOGIC_1164.all;

This is the preferred data type for bits in VHDL. The file should now appear as in Figure 1.17.

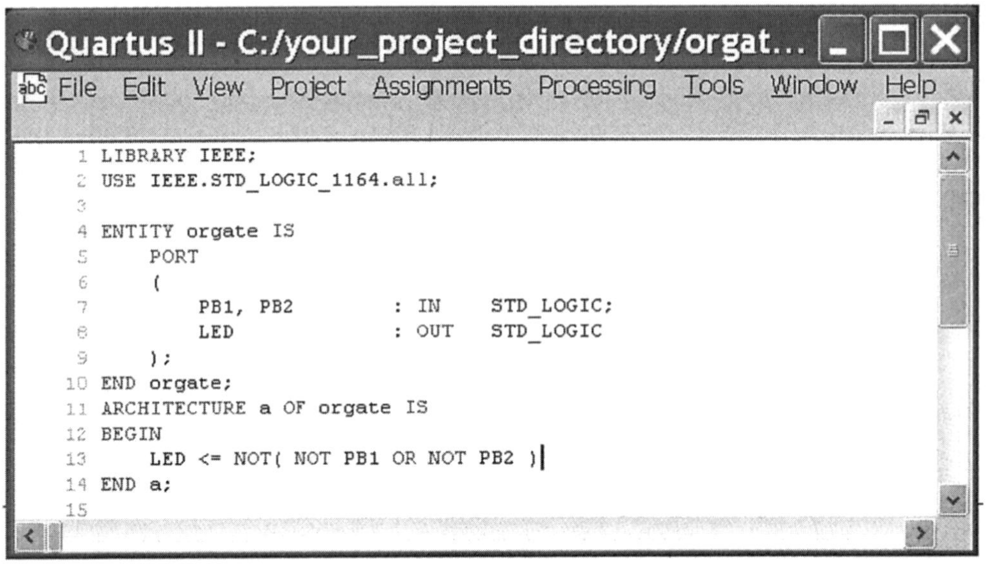

Figure 1.17 VHDL OR-gate model (with syntax error).

Before You Compile

Before you compile the VHDL code, the FPGA device type and pin numbers need to be assigned with **Assignments** ⇨ **Device** and **Assignments** ⇨ **Pin**. If your pins are already defined from the earlier Schematic Entry Tutorial, just confirm the pin assignments. If you did not do this step earlier in the tutorial see the device and pin assignment instructions at the end of section 1.1. At this point, VHDL code is generally ready to be compiled, simulated, and downloaded to the board using steps identical to those used earlier in the schematic entry method. Once pin assignments are made, they are stored in the project's *.qsf file.

1.7 Compiling the VHDL Design

The Compile process checks for syntax errors, synthesizes the logic design, produces timing information for simulation, fits the design on the FPGA, and generates the file required to program the FPGA. After any changes are made to the design files or pin assignments, the project should always be re-compiled prior to simulation or programming.

Select **Project** ⇨ **ADD/Remove Files in Current Project**. Confirm that the new orgrate.vhd file is now part of project and **remove** the tutorial's earlier **orgate.bdf** file that the new VHDL file replaces from the project's file list, if it is present. Click **OK**. Start the compiler with **Processing** ⇨ **Start Compilation**.

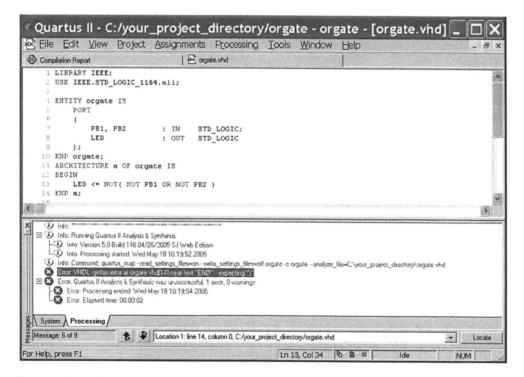

Figure 1.18 VHDL compilation with a syntax error.

Checking for Compile Warnings and Errors

The project should compile with an error. After compiling the VHDL code, a window indicating an error should appear. The result should look something like Figure 1.18.

Double click on the first red error line and note that the cursor is placed in the editor either on or after the line missing the ";" (semicolon). VHDL statements should end with a semicolon. Add the semicolon to the end of the line so that it is now reads:

LED <= NOT (NOT PB1 OR NOT PB2);

Now, recompile, and you should have no errors. You can simulate your VHDL code using steps identical to the tutorial's earlier schematic version of the project.

1.8 The 10 Minute Verilog Entry Tutorial

Verilog is another widely used hardware description language (HDL). Verilog and VHDL have roughly the same capabilites. VHDL is based on a PASCAL style syntax and Verilog is based on the C language. In large designs, HDLs greatly increase productivity and reduce design cycle time. Logic minimization and synthesis are automatically performed by the compiler and synthesis tools.

Just like the previous VHDL example, to perform addition, the Verilog statement:

A = B + C;

will automatically generate an addition logic circuit with the correct number of bits to generate the new value of A.

Using the OR-gate design from the Schematic Entry Tutorial and the VHDL Tutorial, we will now create the same circuit in Verilog.

PRIOR KNOWLEDGE OF VERILOG IS NOT NEEDED TO COMPLETE THIS TUTORIAL.

Using a Template to Begin the Entry Process

Choose **File ➪ New**, select **Verilog HDL File** and **OK**. Place the cursor within the text area, right click the mouse, and select **Insert Template** and then select **Verilog HDL**. (Note the different prewritten templates. These are built to expedite the entry of Verilog.) Select **Module Declaration** – this declaration is the one you will generally start with since it also sets up the input and output declarations. Click **OK** and the template for the module declaration appears in the Text editor. Since the editor knows that it is a Verilog source file, the text will appear in different context-sensitive colors. Verilog keywords appear in blue and strings in green. The coloring information should be used to detect syntax errors while still in the text editor.

Saving the new Verilog File

Select **File** ⇨ **Save As.** Note the automatic extension is **.v** (Verilog) and save the file as **orgate.v** – click **Save.**

Replacing Comments in the Verilog Code

The entire string indicating the position of the entity name, **__module_name**, should be set to the name used for the filename – in this case, **orgate**. There is one occurrence of **__module_name** in the text. Find and change it accordingly. Lines starting with // are comments and these will need to be replaced with the appropriate Verilog code.

Declaring the I/O Pins

The input pins, PB1, PB2 and the output pin LED need to be specified in the arguments of the Module statement and Port declaration. Since there are no inout pins, wire or integer declarations, or Always statements in this design, remove all of these lines. The source file should now look like Figure 1.19.

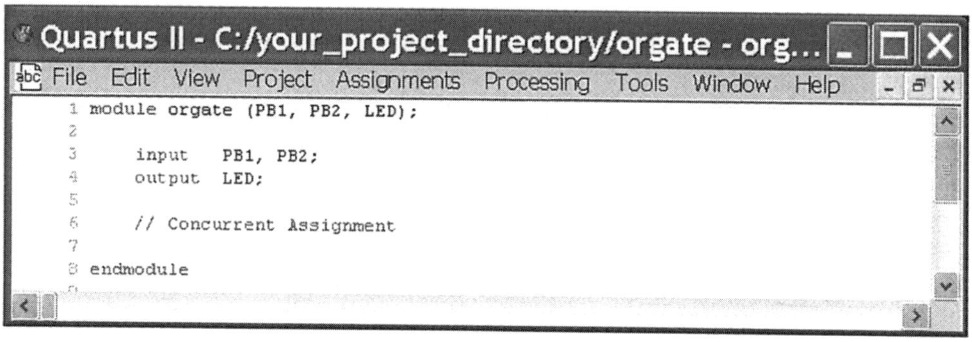

```
 1  module orgate (PB1, PB2, LED);
 2
 3      input   PB1, PB2;
 4      output  LED;
 5
 6      // Concurrent Assignment
 7
 8  endmodule
```

Figure 1.19 Verilog module declaration text.

Setting up the Behavioral Code

Click the mouse to just after the line starting with // Concurrent Assignment. (We will be inserting another template here.) Following the earlier procedure for selecting a Verilog template (start with a right click), select a **Continuous Assignment Statement**. (A single assign statement will specify the internal logic of this design.) The syntax for a **Continuous Assignment Statement** appears in the text window after the other text. (You can now see why the template is left highlighted – had you not placed your cursor first, text would have appeared at your last cursor position. If you do misplace the template, hitting the delete key removes the highlighted text.)

Editing the Continuous Assignment Statement

Change the identify name in the assign statement to "LED" and value to:

! (! PB1 | ! PB2);

Verilog is based on C and "|" (vertical line) is the bit wise OR operator. The "!" (exclamation point) is the NOT operator. Delete the remaining comment lines that start with "//". Delete the ";" at the end of the assign LED statement (This causes a deliberate syntax error that will be detected and fixed later.) The file should now appear as in Figure 1.20.

Before You Compile

Before you compile the Verilog code, the FPGA device type and pin numbers need to be assigned with **Assignments** ⇨ **Device** and **Assignments** ⇨ **Pin**. If your pins are already defined from the earlier Schematic Entry Tutorial, just confirm the pin assignments. If you did not do this step earlier in the tutorial see the device and pin assignment instructions at the end of Section 1.1. At this point, Verilog code is generally ready to be compiled, simulated, and downloaded to the board using steps identical to those used earlier in the schematic entry method. Once pin assignments are made, they are stored in the project's *.qsf file.

```
Quartus II - C:/your_project_directory/orgate - org...
File  Edit  View  Project  Assignments  Processing  Tools  Window  Help
1 module orgate (PB1, PB2, LED);
2
3     input   PB1, PB2;
4     output  LED;
5
6     assign LED = ! ( ! PB1 | ! PB2 )
7
8 endmodule
```

Figure 1.20 Verilog active low OR-gate model (with syntax error).

1.9 Compiling the Verilog Design

The Compile process checks for syntax errors, synthesizes the logic design, produces timing information for simulation, fits the design on the FPGA, and generates the file required to program the FPGA. After any changes are made to the design files or pin assignments, the project should always be re-compiled prior to simulation or programming.

Select **Project** ⇨ **ADD/Remove Files in Current Project**. Confirm that the new **orgate.v** file is now part of project and **remove** the tutorial's earlier **orgate.bdf** or **orgate.vhd** files that the new Verilog file replaces from the project if either file is present. Click **OK**. Start the compiler with **Processing** ⇨ **Start Compilation**.

Checking for Compile Warnings and Errors

The project will compile with an error. After compiling the Verilog code, a window indicating an error should appear. (See Figure 1.21.)

Double click on the first red error line and note that the cursor is placed in the editor either on or after the line missing the ";" (semicolon). Verilog statements should end with a semicolon. Add the semicolon to the end of the line so that it is now reads:

assign LED = ! (! PB1 | ! PB2);

Now, recompile, and you should have no errors. You can simulate your Verilog code using steps identical to the tutorial's earlier schematic version of the project.

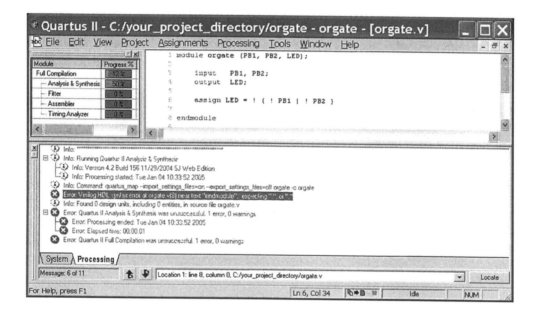

Figure 1.21 Verilog compilation with a syntax error.

1.10 Timing Analysis

With every physical device, there are timing considerations. An FPGA's timing is affected by:

- Input buffer delays,
- Signal routing interconnect delays within the FPGA,
- The internal logic delays (in this case the OR), and
- Output buffer delays.

The timing analysis tool can be used to determine:

- The physical delay times and
- The maximum clock rates in your design.

Starting the Analyzer

At the top menu, select **Tools** ⇨ **Timing Analyzer** and click the **t$_{pd}$** (propagation delay time) tab at the top of the Timing Analyzer window. A matrix of input to output delay times for the project will be computed and displayed as seen in Figure 1.22.

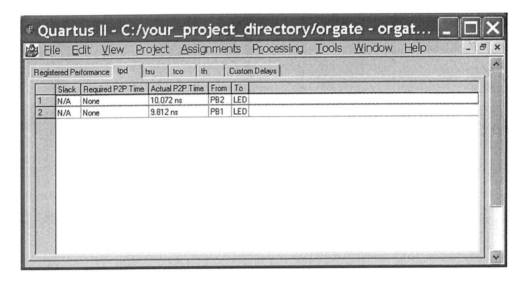

Figure 1.22 Timing analyzer showing input to output timing delays.

Note that this is the same delay time seen in the simulator. These times include the input-to-output buffer delays at the pins and the interconnect delays inside the FPGA. The internal OR logic delay is only around a nanosecond relative to the rest of the device delay. The actual time shown will vary with different versions of the Altera CAD tools and different FPGA chip speed grades. Other timing analysis options include setup times, hold times, and clock rates for sequential circuits.

1.11 The Floorplan Editor

A floorplan editor is a visual tool to assist expert users in manually placing and moving portions of logic circuits to different logic cells inside the FPGA. This is done in an attempt to achieve faster timing or better utilization of the FPGA. Floorplanning is typically used only on very large designs that contain subsections of hardware with critical high-speed timing. Since the interconnect delays are as large as the design's logic delays, logic element and I/O pin placement is very critical in high speed designs. Vertical and horizontal interconnect buses are used through the FPGA to connect Logic Elements.

For all but expert users, the compiler's automatic place-and-route tools should be used. Automatic place-and-route was already performed by the fitter in the compile process of the tutorial. Timing constraints for critical signals can also be specified in some FPGA place and routing tools to help the fitter meet the design's timing goals.

To see the fitter's automatic placement of the design inside the FPGA, select **Assignment ⇨ Back-Annotate Assignments** click **OK** and then **Assignments ⇨ Timing Closure Floorplan.** In the display that opens, zoom in and scroll around to find the yellow logic element and gray shaded I/O pins used in your design. Find and select the yellow LE (blue on UP 2), then **View ⇨ Routing ⇨ show fan in** and then **show fan out,** and a view like Figure 1.23 showing the design will be produced.

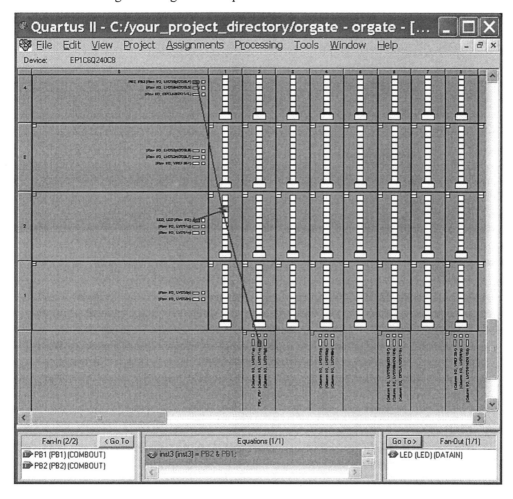

Figure 1.23 Floorplan view showing internal FPGA placement of OR-gate in LE and I/O pins.

There is a lot of empty space since the Cyclone EP1C6 contains 5,980 Logic Elements (LEs) and the larger EP1C12 contains 12,060 LEs. Only 1 LE was used in this design. If you move the logic cell or I/O pins to other locations, it will make small changes to the circuit timing because of changes in the interconnect delays inside the FPGA. Due to the vast number of combinations, FPGA CAD tools cannot explore every possible placement and routing option. The Quartus II Design Space Explorer tool can also be used to search and explore other design options in the design space. Large FPGA designs containing millions of gates can require several hours or even days of CPU time to examine many of the different place and route alternatives in the design space.

1.12 Symbols and Hierarchy

The Symbol Editor is used to edit or create a symbol to represent a logic circuit. Symbols can be created for a design whenever a VHDL or Verilog file is compiled. Create a symbol for your VHDL design by opening the orgate.vhd file, and then select **File ➪ Create/Update ➪ Create Symbol Files for Current File.**

Select **File ➪ Open,** change the file type setting for *.bsf, find and chose **orgate.bsf** to see the new symbol for your VHDL based design as shown in Figure 1.24. Inputs are typically shown on the left side of the symbol and outputs on the right side. Symbols are used for design hierarchy in more complex schematics. This new symbol can be used to add the circuit to a design with the graphic editor just like the BNOR2 symbol that was used earlier in the tutorial. Clicking on a symbol in the graphic editor will take the user to the underlying logic circuit or HDL code that the symbol represents.

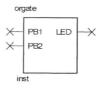

Figure 1.24 ORgate design symbol.

1.13 Functional Simulation

In large designs with long compile and simulation times, another type of simulation that runs faster is commonly used. A functional simulation does not include device delay times and it is used to check for logical errors only. A timing simulation that included device delay times, was used earlier in the tutorial. After fixing logical errors using functional simulation, a timing simulation is still necessary to check for any timing related errors in a design.

Performing a Functional Simulation

To perform a functional simulation, set the simulator for functional simulation with **Assignments** ⇨ **Settings.** Select **Simulator** in the left column and then change the simulation mode from Timing to **Functional.** Run **Processing** ⇨ **Generate Functional Simulation Netlist.** Finally, select **Processing** ⇨ **Start Simulation.** Open the Simulation Report waveform and note that the output changes without any delay in response to an input unlike the earlier timing simulation. To switch back to a timing-mode simulation, change the simulator setting back to timing, recompile, and restart the simulation.

This short tutorial has gone through the basics of a simple design using a common path through the design tools. As you continue to work with the tools, you will want to explore more of the menus, options and shortcuts. Chapter 4 contains a tutorial that will introduce a more complex design example. In Quartus II, **Help** ⇨ **Tutorial** also contains more tutorials. Quartus II video tutorials and reference manuals are also available online at Altera's website, www.altera.com . A number of files such as the *.q* files are maintained in the project directory to support a design. Appendix B contains a list of different file extensions used by Quartus II.

1.14 Laboratory Exercises

1. The tutorials ORed the active low signals from the pushbuttons and produced an output that was required to be low to turn off an LED. This was accomplished with the "negative-logic OR" gate illustrated to the left in Figure 1.25.

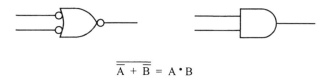

$$\overline{\overline{A} + \overline{B}} = A \cdot B$$

Figure 1.25 Equivalent gates: A negative logic OR and a positive-logic AND.

We know from DeMorgan's Law that the equation in Figure 1.25 represents an equivalence. We should therefore be able to substitute a simple two-input AND gate as illustrated in Figure 1.25 and accomplish the same task as the single gate used in the tutorial. Substitute the AND2 gate for the BNOR2 gate in the schematic capture, then compile, simulate, and download the AND circuit. What can you conclude?

2. Substitute in the VHDL code:

    LED <= PB1 AND PB2;

    Or into the Verilog code:

assign LED = PB1 & PB2;

Compile, simulate, and download and test the new circuit. What can you conclude about gate equivalence using DeMorgan's Theorem?

3. Design a logic circuit to turn on the LED when both pushbuttons are pressed. Compile, simulate, and download the new circuit.

4. Try a different logic function such as XOR. Start at the beginning or edit your existing schematic by deleting and replacing the BNOR2 symbol. Next repeat the tutorial steps to compile, simulate, download and test.

5. Repeat problem 2 for all of the basic gates including, OR, NOR, NAND, XOR, XNOR, and NOT. Try using different LEDs and output your results simultaneously. Look up the pin connections to the Cyclone chip in Appendix C and be sure to give each pinout a different name.

6. Design, enter, simulate and implement a more complex logic gate network. One suggestion is a half adder. You will need two LED outputs.

7. In the schematic editor, try building the design with some 74xx TTL parts from the others maxplus2 symbol library.

8. Draw a schematic and develop a simulation to test the 2-to-1 Mux function in the others maxplus2 symbol library.

9. View the orgate.rpt file and find the device utilization, the pin assignments, and the netlist. A substantial portion of the time delay in this simple logic design is the input and output buffer delays and the internal routing of this signal inside the FPGA. Find this delay time by removing the BNOR2 gate and one of the inputs in the schematic. Connect the input pin to the output pin, recompile and rerun the timing analyzer to estimate this time delay.

10. Use the chip editor to move the logic cell used in the OR-gate design to another location inside the FPGA. For information of the chip editor, use the Quartus II Help function. Try moving the LE used several columns farther away from the pushbutton and LED pins. Not all locations of the logic cell will work and some trial and error will be required. Save the edited design, rerun the timing analyzer, and compare the resulting time delays with the original time delays. See if you are able to achieve a faster implementation than the automatic place-and-route tools.

11. Remove the pin number constraints from the schematic and let the compiler assign the pin locations. Rerun the timing analyzer and compare the time delays. Are they faster or slower than having specified the input pins?

12. If you are using a UP 2 board, retarget the example design to the MAX chip. Pin numbers for the MAX decimal point LED can be found in Appendix C. It will be necessary to connect jumper wires from the MAX header to the pushbuttons. Select pins near the pushbuttons. Pin numbers can be seen on the board's silk-screen. Compare the timing from the MAX implementation to the Flex implementation.

13. If a storage oscilloscope or a fast logic analyzer is available, compare the predicted delay times from the simulation and timing analysis to the actual delays measured on the FPGA board. Force the pins to a header connector so that you can attach probes to the signal wires.

14. Draw a schematic that uses the LPM_ADD_SUB megafunction to add two signed numbers on the Cyclone device. Use **Tools** ⇨ **Megawizard** to start the megawizard to help configure LPM symbols. Verify the proper operation using a simulation with two 4-bit numbers. Do not use pipelining, clock, or carry in. Vary the number of bits in the adder and find the maximum delay time using the timing analyzer. Plot delay time versus number of bits for adder sizes of 4, 8, 16, 32, and 64 bits. Using the LC percentages listed in the compiler's report file, estimate the hardware size in LEs. Plot LEs required versus number of bits.

15. Use the DFF part from the primitives storage library and enter the symbol in a schematic using the graphical editor. Develop a simulation that exercises all of the features of the D flip-flop. Use Help on DFF for more information on this primitive.

16. Use the DFFE part from the primitives storage library and enter the symbol in a schematic using the graphical editor. Develop a simulation that exercises all of the features of the D flip-flop with a clock enable. Use Help on DFFE for more information on this primitive.

17. Use gates and a DFF part from the primitives storage library with graphical entry to implement the state machine shown in the following state diagram. Verify correct operation with a simulation using the Altera CAD tools. The simulation should exercise all arcs in the state diagram. A and B are the two states, X is the output and Y is the input. Use the timing analyzer's **Tools** ⇨ **Timing Analysis** ⇨ **Registered performance** option tab to determine the maximum clock frequency on the Cyclone device. Reset is asynchronous and the DFF Q output should be high for state B.

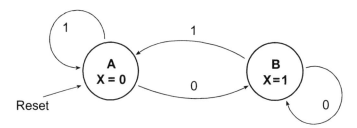

18. Repeat the previous problem but use one-hot encoding on the state machine. For one-hot encoding use two flip-flops with only one active for each state. For state A the flip-flop outputs would be "10" and for state B "01". One-hot encoding is common in FPGAs.

CHAPTER 2

The Altera UP 3 Board

Photo: The Altera UP 3 board contains a Cyclone FPGA, external SRAM, SDRAM &
Flash memory, and a wide assortment of I/O devices and connectors.

2 The Altera UP 3 Board

The Altera University Program 3 (UP 3), FPGA design laboratory board is shown in Figures 2.1 and 2.2. This board contains a Cyclone FPGA, several external memory devices and a wide range of I/O features. Two versions of the UP3 board are currently available with either a Cyclone EP1C6Q240 or the larger EP1C12Q240 FPGA. The FPGA and memory devices can be programmed using a JTAG ByteBlasterII cable attached to a PC printer port. An external 6V DC power supply or an AC to DC wall adapter is used to provide power. An on-board clock oscillator and clock chip provides several clock signals that are selectable with the board's jumpers.

Note the orientation of the LCD Display module in Figure 2.1 and how it extends beyond to edge of the board. Do not plug the module in backwards as it may damage the LCD by reversing its power connections.

A ByteBlasterII programming adapter cable is supplied with the UP 3 board and it connects the UP 3 board to the PC's parallel port (LPT) for device programming. The printer port mode of the PC should be set in the PC's BIOS to ECP or EPP.

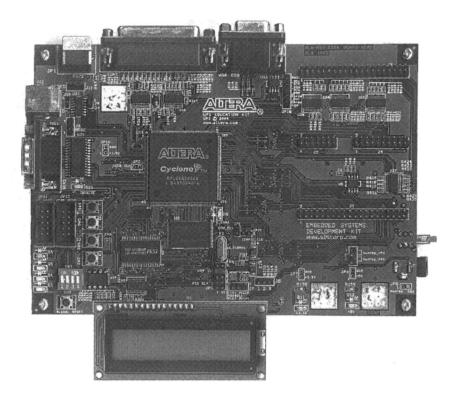

Figure 2.1 The Altera UP 3 board.

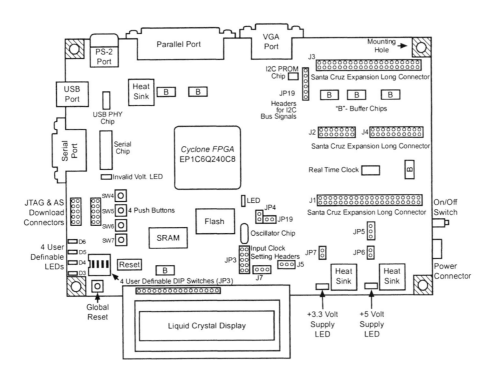

Figure 2.2 The Altera UP 3 board's features.

2.1 The UP 3 Cyclone FPGA Features

The Cyclone FPGA is the large square chip located in the center of the UP 3 board as seen in Figure 2.1. Two versions of the UP 3 are currently available. The 312 model has the larger EP1C12 and the 306 model uses the EP1C6 as seen in Table 2.1. There are some minor pin changes between the EP1C6 and EP1C12 board's LCD and memory devices. You must also compile your design for the correct Cyclone device assignment or it will not download the device.

Table 2.1 UP 3 Board's Cyclone FPGA Features

Cyclone FPGA Feature	EP1C6Q240	EP1C12Q240
Logic Elements (LEs)	5,980	12,060
4K bit RAM blocks (M4Ks)	20	52
Total Internal RAM bits	92,160	239,616
Phase Locked Loops (PLLs)	2	2
User I/O pins	185	173

2.2 The UP 3 Board's Memory Features

In addition to the Cyclone FPGA's internal memory, the UP 3 board provides several external ROM and RAM memory devices as seen in Table 2.2. Capacities of external memory are much larger than the internal memory, but they will have a longer access time. FPGA processor cores such as the Nios use external memory for program and data memory and the FPGA's internal memory for register files and cache. Flash and EEPROM are used to provide non-volatile memory storage. The EPCS1 serial Flash chip is used to automatically load the FPGA's serial configuration data at power up in systems where you do not want to download the board with the ByteBlaster II each time power is applied. Links to datasheets for all of the UP 3 board's chips can be found at the book's website.

Table 2.2 UP 3 Board's Memory Features

Memory Device	Size	Part Number
SRAM	64K by 16 bits	ISSI IS61C6416
SDRAM	1M by 16 bits	ISSI IS42S16400B
Flash Memory	1M by 16 bits	Toshiba TC58FVB106AFT-70
I^2C EEPROM	16K by 1bit	ISSI IS24C16
Serial Flash Memory	1M by 1bit	Altera EPCS1

2.3 The UP 3 Board's I/O Features

The UP 3 board provides a wide variety of I/O features as summarized in Table 2.3. For most devices, the UP 3 board's hardware provides only an electrical interface to the FPGA's I/O pins. Logic that provides a device interface circuit or controller will need to be constructed using the FPGA's internal logic. Many design examples of interfacing these various I/O devices can be found in the following chapters of this book.

The Cyclone FPGA is a surface-mount chip that it is soldered directly to the board. It is difficult if not impossible to replace the Cyclone chip without expensive surface mount soldering equipment, so extreme care should be exercised when interfacing the Cyclone I/O pins to any external devices.

Also, remember to assign pins as shown in the tutorials to avoid randomly turning on several of the memory devices at the same time. A tri-state bus conflict occurs when several tri-state outputs are turned on and they attempt to drive a single signal line to different logic levels. It is possible that such a tri-state bus conflict on the memory data bus could damage the devices by overheating them after several minutes of operation.

Table 2.3 Overview of the UP 3 Board's I/O Features

I/O Device	Description	Hardware Interface Needed
USB 1.1	Full Speed and Low Speed	Processor & USB SIE engine core
Serial Port	RS 232 Full Modem	UART to send and receive data
Parallel Port	IEEE 1284	State machine or Proc. for handshake
PS/2 Port	PC Keyboard or Mouse	Serial Data - PS/2 state machine
VGA Port for Video Display on Monitor	RGB three 1-bit signals provide 8 colors	State machine for sync signals & user logic to generate RGB color signals
IDE Port	Connector	Processor & IDE Device Driver
Reset Switch	Use for Global Reset	Must use a reset in design
Pushbutton Switches	4 Non-debounced (0=HIT)	Most applications will need a switch debounce Circuit
Expansion Card	Santa Cruz Long 72 I/O	Depends on expansion card used
LEDs	4 User Definable (1=ON)	None
LCD Display	16 Character by 2 line ASCII Characters	State machine or Processor to send ASCII characters and LCD commands
Real Time Clock	I^2C clock chip	Serial Data - I^2C state machine
DIP Switch	4 Switches (1=ON)	None or Synchronizer Circuit

WHEN CONNECTING EXTERNAL HARDWARE, ADDITIONAL PINS ARE AVAILABLE FOR USE ON THE SANTA CRUZ EXPANSION CONNECTORS ON THE BOARD. REFER TO THE UP 3 EDUCATION KIT REFERENCE MANUAL FOR DETAILS. THIS MANUAL COMES WITH THE UP 3 BOARD AND IS AVAILABLE FREE AT HTTP://WWW.ALTERA.COM OR AT HTTP://WWW.SLSCORP.COM.

Table 2.4 contains the pin assignments and names used for the UP 3 board's most commonly used I/O devices. Pin differences for the larger 1C12 UP 3 board are listed in parenthesis. The larger 1C12 has twelve additional power and ground pins, so fewer pins are left for general purpose I/O.

NOTE: If you ever switch between 1C6 and 1C12 boards, you will need to change the device, fix the pin assignments, and then recompile for the new FPGA device. A complete table including all I/O devices can be found in Appendix C. The voltage levels on FPGA pins can vary, so be sure to check for the proper voltage levels when selecting an I/O pin to interface new external hardware to the board. 5V logic levels are available on J3.

Do not connect high current devices such as motors or relay coils directly to FPGA I/O pins. These pins cannot provide the high current levels needed, and it may damage the FPGA's output circuit.

Table 2.4 UP 3 Board's most commonly used FPGA I/O pin names and assignments

Pin Name	Pin#	Pin I/O Type	Function of Pin
PS2_CLK	12	Bidirectional	PS2 Connector
PS2_DATA	13	Bidirectional	PS2 Connector
RESET	23	Input	Power on or SW8 pushbutton reset (Reset = 0)
USB_CLK	29	Input	USB 48MHz Clock - jumper
USER_CLOCK	38	Input	External Clock from J2 Pin 2
PBSWITCH_4	48	Input	Pushbutton SW4 (non-debounced, 0 = button hit)
PBSWITCH_5	49	Input	Pushbutton SW5 (non-debounced, 0 = button hit)
LCD_E	50	Output	LCD Enable line
LED_D6	53	Output	LED D3 (0 = LED ON, 1= LED OFF)
LED_D5	54	Output	LED D4 (0 = LED ON, 1= LED OFF)
LED_D4	55	Output	LED D5 (0 = LED ON, 1= LED OFF)
LED_D3	56	Output	LED D6 (0 = LED ON, 1= LED OFF)
PBSWITCH_6	57	Input	Pushbutton SW6 (non-debounced, 0 = hit)
DIPSWITCH_1	58	Input	DIP Switch SW3 #1 (ON = 1, OFF = 0)
DIPSWITCH_2	59	Input	DIP Switch SW3 #2 (ON = 1, OFF = 0)
DIPSWITCH_3	60	Input	DIP Switch SW3 #3 (ON = 1, OFF = 0)"
DIPSWITCH_4	61	Input	DIP Switch SW3 #4 (ON = 1, OFF = 0)
PBSWITCH_7	62	Input	Pushbutton SW7 (non-debounced, 0 = hit)
LCD_RW	73	Output	LCD R/W control line
MEM_DQ[0]	94	Bidirectional	Memory/LCD Data Bus
MEM_DQ[1]	96(133)	Bidirectional	Memory/LCD Data Bus
MEM_DQ[2]	98	Bidirectional	Memory/LCD Data Bus
MEM_DQ[3]	100	Bidirectional	Memory/LCD Data Bus
MEM_DQ[4]	102(128)	Bidirectional	Memory/LCD Data Bus
MEM_DQ[5]	104	Bidirectional	Memory/LCD Data Bus
MEM_DQ[6]	106	Bidirectional	Memory/LCD Data Bus
LCD_RS	108	Output	LCD Register Select Line
MEM_DQ[7]	113	Bidirectional	Memory/LCD Data Bus
VGA_GREEN	122	Output	VGA Connector Green Video Signal
CPU_CLOCK	153	Input	CPU Clock 100 or 66MHz - jumper
VGA_BLUE	170	Output	VGA Connector Blue Video Signal
VGA_VSYNC	226	Output	VGA Connector Vert Sync Signal
VGA_HSYNC	227	Output	VGA Connector Horiz Sync Signal
VGA_RED	228	Output	VGA Connector Red Video Signal

2.4 Obtaining a UP 3 Board and Cables

UP 3 boards are available for purchase from Altera's University Program at special educational pricing for schools and students (www.altera.com). UP 3 education kits come with an AC to 6V DC power supply, a special serial cable, and a ByteBlasterII cable.

A Longer Cable for the ByteBlaster

For use with the UP 3 or UP 2 board, a longer 25pin to 25pin PC M/F parallel printer cable is useful since the 1 foot Byteblaster II cable provided with the boards is often too short to reach the PC's printer port. All 25 wires must be connected in the printer extension cable. Any computer store should have these cables. A three-foot well-shielded cable works best. Avoid using extra long cables or very low-cost cables without good shielding as they can cause problems.

ADDITIONAL UP 2 RELATED MATERIALS CAN BE FOUND IN THE CD-ROM CHAPTER DIRECTORIES IN THE \UP2 SUBDIRECTORY.

CHAPTER 3

Programmable Logic Technology

Photo: An Altera Flex 10K100 FPGA containing 10,000,000 Transistors and 100,000 gates. The FPGA is in a pin grid array (PGA) package. The cover has been removed so that the chip die is visible in the center of the package.

3 Programmable Logic Technology

A wide spectrum of devices is available for the implementation of digital logic designs as shown in Figure 3.1. Traditional off-the-shelf integrated circuit chips, such as SSI and MSI TTL, perform a fixed operation defined by the device manufacturer. A user must connect different chip types to build a circuit. Application specific integrated circuits (ASICs), complex programmable logic devices (CPLDs), and field programmable gate arrays (FPGAs) are integrated circuits whose internal functional operation is defined by the user. ASICs require a final customized manufacturing step for the user-defined function. A CPLD or FPGA requires user programming to perform the desired operation.

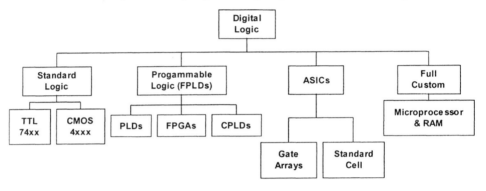

Figure 3.1 Digital logic technologies.

The design tradeoffs of the different technologies are seen in Figure 3.2. Full custom VLSI development of a design at the transistor level can require several years of engineering effort for design and testing. Such an expensive development effort is warranted only for the highest volume devices. This approach can generate the highest performance devices. Examples of full custom devices include the microprocessor and RAM chips used in PCs.

ASICs can be divided into three categories, Gate Arrays, Standard Cell and Structured. Gate Arrays are built from arrays of pre-manufactured logic cells. A single logic cell can implement a few gates or a flip-flop. A final manufacturing step is required to interconnect the sea of logic cells on a gate array. This interconnection pattern is created by the user to implement a particular design. Standard Cell devices contain no fixed internal structure. For standard cell devices, the manufacturer creates a custom photographic mask to build the chip based on the user's selection of devices, such as controllers, ALUs, RAM, ROM, and microprocessors from the manufacturer's standard cell library. New Structured ASICs are similar to gate arrays but each array element contains more logic. They offer tradeoffs somewhere between other ASICs and FPGAs.

Since ASICs require custom manufacturing, additional time and development costs are involved. Several months are normally required and substantial setup fees are charged. ASIC setup fees can be as high as a few million dollars. Additional effort in testing must be performed by the user since chips are tested after the final custom-manufacturing step. Any design error in the chip will lead to additional manufacturing delays and costs. For products with long lifetimes and large volumes, this approach has a lower cost per unit than CPLDs or FPGAs. Economic and performance tradeoffs between ASICs, CPLDs, and FPGAs are changing with each new generation of devices and design tools.

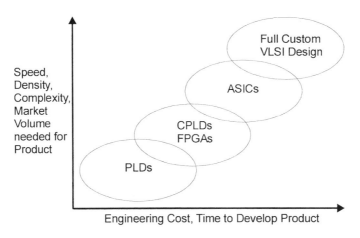

Figure 3.2 Digital logic technology tradeoffs.

Simple programmable logic devices (PLDs), such as programmable array logic (PALs), and programmable logic arrays (PLAs), have been in use for over thirty years. An example of a small PLA is shown in Figure 3.3. First, the logic equation is minimized and placed in sum of products (SOP) form. The PLA has four inputs, A, B, C, and D shown in the upper left corner of Figure 3.3. Every input connects to an inverter, making the inverted values of A, B, C, and D available for use. Each product term is implemented using an AND gate with several inputs. Outputs from the two product term's AND gates then feed into an OR gate.

A special shorthand notation is used in PLAs and PALs to represent the large number of inputs present in the AND and OR gate arrays. A gate input is present at each point where the vertical and horizontal signal lines cross in Figure 3.3. Note that this means that the two AND gates actually have eight inputs and the OR gate has two inputs in the PLA. Every input signal and its complement is available as an input to the AND gates. Each gate input in the PLA is controlled by a fuse. Initially all fuses are intact. By blowing selected fuses, or programming the PLA, the desired SOP equation is produced. The top AND gate in Figure 3.3 has fuses intact to the A and B inputs, so it produces the AB product term. The lower AND gate has fuses set to produce $C\overline{D}$. The OR gate has both fuses intact, so it ORs both product terms from the AND gates to produce the final output, $F = AB + C\overline{D}$.

Small PLDs can replace several older fixed function TTL-style parts in a design. Most PLDs contain a PLA-like structure in which a series of AND gates with selectable or programmable inputs, feed into an OR gate. In PALs, the OR gate has a fixed number of inputs and is not programmable. The AND gates and OR gate are programmed to directly implement a sum-of-products Boolean equation. On many PLDs, the output of the OR gate is connected to a flip-flop whose output can then be feed back as an input into the AND gate array. This provides PLDs with the capability to implement simple state machines. A PLD can contain several of these AND/OR networks.

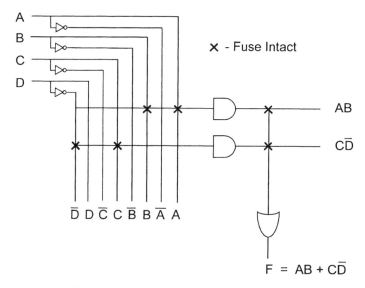

Figure 3.3 Using a PLA to implement a Sum of Products equation.

In more recent times, higher densities, higher speed, and cost advantages have enabled the use of programmable logic devices in a wider variety of designs. CPLDs and FPGAs are the highest density and most advanced programmable logic devices. Designs using a CPLD or FPGA typically require several weeks of engineering effort instead of months. These devices are also sometimes collectively called field programmable logic devices (FPLDs).

ASICs and full custom designs provide faster clock times than CPLDs or FPGAs since they are hardwired and do not have programmable interconnect delays. Since ASICs and full custom designs do not require programmable interconnect circuitry they use less chip area, less power, and have a lower per unit manufacturing cost in large volumes. Initial engineering and setup costs for ASICs and full custom designs are much higher.

For all but the most time critical design applications, CPLDs and FPGAs have adequate speed with maximum clock rates typically in the range of 50-400MHz; however, clock rates up to 1GHz have been achieved on new generation FPGAs and many have a few high-speed 1-10 GHz output pins.

3.1 CPLDs and FPGAs

Internally, CPLDs and FPGAs typically contain multiple copies of a basic programmable logic element (LE) or cell. The logic element can implement a network of several logic gates that then feed into 1 or 2 flip-flops. Logic elements are arranged in a column or matrix on the chip. To perform more complex operations, logic elements can be automatically connected to other logic elements on the chip using a programmable interconnection network. The interconnection network is also contained in the CPLD or FPGA. The interconnection network used to connect the logic elements contains row and/or column chip-wide interconnects. In addition, the interconnection network often contains shorter and faster programmable interconnects limited only to neighboring logic elements.

When a design approaches the device size limits, it is possible to run out of either gate, interconnect, or pin resources when using a CPLD or FPGA. CPLDs tend to have faster and more predictable timing properties while FPGAs offer the highest gate densities and more features.

Clock signals in large FPGAs normally use special low-skew global clock buffer lines. These are dedicated pins connected to an internal high-speed bus. This special bus is used to distribute the clock signal to all flip-flops in the device at the same time to minimize clock skew. If the global clock buffer line is not used, the clock is routed through the chip just like a normal signal. The clock signal could arrive at flip-flops at widely different times since interconnect delays will vary in different parts of the chip. This delay time can violate flip-flop setup and hold times and can cause metastability or unpredictable operation in flip-flops. Most large designs with a common clock that is used throughout the FPGA will require the use of the global clock buffer.

The size of CPLDs and FPGAs is typically described in terms of useable or equivalent gates. This refers to the maximum number of two input NAND gates available in the device. This should be viewed as a rough estimate of size only.

Figure 3.4 Examples of FPGAs and advanced high pin count package types.

The internal architecture of three examples of CPLD and FPGA device technologies, the Altera MAX 7000, the Altera Cyclone, and the Xilinx 4000 family will now be examined. An example of each of these devices is shown in Figure 3.4. From left to right the chips are an Altera MAX 7128S CPLD in a Plastic J-Lead Chip Carrier (PLCC), an Altera Cyclone 10K70 FPGA in a Plastic Quad Flat Pack (PQFP), and a Xilinx XC4052 FPGA in a ceramic Pin Grid Array Package (PGA). The PGA package has pins on .1" centers while the PQFP has pins on .05" centers at the edges of the package. Both Altera and Xilinx devices are available in a variety of packages.

Packaging can represent a significant portion of the FPGA chip cost. The number of I/O pins on the FPGA package often limits designs. Larger ceramic packages such as a PGA with more pins are more expensive than plastic.

3.2 Altera MAX 7000S Architecture – A Product Term CPLD Device

The multiple array matrix (MAX) 7000S is a CPLD device family with 600 to 20,000 gates. This device is configured by programming an internal electrically erasable programmable read only memory (EEPROM). Since an EEPROM is used for programming, the configuration is retained when power is removed. This device also allows in-circuit reprogrammability.

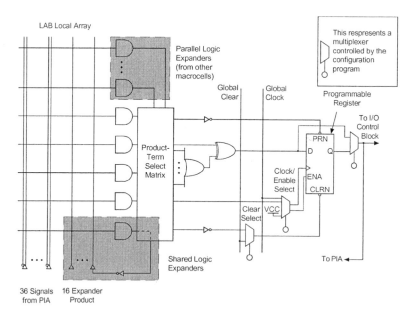

Figure 3.5 MAX 7000 macrocell.

The 7000 device family contains from 32 to 256 macrocells of the type seen in Figure 3.5. Similar to the early PALs, an individual macrocell contains five programmable AND gates with wide inputs that feed into an OR gate with a programmable inversion at the output. Just like a PAL, the AND/OR network is designed to implement Boolean equations expressed in sum-of-products form. Inputs to the wide AND gate are available in both normal and inverted forms. Parallel expanders are included that are used to borrow extra product terms from adjacent macrocells for logic functions needing more than five product terms.

The output from the AND/OR network can then be fed into a programmable flip-flop. Inputs to the AND gates include product terms from other macrocells in the same local block or signals from the chip-wide programmable interconnect array (PIA). The flip-flop contains Bypass, Enable, Clear and Preset functions and can be programmed to act as a D flip-flop, Toggle flip-flop, JK flip-flop, or SR latch.

Macrocells are combined into groups of 16 and called logic array blocks (LABs), for the overall device architecture as shown in Figure 3.6. The PIA can be used to route data to or from other LABs or external pins on the device. Each I/O pin contains a programmable tri-state output buffer. An FPGA's I/O pin can thus be programmed as input, output, output with a tri-state driver, or even tri-state bi-directional.

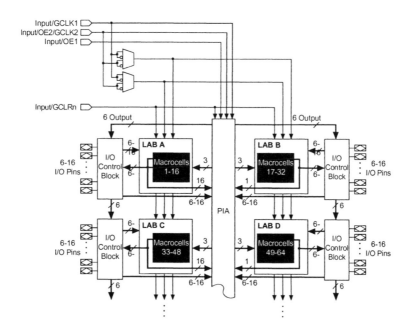

Figure 3.6 MAX 7000 CPLD architecture.

3.3 Altera Cyclone Architecture – A Look-Up Table FPGA Device

The Cyclone device is configured by loading internal static random access memory (SRAM). Since SRAM is used in FPGAs, the configuration will be lost whenever power is removed. In actual systems, a small external low-cost serial flash memory or programmable read only memory (PROM) is normally used to automatically load the FPGA's programming information when the device powers up.

FPGAs contain a two-dimensional row and column-based architecture to implement user logic. A column and row interconnection network provides signal connections between Logic Array Blocks (LABs) and embedded memory blocks. Interconnect delay times are on the same order of magnitude as logic delays.

The Cyclone FPGA's logic array consists of LABs, with 10 Logic Elements (LEs) in each LAB. An LE is a small unit of logic providing efficient implementation of user logic functions. LABs are grouped into rows and columns across the device. Cyclone devices range from 2,910 to 20,060 LEs.

M4K RAM embedded memory blocks are dual-port memory blocks with 4K bits of memory plus parity (4,608 bits). These blocks provide dual-port or single-port memory from 1 to 36-bits wide at up to 200 MHz. These blocks are grouped into columns across the device in between certain LABs. The Cyclone EP1C6 and EP1C12 contain 92K and 239K bits of embedded RAM.

Each of the Cyclone device's I/O pins is fed by an I/O element (IOE) located at the ends of LAB rows and columns around the periphery of the device. I/O pins support various single-ended and differential I/O standards. Each IOE contains a bidirectional I/O buffer and three registers for registering input, output, and output-enable signals.

Cyclone devices also provide a global low-skew clock network and up to two Phase Locked Loops (PLLs). The global clock network consists of eight global clock lines that drive throughout the entire device. The global clock network can provide clocks for all resources within the device, such as IOEs, LEs, and memory blocks. Cyclone PLLs provide general-purpose clocking with clock multiplication/division and phase shifting as well as external outputs for high-speed differential I/O support.

Figure 3.7 shows a Cyclone logic element. Logic gates are implemented using a look-up table (LUT). The LUT is a high-speed 16 by 1 SRAM. Four inputs are used to address the LUT's memory. The truth table for the desired gate network is loaded into the LUT's SRAM during programming. A single LUT can therefore model any network of gates with four inputs and one output. The multiplexers seen in Figure 3.7 are all controlled by bits in the FPGA's SRAM configuration memory.

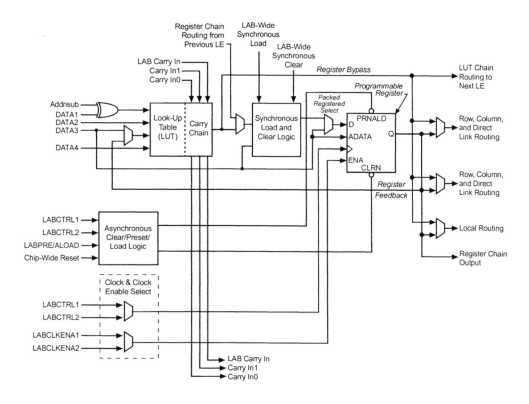

Figure 3.7 Cyclone Logic Element (LE).

An example showing how a LUT can model a gate network is shown in Figure 3.8. First, the gate network is converted into a truth table. Since there are four inputs and one output, a truth table with 16 rows and one output is needed. The truth table is then loaded into the LUT's 16 by 1 high-speed SRAM when the FPGA is programmed.

Note that the four gate inputs, A, B, C, and D, are used as address lines for the RAM and that F, the output of the truth table, is the data that is stored in the LUT's RAM. In this manner, the LUT's RAM implements the gate network by performing a RAM based table lookup instead of using actual logic gates.

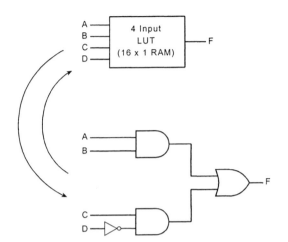

RAM Contents				
Address				Data
A	B	C	D	F
0	0	0	0	0
0	0	0	1	0
0	0	1	0	1
0	0	1	1	0
0	1	0	0	0
0	1	0	1	0
0	1	1	0	1
0	1	1	1	0
1	0	0	0	0
1	0	0	1	0
1	0	1	0	1
1	0	1	1	0
1	1	0	0	1
1	1	0	1	1
1	1	1	0	1
1	1	1	1	1

Figure 3.8 Using a look-up table (LUT) to model a gate network.

More complex gate networks require interconnections with additional neighboring logic elements. The output of the LUT can be fed into a D flip-flop and then to the interconnection network. The clock, Clear, and Preset can be driven by internal logic or an external I/O pin. The flip-flop can be programmed to act as a D flip-flop, T flip-flop, JK flip-flop, or SR latch. Carry and Cascade chains connect to all LEs in the same row.

Figure 3.9 shows a logic array block (LAB). A logic array block is composed of ten logic elements (LEs). Both programmable local LAB and chip-wide row and column interconnects are available. Carry chains are also provided to support faster addition operations.

Input-output elements (IOEs) are located at each of the device's I/O pins. IOEs contain a programmable tri-state driver and an optional 1-bit flip-flop register. Each I/O pin can be programmed as input, output, output with a tri-state driver, or even tri-state bi-directional with or without a register. Four clock I/O pins connect to the eight low-skew global clock buffer lines that are provided in the device.

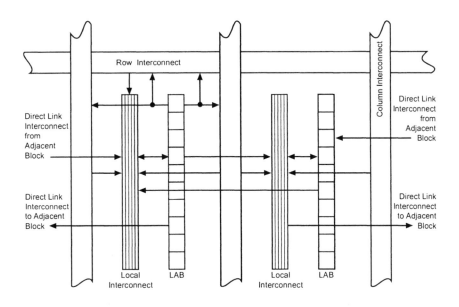

Figure 3.9 Cyclone Logic Array Blocks (LAB) and Interconnects.

3.4 Xilinx 4000 Architecture – A Look-Up Table FPGA Device

The Xilinx 4000 Family was a popular first generation FPGA device family with 2,000 to 180,000 usable gates. It is configured by programming internal SRAM. Figure 3.10 is a photograph of a six-inch silicon wafer containing several XC4010E 10,000 gate FPGA chip dice. Figure 3.11 is a contrast-enhanced view of a single XC4010E die. If you look closely, you can see the 20 by 20 array of logic elements and the surrounding interconnect lines. Die that pass wafer-level inspection and testing are sliced from the wafer and packaged in a chip. FPGA yields are typically 90% or higher after the first few production runs.

As seen in Figure 3.12, this device contains a more complex logic element called a configurable logic block (CLB). Each CLB contains three SRAM-based lookup tables. Outputs from the LUTs can be fed into two flip-flops and routed to other CLBs. A CLB's lookup tables can also be configured to act as a 16 by 2 RAM or a dual-port 16 by 1 RAM. High-speed carry logic is provided between adjacent CLBs.

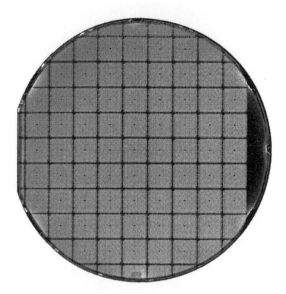

Figure 3.10 Silicon wafer containing XC4010E 10,000 gate FPGAs.

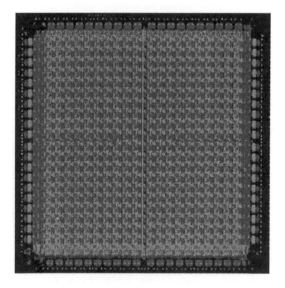

Figure 3.11 Single XC4010E FPGA die showing 20 by 20 array of logic elements and interconnect.

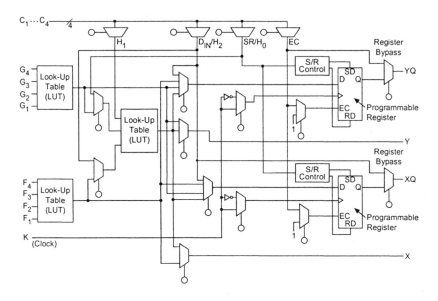

Figure 3.12 Xilinx 4000 Family Configurable Logic Block (CLB).

CLBs are arranged in a square matrix with a programmable hierarchical interconnection network. Devices in the family contain from 100 to 3,136 CLBs. The multiplexers seen in Figure 3.12 are all controlled by bits in the FPGA's SRAM configuration memory.

The complex hierarchical interconnection network contains varying length row, column, and neighboring CLB interconnect structures. Eight low-skew global clock buffers are also provided. Input-output blocks (IOBs), contain programmable tri-state drivers and optional registers. Each I/O pin can be programmed as input, output, output with a tri-state driver, or tri-state bi-directional with or without a register. In the more recent Xilinx Virtex 4 FPGAs, each CLB now contains four circuits similar to the earlier 4000 CLBs.

3.5 Computer Aided Design Tools for Programmable Logic

Increasing design complexity and higher gate densities are forcing digital designs to undergo a paradigm shift. Old technology, low-density logic families, such as the TTL 7400 or simple PLD families are rarely if ever used in new designs. With logic capacities of an individual FPGA chip approaching 10,000,000 gates, manual design at the gate level is no longer a viable option in complex systems. Rapid prototyping using hardware description languages (HDLs), IP cores, and logic synthesis tools has all but replaced traditional gate-level design with schematic capture entry. These new HDL-based logic synthesis tools can be used for both ASIC and FPGA-based designs. The two most widely used HDLs at the present time are VHDL and Verilog.

The typical FPGA CAD tool design flow is shown in Figure 3.13. After design entry using an HDL or schematic, the design is automatically translated, optimized, synthesized, and saved as a netlist. (A netlist is a text-based representation of a logic diagram.) A functional simulation step is often added prior to the synthesis step to speed up simulations of large designs.

An automatic tool then fits the design onto the device by converting the design to use the FPGA's logic elements, first by placing the design in specific logic element locations in the FPGA and then by selecting the interconnection network routing paths. The place and route process can be quite involved and can take several minutes to compute on large designs. On large devices, combinatorial explosion (exponential growth) will prevent the tool from examining all possible place and route combinations. When designs require critical timing, some tools support timing constraints that can be placed on critical signal lines. These optional constraints are added to aid the place and route tool in finding a design placement with improved performance.

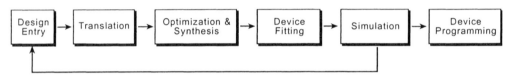

Figure 3.13 CAD tool design flow for FPGAs.

After place and route, simulation can be performed using actual gate and interconnect time delays from a detailed timing model of the device. Although errors can occur at any step, the most common path is to find errors during an exhaustive simulation. The final step is device programming and hardware verification on the FPGA.

3.6 Next Generation FPGA CAD tools

A few HDL synthesis tools now support behavioral synthesis. Unlike the more widely used register transfer level (RTL) models contained in this book, behavioral synthesis models do not specify the exact states and sequence of register transfers. A separate constraint file specifies the number of clocks needed to obtain selected signals and the tool automatically generates the state machines, logic, and register transfers needed.

Although not currently in widespread use for current designs, newer FPGA CAD tools are also appearing based on other languages such as C and Java. Some of these system-level tools output VHDL or Verilog models as an intermediate step. New HDLs such as SystemVerilog (www.systemverilog.org) and SystemC (www.systemC.org) provide enhanced support for verification.

Tools that automatically generate an FPGA design from other engineering tools such as MATLAB-Simulink or LabVIEW have also been introduced. These

graphical based tools are primarily aimed at DSP application development for
FPGAs using a library of specialized DSP blocks.

3.7 Applications of FPGAs

The last decade has seen ever increasing application areas for FPGAs. A recent
market study found over twelve times as many new FPGA-based designs as
ASIC-based designs and ASIC costs continue to increase. New generation
FPGAs can have nearly ten million gates with clock rates approaching 1GHz.
Example application areas include single chip replacements for old multichip
technology designs, Digital Signal Processing (DSP), image processing,
multimedia applications, high-speed communications and networking
equipment such as routers and switches, the implementation of bus protocols
such as peripheral component interconnect (PCI), microprocessor glue logic,
co-processors, and microperipheral controllers.

Several large FPGAs with an interconnection network are used to build
hardware emulators. Hardware emulators are specially designed commercial
devices used to prototype and test complex hardware designs that will later be
implemented on gate arrays or custom devices. Hardware emulators are
commonly used to build a prototype quickly during the development and
testing of microprocessors. Several of the recent Intel and AMD processors
used in PCs were tested on FPGA-based hardware emulators before the full
custom VLSI processor chip was produced.

A newer application area is reconfigurable computing. In reconfigurable
computing, FPGAs are quickly reprogrammed or reconfigured multiple times
during normal operation to enable them to perform different computations at
different times for a particular application.

3.8 Features of New Generation FPGAs

Each new generation of FPGAs increases in size and performance. In addition
to more logic elements, embedded memory blocks, and interconnect other new
features are appearing. Some FPGAs contain a mix of both product term and
lookup tables to implement logic. Such product term structures typically
require less chip area to implement the complex gating logic present in large
state machines and address decoders. Many FPGAs include several phase-
locked loops (PLLs). These PLLs are used to multiply, divide, and adjust high-
speed clock signals. Similar to microprocessors used in PCs, many new FPGAs
use a lower 1.5 to 3 Volt internal core power supply. To easily interface to
external processor and memory chips, new FPGAs feature selectable I/O
standards on I/O pins.

High-speed hardware multipliers and multiply accumulators (MACs) are also
available in FPGA families targeted for multiply intensive DSP and graphics
applications. Several FPGAs from Altera and Xilinx are available with
commercial internal RISC microprocessor intellectual property (IP) cores.
These include the Nios, ARM, Microblaze, and PowerPC. The Nios and
Microblaze processors are an HDL model that is synthesized using the FPGA's
standard logic elements. The ARM, and PowerPC are commercial IP cores with

custom VLSI layouts. These new devices are a hybrid that contains both ASIC and FPGA features. Several processors can be implemented in a single FPGA.

These FPGAs come with additional software tools for the processor, including C/C++ compilers. Some processor cores are available with a small operating system kernel. These new large FPGAs with a microprocessor IP core are targeted for system on-a-chip (SOC) applications. When an FPGA is used for SOC applications it is also called system on-a-programmable chip (SOPC).

On many of the largest FPGAs, redundant rows of logic elements are included to increase yields. As any VLSI device gets larger the probability of a manufacturing defect increases. If a defective logic element is found during initial testing, the entire row is mapped out and replaced with a spare row of logic elements. This operation is transparent to the user.

3.9 For additional information

This short overview of programmable logic technology has provided a brief introduction to FPGA architectures. Altera and Xilinx have the largest market share of current FPGA vendors. Additional CPLD and FPGA manufacturers include Lattice, Actel, Quicklogic, and Cypress. Actel, Quicklogic, and Cypress have one-time programmable FPGA devices. These devices utilize antifuse programming technology. Antifuses are open circuits that short circuit or have low impedance only after programming. Trade publications such as *Electronic Design News* periodically have a comparison of the available devices and manufacturers.

Altera MAX 7000, Cyclone, and the new Cyclone II and Stratix II family data manuals with a more in-depth explanation of device hardware details are available free online at Altera's website, http://www.altera.com.

For other examples of FPGA architectures, details on the newer Xilinx Spartan and Virtex families can be found at http://www.xilinx.com.

An introduction to the mathematics and algorithms used internally by digital logic CAD tools can be found in *Synthesis and Optimization of Digital Circuits* by Giovanni De Micheli, McGraw-Hill, 1994 and *Logic Synthesis and Verification Algorithms* by Hactel and Somenzi, Springer Publishers, 1996. *The Design Warrior's Guide to FPGAs* by Clive Maxfield, Elsevier, 2004 contains an overview of commercial FPGA devices and commercial EDA tool flows for FPGA design.

3.10 Laboratory Exercises

1. Show how the logic equation (A AND NOT(B)) OR (C AND NOT(D)) can be implemented using the following:

    A. The PLA in Figure 3.3
    B. The LUT in Figure 3.9

Be sure to include the PLA fuse pattern and contents of the LUT.

2. Examine the compiler report file and use the chip editor to explain how the OR-gate design in the tutorial in Chapter 1 was mapped into the Cyclone device.

3. Retarget the design from Chapter 1 to a MAX 7000S device. Examine the compiler report file and use the chip editor to explain how the OR-gate design in the tutorial in Chapter 1 was mapped into the MAX device.

4. Show how the logic equation (A AND NOT(B)) OR (C AND NOT(D)) can be implemented in the following:

> A. A MAX Logic Element
> B. A Cyclone Logic Element
> C. An XC4000 CLB

Be sure to include the contents of any LUTs required and describe the required mux settings.

5. Using data sheets available on the web, compare and contrast the features of newer generation FPGAs such as Altera's Cyclone II and Stratix II and Xilinx's Virtex II and Virtex 4 families.

CHAPTER 4

Tutorial II: Sequential Design and Hierarchy

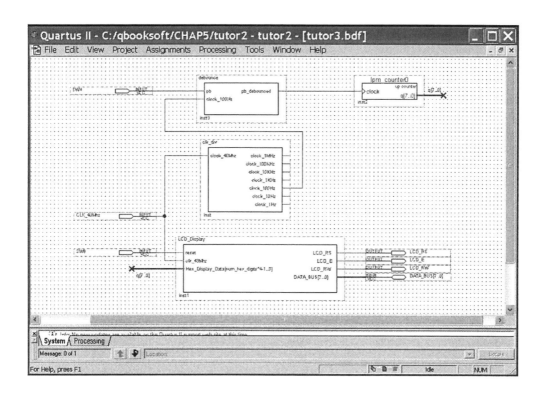

4 Tutorial II: Sequential Design and Hierarchy

The second tutorial contains a more complex design containing sequential logic and hierarchy with a counter and an LCD display. To save time, much of the design has already been entered. The existing design will require some modifications.

Once completed, you will:

- Understand the fundamentals of hierarchical design tools,
- Complete an example of a sequential logic design,
- Use the UP3core library designed for the UP 3 board,
- Use the LCD display, pushbuttons, and the onboard clock,
- Use buses in a schematic, and
- Be able to perform automatic timing analysis of sequential circuits.

4.1 Install the Tutorial Files and UP3core Library

Locate the **booksoft\chap4** directory on the CD-ROM that came with the book. For the UP 3 board, copy all of the Chapter 4 tutorial files in this directory to your ***drive:\mydesigns*** directory or another subdirectory. A special version of the files for the larger 1C12 UP 3 board is in the subdirectory **\1C12**. If you are using the UP 2, a version of the files for the UP 2 board is in the subdirectory **\UP2** along with special instructions for UP 2 users.

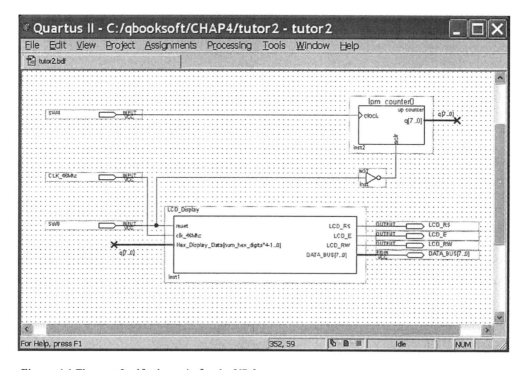

Figure 4.1 The tutor2.gdf schematic for the UP 3.

4.2 Open the tutor2 Schematic

After setting up the files in your directory, select **File** ⇨ **Open Project** ⇨ *drive*:*mydesigns***tutor2.qpf.** Open the top-level schematic by selecting **File** ⇨ **Open** ⇨ *drive*:*mydesigns***tutor2.bdf** and the schematic in Figure 4.1 should be displayed. This design has been partially entered to save time. This is an 8-bit counter design that outputs the counter value to the UP 3's LCD display panel. On the UP 2 version of the tutorial, you will see the counter value in the seven segment LED displays.

Click on the lpm_counter0 symbol to activate the MegaWizard Plug-In Manager. The MegaWizard can be used to create and edit megafunctions. In this case, you can see that lpm_counter0 is an 8-bit binary counter that counts up. You can click on the documentation button and then generate sample waveforms to view more details about the counter's operation. You can create new functions with the MegaWizard using **Tools** ⇨ **MegaWizard Plug-In Manager.** Close the MegaWizard windows to continue.

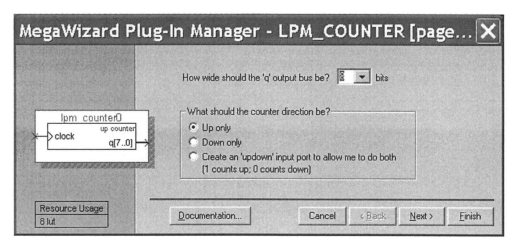

Figure 4.2 Lpm_counter0 MegaWizard edit window.

Special hardware blocks have been designed to support the easy use of the advanced I/O features found on the UP 3 board. They include pushbuttons, LCD displays, keyboard, mouse, and video output. More details on all of these UP3core functions are provided in Chapter 5. The UP3core function needed for this tutorial have already been placed in this project's directory. Several symbols from the UP3core library appear in the project library and are available to be entered in a design. As an alternative to copying these files to each project's directory, under **Project** ⇨ **Add/Remove Files in Project,** you can click on **User Libraries** and enter another path to an external library such as the UP3core library.

4.3 Browse the Hierarchy

In engineering, the principle of functional decomposition is normally used in large designs. Complex designs are typically broken into smaller design units.

The smaller design units are then more easily understood and implemented. The smaller designs are interconnected to form the complex system. The overall design is a hierarchy of interconnected smaller design units. This also promotes the re-use of portions of the design.

The current schematic is a view of the top level of the design. In this design, the problem was decomposed into a design unit or symbol with logic for a counter and another design unit to display the count. Each symbol also has an internal design that can be any combination of another schematic, megafunction, VHDL, or Verilog file.

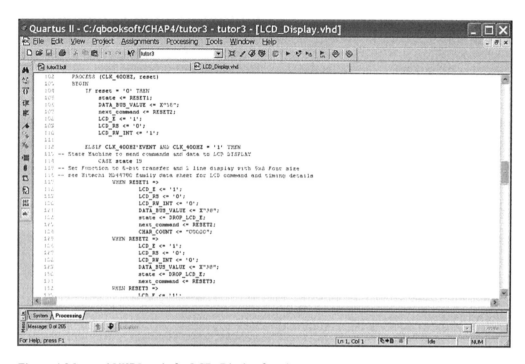

Figure 4.3 Internal VHDL code for LCD_Display function.

On scehamtic for the UP 3, double click on the **LCD_Display** symbol to see the underlying VHDL code that describes the internal operation of the LCD_Display block. UP 2 users should click on the **DEC_7SEG** symbol. As shown in Figure 4.3, it contains a complex state machine that sends commands and ASCII character data to the LCD controller. As an alternative, it could be designed in Verilog or even at the gate level using basic logic symbols (if you had infinite time and patience to work at that low of a level!). Close the VHDL text editor and return to the graphic editor.

To see the overall hierarchy of the design, select **View** ⇨ **Utility Windows Project Navigator** and make sure the **Hierarchy** tab is selected. After expanding this window as seen in Figure 4.4, note that the tutor2 schematic is comprised of two symbols. For the UP 3, the LCD_Display symbol is used in the design to output the count to the LCD display. The lpm_counter0 symbol contains the 8-bit binary counter. If you click on the "+" block on each symbol, you will see that they each contain lpm_counter megafunctions. In this case, the design hierarchy is three levels deep. After examining the hierarchy display window, close it and return to the graphic editor window that contains the tutor2 schematic.

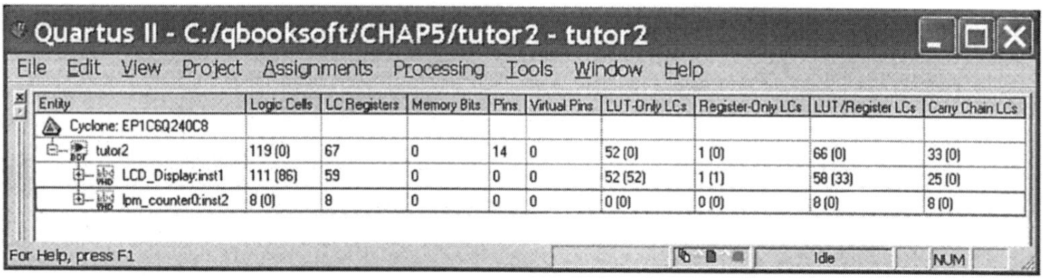

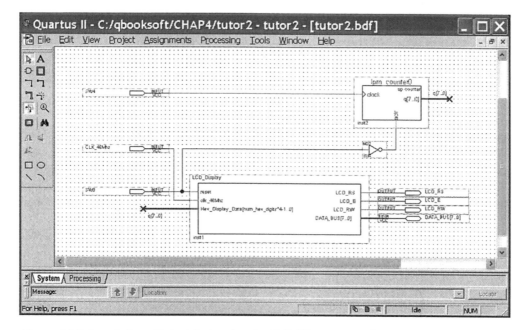

Figure 4.4 Hierarchy display window for the tutor2 design.

4.4 Using Buses in a Schematic

In Figure 4.5, find the heavy purple lines flowing out of the lpm_counter0 symbol in the upper right corner and into the LCD_Display symbol in the lower left corner. This is an example of a bus. A bus is just a parallel collection of bits. The bus is labeled Q[7..0] indicating the bus has eight signals (bits) named Q[7], Q[6] ... Q[0]. This bus sends the counter's eight output bits to the LCD display function.

Figure 4.5 Enlarged view of tutor2 design showing Q[7..0] bus connnections.

To connect single node lines to a bus, it is necessary to assign a name such as Q[7..0] to the bus. Then the node line that needs to connect to a bus line is given the name of one of the bus elements. As an example, the counter output MSB signal line is labeled Q[7]. To label a bus or node, right click on the node or bus line and select **Properties**. You can then type in or edit the name. When signal lines have the same name, they are automatically connected in the graphic editor. A physical node line connecting a node and a bus with the same name is optional. Leaving it out often times makes a complex schematic easier to follow since there will be fewer lines crossing on the schematic. Node and bus names must be assigned first when connecting a node to a bus.

4.5 Testing the Pushbutton Counter and Displays

Compile the design with **Processing** ⇨ **Start Compilation.** Wait a few seconds for the "Full Compilation was successful" message to appear. Select **Tools** ⇨ **Timing Analyzer Tool**. This counter circuit is a sequential design. The primary timing issue in sequential circuits is the maximum clock rate. Whenever you compile, a timing analysis tool automatically runs that will determine the maximum clock frequency of the logic circuit.

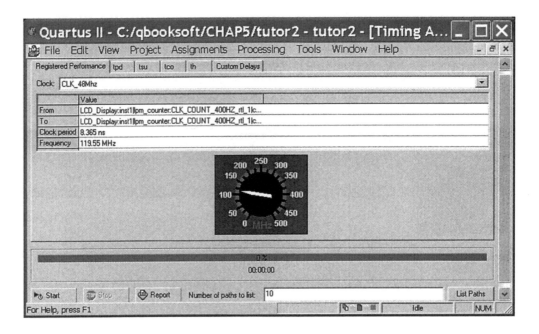

Figure 4.6 Timing analysis of a Sequential Circuit

The Timing Analyzer shows the maximum clock frequency of this logic circuit to be approximately 120 MHz. Clock rates you may obtain will vary depending on the complexity and size of the logic circuits, the speed grade of the chip, and the CAD tool version and settings. In this design, the clock is supplied by a pushbutton input so a clock input of only a few hertz will be used for the counter. Since the UP 3's clock input is only 48 MHz for the LCD_Display core this simple counter cannot be overclocked. Close and exit the timing analyzer.

4.6 Testing the Initial Design on the Board

Download the design to the FPGA board. If you need help downloading to the board, refer to Section 1.4 for the UP 3 or Section 1.5 for the UP 2 board.

Press the top UP 3 pushbutton (SW4) several times to clock the counter and watch the count display as it counts up. On the UP 2, use the left FLEX pushbutton. When the pushbutton is pressed, it will occasionally count up by more than one. This is a product of mechanical bounce in the pushbutton. The pushbutton contains a metal spring that actually forces contact several times before stabilizing. The high-speed logic circuits will react to the switch contact bounce as if several clock signals have occured. This makes the counter count up by more than one.

The actual output of the pushbutton as it appears on a digital oscilloscope is shown in Figure 4.7. When the pushbutton is hit, a random number of pulses appear as the switch contacts bounce several times and then finally stabilize. Several of the pulses will have a voltage and duration long enough to generate extra clock pulses to the counter. An FPGA will respond to pulses in the ns range, and these pulses are in the μs range.

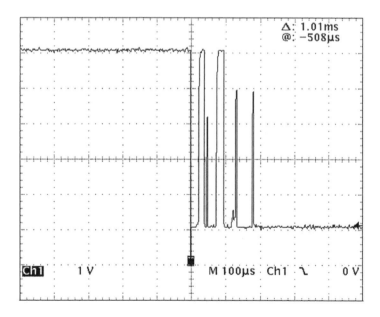

Figure 4.7 Oscillosope display of pushbutton switch contact bounce.

This problem occurs with all pushbuttons in digital designs. If the pushbutton is a DPDT, double-pole double-throw (i.e., has both an ON and an OFF contact), an SR latch is commonly used to remove the contact bounce. The pushbutton on the UP 3 is SPST, single pole single throw, so a time averaging filter is used. This example demonstrates why designs must be tested on actual hardware after simulation. This problem would not have shown up in a simulation.Verify that the UP 3 board's global reset pushbutton on the lower left corner of the board (SW8) resets the display and the counter. On the UP 2, the right FLEX pushbutton should reset the count.

4.7 Fixing the Switch Contact Bounce Problem

For the hardware implementation to work correctly, the switch contact bounce must be removed. A logic circuit that filters the pushbutton output using a small shift register can be added to filter the output. This process is called switch debouncing. Add the symbol **debounce** from the project library to the schematic.

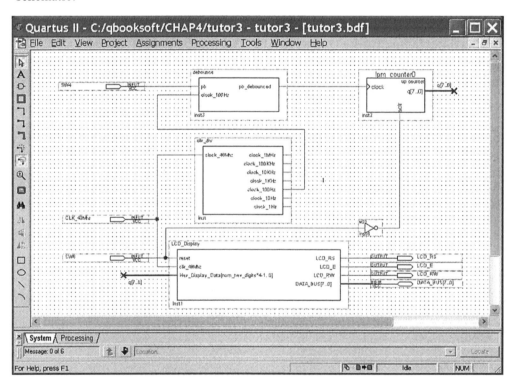

Figure 4.8 Modified tutor2 design schematic.

Disconnect the pushbutton from the lpm_counter0's clock pin and connect it to the pushbutton input pin, PB, on the debounce symbol. Now connect the PB_DEBOUNCED pin to the lpm_counter0's clock pin. The debounce circuit needs a 100Hz clock signal for the time averaging filter. The clock needed is much slower than the 48MHz system clock, so a clock prescalar is needed. A clock prescalar is a logic circuit that divides a clock signal.

Add the **clk_div** symbol from the project library to the schematic. Connect the 100Hz input pin on the debounce symbol to the 100Hz output pin on the clock prescalar. Connect the 48MHz clock input on the **clk_div** symbol to an the clk_48MHz input pin.

The internal VHDL design in the debounce module generates the switch debounce circuit. The debounce circuit contains a 4-bit shift register that is clocked at 100Hz. The shift register shifts in the inverted pushbutton output. When any of the four bits of the shift register (i.e., four 10 ms time-spaced samples of the pushbutton's output) are High the output of the debounce circuit changes to High. When all four bits of the shift register are Low the output goes Low. This delays the High to Low change until after the switch contact bounce stops.

4.8 Testing the Modified Design on the UP 3 Board

Verify that your schematic has the same connections as seen in Figure 4.8. Compile the design and download the design to the board again. Hit the count pushbutton several times to clock the counter and watch the LCD display as it counts up. It should now count up reliably by one whenever the pushbutton is hit. The UP3core functions LCD_Display, clk_div, and debounce will be useful in future design projects using the UP 3 board. They can be used in any VHDL, Verilog, or schematic designs by using the graphical editor and UP3core symbols or by using an HDL component instantiation statement.

ALL SOURCE CODE FOR THE BOOK'S DESIGNS IS AVAILABLE ON THE CD-ROM.

ADDITIONAL UP 2 RELATED MATERIALS CAN BE FOUND IN EACH CD-ROM

CHAPTER IN THE \UP2 SUBDIRECTORY.

4.9 Laboratory Exercises

1. Simulate the initial design without the switch debounce circuit by setting up an initial reset pulse and a periodic 200 ns clock input in the simulator. In sequential simulations, turn on the **setup and hold time violation detection** simulator setting option before

running the simulator. This will check for flip-flop timing problems that would otherwise go undetected in the simulation. Adjust the reset pulse so that it changes right before the clock edge and run another simulation to see if you can produce a setup or hold violation.

2. Modify the counter circuit so that it counts down or up depending on the state of UP 3's DIP switch 1.

3. Modify the counter circuit so that it parallel loads a count value from the four DIP switches on the UP 3 board when PB2 is pushed. Zero out the low four counter bits during a load. Since the Cyclone DIP switch inputs are only used when PB2 is hit, they do not need to be debounced. Here are the pin connections for the Cyclone DIP switches.

Input Pin	Pin
DIPswitch_1	58
DIPswitch_2	59
DIPswitch_3	60
DIPswitch_4	61
(1 = Open, 0 = Closed)	

4. Build a stopwatch with the following modifications to the design. Disconnect the counter clk line and connect it to the clock_10hz pin on the clock_div symbol. Clock a toggle flip-flop with the pb_debounced output. A toggle flip-flop, tff, can be found in the prim symbol library. A toggle flip-flop's output changes state every time it is clocked. Connect the output of the toggle flip-flop to a new count enable input added to the counter with the megawizard. The count should start and stop when PB1 is hit. Elapsed time in tenths of seconds should be displayed in hexadecimal. Pushing PB2 should reset the stopwatch.

5. The elapsed time in the stopwatch from problem 3 is displayed in hexadecimal. Replace the counter with two cascaded binary-coded-decimal (BCD) counters so that it displays the elapsed time as two decimal digits.

6. Build a watch by expanding the counter circuit to count seconds, hours, and minutes. The two pushbuttons reset and start the watch.

7. Replace the lpm_counter0 logic with a VHDL counter design, simulate the design, and verify operation on the UP 3 board. Read Chapter 5 and note the example counter design in section 6.10.

8. Draw a schematic, develop a simulation, and download a design to the UP 3 board that uses the LCD displays for outputs and the DIP switch for input, to test the 74161 4-bit TTL counter function found in the /others/maxplus2 symbol library. Use the Cyclone DIP switch to provide four inputs for a parallel load of the count. Use a debounced pushbutton input for the clock. Use the second pushbutton for the load input.

9. Draw a schematic, develop a simulation, and download a design to the UP 3 board to test the following functions that can be created with the MegaWizard:

LPM_ADD_SUB:	a 2-bit adder/subtractor; test the add operation
LPM_ADD_SUB:	a 2-bit adder/subtractor; test the subtract operation
LPM_COMPARE:	compare two 2-bit unsigned numbers
LPM_DECODE:	a 4 to 16-bit decoder
LPM_CLSHIFT:	a 4-bit shift register
LPM_MULT:	a 2-bit unsigned multiply

The LPM megafunctions require several parameters to specify bus size and other various options. For this problem, do not use pipelining and use the unregistered input options. Refer to the online help files for each LPM function for additional information. In the **enter symbol** window, use the megawizard button to help configure LPM symbols. Use the UP 3's DIP switches for four inputs as needed and display the output in hex on the two seven-segment displays. Use a debounced pushbutton input for the clock, if one is required. Use the second pushbutton for a Clear or Reset input. Use the timing analyzer to determine the worst-case delay time for each function.

10. Draw a schematic and develop a simulation to test the LPM_ROM megafunction. Create a sixteen word ROM with eight data bits per word. Specify initial values in hex for the ROM in a memory initialization file (*.mif) file. The contents of each memory location should be initialized to four times its address. See MIF in the online help for details on the syntax of a MIF file. Enter the address in four Cyclone DIP switches and display the data from the ROM in the two seven-segment LEDs. Determine the access time of the ROM.

11. Using gates and the DFF part from the primitives/storage library, design a circuit that implements the state machine shown below. Use two D flip-flops with an encoded state.

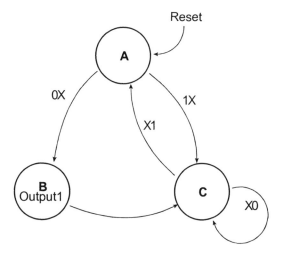

For the encoded states use A = "00", B = "01", and C = "10". Ensure that the undefined "11" state enters a known state. Enter the design using the graphical editor. Develop a simulation that tests the state machine for correct operation. The simulation should test all states and arcs in the state diagram and the "11" state. Use the **Tools** ⇨ **Timing Analyzer Tool** option to determine the maximum clock frequency on the Cyclone device. Use an asynchronous reset.

12. Repeat the previous problem using one-hot encoding. Recall that one-hot encoding uses one flip-flop per state, and only one flip-flop is ever active at any given time in valid states. The state encoding for the one-hot state machine would be A = "100", B = "010", and C = "001". Start with a reset in the simulation. It is not necessary to test illegal states in the one-hot simulation. One-hot state machine encoding is recommended by many FPGA device manufacturers.

CHAPTER 5

UP3core Library Functions

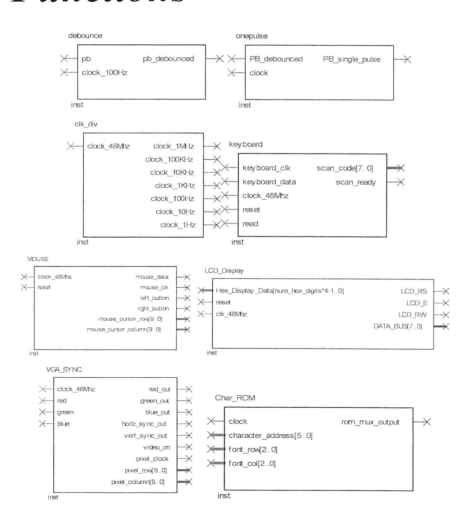

5 UP3core Library Functions

In complex hierarchical designs, intellectual property (IP) cores are frequently used. An IP core is a previously developed synthesizable hardware design that provides a widely used function. Commercially licensed IP cores include functions such as microprocessors, microcontrollers, bus interfaces, multimedia and DSP functions, and communications controllers. IP cores promote design reuse and reduce development time by providing common hardware functions for use in a new design.

The UP3core functions listed in Table 5.1 are designed to simplify use of the UP 3 board's pushbuttons, keyboard, mouse, LCD display, and video output features. They can be used in schematic capture, VHDL, or Verilog based designs. Full source code is provided on the CD-ROM.

Table 5.1 The UP3core Functions.

UP3core Name	Description
LCD_Display	Displays ASCII Characters and Hex Data on the UP3's LCD Panel
Debounce	Pushbutton Debounce Circuit
OnePulse	Pushbutton Single Pulse Circuit
Clk_Div	48MHz Clock Prescaler with 7 frequency outputs (1MHz to 1hz)
VGA_Sync	VGA Sync signal generator for UP 3 that outputs pixel addresses
Video_PLL	Used by VGA Sync to generate the video pixel clock using a PLL
Char_ROM	Small Character Font ROM for video character generation
Keyboard	Reads keyboard scan codes from the UP 3's PS/2 connector
Mouse	Reads PS/2 mouse data and outputs cursor row and column address

UP3cores can be used as symbols from the UP3core library, accessed via a VHDL package, or used as a component in other VHDL or Verilog files. An example of using the UP3core package in VHDL can be found in the file \booksoft\chap5\UP3pack.vhd available on the CD-ROM. The use of UP3pack's VHDL package saves retyping lengthy component declarations for the core functions in each VHDL-based design.

This section contains a one-page summary of each UP3core interface. VHDL source code is provided for all UP3cores on the CD-ROM. Additional documentation, examples, and interface details can be found in later chapters on video signal generation, the keyboard, and the mouse. The Clk_Div, LCD_Display, and Debounce functions were already used in the tutorial design example in Chapter 4.

For correct operation of the UP3core functions, I/O pin assignments must be made as shown in the description of each UP3core function. Clock inputs are also required on several of the UP3core functions. The Clk_Div UP3core is setup to provide the slower clock signals needed by some of the core functions.

Source code for the UP3core functions must be in the project directory or in the user's library search path. Review Section 4.2 for additional information on checking the library path. The VGA_Sync, Video_PLL, and LCD_Display core functions are often modified by the user to support different display resolutions and message options for each design. Be sure to select the right version of these core functions when adding file paths for a new project.

For UP 2 users, the same functionality is provided in the UP2core library functions found on the CD-ROM in the \UP2 subdirectory. Since the UP 2 does not have an LCD display, the seven segment display core, DEC_7SEG, must be used instead. On the UP 2, only the 640 by 480 video mode is possible since the FPGA does not have a PLL to generate the higher clock frequencies needed for other video display resolutions.

SPECIAL CORE FUNCTIONS FOR THE UP 1 & UP 2's LED AND VIDEO DISPLAY

CAN BE FOUND ON THE CD-ROM IN THE CHAP5 \UP2 SUBDIRECTORY.

\UP2\VGA_SYNC SHOULD BE USED FOR UP 2 VIDEO AND\UP2\ DEC_7SEG IS

USED TO DISPLAY HEX VALUES IN THE UP 2's SEVEN SEGMENT DISPLAYS.

5.1   UP3core LCD_Display: LCD Panel Character Display

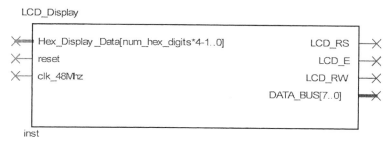

Figure 5.1 Symbol for LCD_Display UP3core.

The LCD_Display core is used to display static ASCII characters and changing hex values from hardware on the UP 3's 16 by 2 line LCD display panel. The core's VHDL code can be configured internally by the user to display different ASCII strings and hex data fields. Instructions can be found in comments in the core's VHDL code. A Generic, Num_Hex_Digits, is used to set the size of the Hex_Display_Data input (Each hex digit displayed requires a 4-bit signal). The LCD controller datasheet contains information on graphics characters and LCD commands. A state machine is used to send data and commands to the LCD controller and to generate the required handshake signals. An ASCII to hex table can be found in Appendix D. See LCD_Display.vhd for more information.

5.1.1   VHDL Component Declaration

```
COMPONENT LCD_Display
  PORT(Hex_Display_Data:
              IN    STD_LOGIC_VECTOR((Num_Hex_Digits*4)-1 DOWNTO 0);
       reset, clock_48MHz:   IN STD_LOGIC;
       LCD_RS, LCD_E:        OUT    STD_LOGIC;
       LCD_RW:               INOUT STD_LOGIC;
       DATA_BUS:             INOUT  STD_LOGIC_VECTOR(7 DOWNTO 0));
  END COMPONENT;
```

5.1.2   Inputs

Hex_Display_Data contains the 4-bit hexadecimal hardware signal values to convert to ASCII hex digits and send to the LED display. The Generic, Num_Hex_Digits, adjusts the size of the input hex data. Generics can be assigned a value in an HDL file or with a block's parameter assignment in a schematic. In a schematic, use **View ⇨ Parameter assignment** to see the generic value and the symbol's properties parameters tab to set it.

5.1.3   Outputs

Outputs control a tri-state bidirectional data bus on the LCD panel.

LCD_RS is pin 108, LCD_E is pin 50, and LCD_RW is pin 73.

Data_Bus(7 DOWNTO 0) is pins 113, 106, 104, 102(128-1C12), 100, 98, 96(133-1C12), and 94.

5.2 UP3core Debounce: Pushbutton Debounce

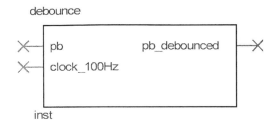

Figure 5.2 Symbol for Debounce UP3core.

The UP3core Debounce shown in Figure 5.2 is a pushbutton debounce circuit. This function is used to filter mechanical contact bounce in the UP 3's pushbuttons. A shift register is used to filter out the switch contact bounce. The shift register takes several time spaced samples of the pushbutton input and changes the output only after several sequential samples are the same value.

5.2.1 VHDL Component Declaration

```
COMPONENT  debounce
      PORT(  pb, clock_100Hz      : IN          STD_LOGIC;
              pb_debounced         : OUT         STD_LOGIC;
END COMPONENT;
```

5.2.2 Inputs

PB is the raw pushbutton input. It should be tied to an input pin connected to a UP 3 pushbutton. See Chapter 2 for pushbutton pin numbers.

Clock is a clock signal of approximately 100Hz that is used for the internal 50ms switch debounce filter circuits.

5.2.3 Outputs

PB_debounced is the debounced pushbutton output. The output will remain Low until the pushbutton is released. If a pulse is needed to be only 1 clock period long, add the OnePulse core function to the debounced switch output.

5.3 UP3core OnePulse: Pushbutton Single Pulse

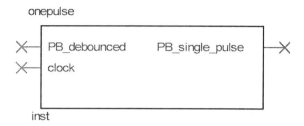

Figure 5.3 Symbol for OnePulse UP3core.

The UP3core OnePulse shown in Figure 5.3 is a pushbutton single-pulse circuit. Output from the pushbutton is High for only one clock cycle no matter how long the pushbutton is pressed. This function is useful in state machines that read external pushbutton inputs. In general, fewer states are required when it is known that inputs only activate for one clock cycle. Internally, an edge-triggered flip-flop is used to build a simple state machine.

5.3.1 VHDL Component Declaration

```
COMPONENT  onepulse
       PORT( PB_debounced, clock   : IN     STD_LOGIC;
              PB_single_pulse       : OUT  STD_LOGIC );
END COMPONENT;
```

5.3.2 Inputs

PB_debounced is the debounced pushbutton input. It should be connected to a debounced pushbutton.

Clock is the user's state-machine clock. It can be any frequency. In some deisgns, the user may want to edit the VHDL code to add a reset input.

5.3.3 Outputs

PB_single_pulse is the output, which is High for only one clock cycle when a pushbutton is hit.

5.4 UP3core Clk_Div: Clock Divider

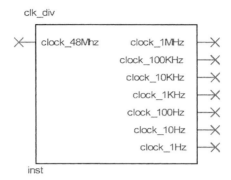

Figure 5.4 Symbol for Clk_Div UP3core.

The UP3core Clk_Div shown in Figure 5.4 is used to provide clock signals slower than the on-board 48MHz oscillator. These signals are obtained by dividing down the 48MHz clock input signal. Multiple output taps provide clock frequencies in powers of ten.

5.4.1 VHDL Component Declaration

```
COMPONENT clk_div
        PORT(   clock_48MHz     : IN       STD_LOGIC;
                clock_1MHz      : OUT      STD_LOGIC;
                clock_100kHz    : OUT      STD_LOGIC;
                clock_10kHz     : OUT      STD_LOGIC;
                clock_1kHz      : OUT      STD_LOGIC;
                clock_100Hz     : OUT      STD_LOGIC;
                clock_10Hz      : OUT      STD_LOGIC;
                clock_1Hz       : OUT      STD_LOGIC );
END COMPONENT;
```

5.4.2 Inputs

Clock_48MHz is an input pin that should be connected to the UP 3 on-board 48MHz USB clock. The pin number for the UP 3's 48MHz USB clock is 29. Make sure the JP3 jumper selects the 48MHz USB clock (default setting).

5.4.3 Outputs

Clock_1MHz through clock_1Hz provide output signals of the specified frequency. The actual frequency is $1.007 \pm .005\%$ times the listed value.

5.5 UP3core VGA_Sync: VGA Video Sync Generation

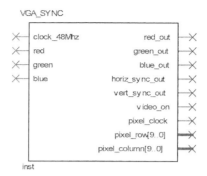

VGA_SYNC

clock_48Mhz	red_out
red	green_out
green	blue_out
blue	horiz_sync_out
	vert_sync_out
	video_on
	pixel_clock
	pixel_row[9..0]
	pixel_column[9..0]

inst

Figure 5.5 Symbol for VGA_Sync UP3core.

The UP3core VGA_Sync shown in Figure 5.5 provides horizontal and vertical sync signals to generate an 8-color 640 by 480 pixel VGA video image. For more detailed information on video signal generation see Chapter 9.

A table of the common screen resolutions and refresh rates along with the required pixel clocks and sync count values can be found at the end of the VGA_Sync IP core. When changing resolutions or refresh rates, use the MegaWizard edit feature to adjust the video_pll.vhd code to output a different pixel clock rate and change the horizontal and vertical sync counter limits to the six new values found in the table. Video_pll.vhd must be present to compile VGA_Sync since it uses this component for the clock.

5.5.1 VHDL Component Declaration

```
COMPONENT vga_sync
   PORT( clock_48MHz, red, green, blue:IN STD_LOGIC;
   red_out, green_out, blue_out,
   horiz_sync_out, vert_sync_out:        OUT  STD_LOGIC;
   pixel_row, pixel_column:              OUT STD_LOGIC_VECTOR( 9 DOWNTO 0 ) );
END COMPONENT;
```

5.5.2 Inputs

Clock_48MHz is an input pin that must be connected to the on-board 48MHz USB clock. One of the Cyclone's two Phase Locked Loops (PLL) is used to generate a video clock. Red, Green, and Blue inputs provide the color information for the video signal. External user logic must generate the RGB signals. One of the Cyclone's two PLLs is used to generate the required pixel clock on the UP 3. An external reference clock of 48MHz (USB clock) is needed by the PLL for the input clock.

5.5.3 Outputs

Horiz_sync is an output pin that should be tied to the horizontal sync, pin 226, on the Cyclone chip.

Vert_sync is an output pin that should be tied to the vertical sync, pin 227, on the Cyclone chip.

Red_out is an output pin that should be tied to the red RGB signal, pin 228, on the Cyclone chip.

Green_out is an output pin that should be tied to the green RGB signal, pin 122, on the Cyclone chip.

Blue_out is an output pin that should be tied to the blue RGB signal, pin 170, on the Cyclone chip.

An interface circuit on the UP 3 board converts the digital red, green, and blue video color signals to the appropriate analog voltage for the monitor. Eight colors are possible using the three digital color signals.

Pixel_clock, pixel_row, and pixel_column are outputs that provide the current pixel clock and the pixel address. Video_on indicates that pixel data is being displayed and a retrace cycle is not presently occurring. These outputs are used by user logic to generate RGB color input data.

5.6 UP3core Char_ROM: Character Generation ROM

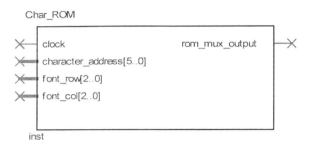

The UP3core Char_ROM shown in Figure 5.6 is a character generation ROM used to generate text in a video display. Each character is represented by an 8 by 8 pixel font. For more information on video character generation see Chapter 9. Character codes are listed in Table 9.1 of Section 9.9. Font data is contained in the memory initialization file, tcgrom.mif. One Cyclone M4K memory block is required for the ROM that holds the font data.

5.6.1 VHDL Component Declaration

```
COMPONENT char_rom
    PORT( clock                  : IN STD_LOGIC;
          character_address      : IN STD_LOGIC_VECTOR( 5 DOWNTO 0 );
          font_row, font_col     : IN STD_LOGIC_VECTOR( 2 DOWNTO 0 );
          rom_mux_output         : OUT STD_LOGIC);
END COMPONENT;
```

5.6.2 Inputs

Character_address is the address of the alphanumeric character to display. Font_row and font_col are used to index through the 8 by 8 font to address the single pixels needed for video signal generation. Clock loads the address register and should be tied to the video pixel_clock.

5.6.3 Outputs

Rom_mux_output is the pixel font value indexed by the address inputs. It is used by user logic to generate the RGB pixel color data for the video signal.

5.7 UP3core Keyboard: Read Keyboard Scan Code

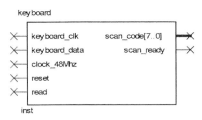

keyboard

keyboard_clk	scan_code[7..0]
keyboard_data	scan_ready
clock_48Mhz	
reset	
read	

inst

Figure 5.7 Symbol for Keyboard UP3core.

The UP3core Keyboard shown in Figure 5.7 is used to read the PS/2 keyboard scan code from a keyboard attached to the UP 3's PS/2 connector. This function converts the serial data from the keyboard to parallel format to produce the scan code output. For detailed information on keyboard applications and scan codes see Table 11.2 in Chapter 11.

5.7.1 VHDL Component Declaration

```
COMPONENT keyboard
      PORT( keyboard_clk, keyboard_data, clock_48MHz ,
            reset, read      : IN STD_LOGIC;
            scan_code        : OUT STD_LOGIC_VECTOR( 7 DOWNTO 0 );
            scan_ready       : OUT STD_LOGIC);
END COMPONENT;
```

5.7.2 Inputs

Clock_48MHz is an input pin that must be connected to the on-board 48MHz oscillator. The pin number for the 48MHz clock is 29 on the Cyclone chip.

Keyboard_clk and keyboard_data are PS/2 input data lines from the keyboard. Keyboard_clk is pin 12 and keyboard_data is pin 13.

Read is a handshake input signal. The rising edge of the read signal clears the scan ready signal. Reset is an input that clears the internal registers and flags used for serial-to-parallel conversion.

5.7.3 Outputs

Scan_code contains the bytes transmitted by the keyboard when a key is pressed or released. See Table 11.2 in Chapter 11 for a listing of scan codes. Scan codes for a single key are a sequence of several bytes. A make code is sent when a key is hit, and a break code is sent whenever a key is released.

Scan_ready is a handshake output signal that goes High when a new scan code is sent by the keyboard. The read input clears scan_ready. The scan_ready handshake line should be used to ensure that a new scan code is read only once.

5.8 UP3core Mouse: Mouse Cursor

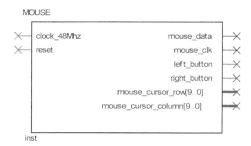

Figure 5.8 Symbol for Mouse UP3core.

The UP3core Mouse shown in Figure 5.8 is used to read position data from a mouse attached to the UP 3's PS/2 connector. It outputs a row and column cursor address for use in video applications. The mouse must be attached to the UP 3 board prior to downloading for proper initialization. Detailed information on mouse applications, commands, and data formats can be found in Chapter 11.

5.8.1 VHDL Component Declaration

```
COMPONENT mouse
    PORT( clock_48MHz, reset    : IN        STD_LOGIC;
          mouse_data            : INOUT    STD_LOGIC;
          mouse_clk             : INOUT    STD_LOGIC;
          left_button, right_button : OUT  STD_LOGIC;
          mouse_cursor_row      : OUT STD_LOGIC_VECTOR( 9 DOWNTO 0 ) );
          mouse_cursor_column   : OUT STD_LOGIC_VECTOR( 9 DOWNTO 0 ) );
END COMPONENT;
```

5.8.2 Inputs

Clock_48MHz is an input pin that must be connected to the on-board 25.175MHz oscillator. The pin number for the 48MHz USB clock is 29 on the Cyclone chip.

Mouse_clk and mouse_data are bi-directional data lines from the mouse. Mouse_clk is pin 12 and mouse_data is pin 13.

5.8.3 Outputs

Mouse_cursor_row and mouse_cursor_column are outputs that contain the current address of the mouse cursor in the 640 by 480 screen area. The cursor is initialized to the center of the screen. Left_button and right_button outputs are High when the corresponding mouse button is pressed.

5.9 For additional information

The FPGA cores summarized in this chapter are used extensively in the textbook's design examples, and complete source code is provided on the CD-ROM. They are provided to support any new FPGA designs that you may develop. Extensive lists of more complex commercial third-party IP cores available for purchase can be found at the major FPGA vendor web sites, www.altera.com and www.xilinx.com. Pricing on commercial cores can be expensive and access to source code may not be provided. An assortment of free open source IP cores for FPGAs is available at www.opencores.org.

SOURCE CODE FOR THE UP3CORE FUNCTIONS IS ON THE CD-ROM.

IN THE BOOK'S DESIGN EXAMPLES, ADDITIONAL UP 2 RELATED MATERIALS

CAN BE FOUND IN EACH CD-ROM CHAPTER IN THE \UP2 SUBDIRECTORY.

CHAPTER 6

Using VHDL for Synthesis of Digital Hardware

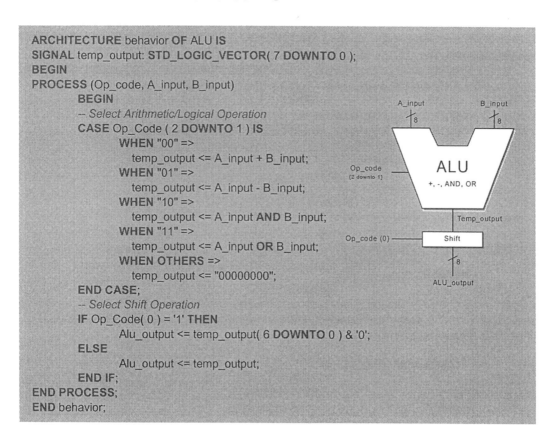

```
ARCHITECTURE behavior OF ALU IS
SIGNAL temp_output: STD_LOGIC_VECTOR( 7 DOWNTO 0 );
BEGIN
PROCESS (Op_code, A_input, B_input)
        BEGIN
        -- Select Arithmetic/Logical Operation
        CASE Op_Code ( 2 DOWNTO 1 ) IS
                WHEN "00" =>
                    temp_output <= A_input + B_input;
                WHEN "01" =>
                    temp_output <= A_input - B_input;
                WHEN "10" =>
                    temp_output <= A_input AND B_input;
                WHEN "11" =>
                    temp_output <= A_input OR B_input;
                WHEN OTHERS =>
                    temp_output <= "00000000";
        END CASE;
        -- Select Shift Operation
        IF Op_Code( 0 ) = '1' THEN
                Alu_output <= temp_output( 6 DOWNTO 0 ) & '0';
        ELSE
                Alu_output <= temp_output;
        END IF;
END PROCESS;
END behavior;
```

6 Using VHDL for Synthesis of Digital Hardware

In the past, most digital designs were manually entered into a schematic entry tool. With increasingly large and more complex designs, this is a tedious and time-consuming process. Logic synthesis using hardware description languages is becoming widely used since it greatly reduces development time and cost. It also enables more exploration of design alternatives, more flexibility to changes in the hardware technology, and promotes design reuse.

VHDL is a language widely used to model and design digital hardware. VHDL is the subject of IEEE standards 1076 and 1164 and is supported by numerous CAD tool and programmable logic vendors. VHDL is an acronym for VHSIC Hardware Description Language. VHSIC, Very High Speed Integrated Circuits, was a USA Department of Defense program in the 1980s that sponsored the early development of VHDL. VHDL has syntax similar to ADA and PASCAL.

Conventional programming languages are based on a sequential operation model. Digital hardware devices by their very nature operate in parallel. This means that conventional programming languages cannot accurately describe or model the operation of digital hardware since they are based on the sequential execution of statements. VHDL is designed to model parallel operations.

> IT IS **CRUCIAL** TO REMEMBER THAT VHDL MODULES, CONCURRENT STATEMENTS, AND PROCESSES ALL OPERATE IN PARALLEL.

In VHDL, variables change without delay and signals change with a small delay. For VHDL synthesis, signals are normally used instead of variables so that simulation works the same as the synthesized hardware.

A subset of VHDL is used for logic synthesis. In this section, a brief introduction to VHDL for logic synthesis will be presented. It is assumed that the reader is already familiar with basic digital logic devices and PASCAL, ADA, or VHDL.

Whenever you need help with VHDL syntax, VHDL templates of common statements are available in the Quartus II online help. In the text editor, just click the right mouse button and Insert ⇨Templates and select VHDL.

6.1 VHDL Data Types

In addition to the normal language data types such as Boolean, integer, and real, VHDL contains new types useful in modeling digital hardware. For logic synthesis, the most important type is standard logic. Type standard logic, STD_LOGIC, is normally used to model a logic bit. To accurately model the operation of digital circuits, more values than "0" or "1" are needed for a logic bit. In the logic simulator, a standard logic bit can have nine values, U, X, 0, 1, Z, W, L, H, and "-". U is uninitialized and X is forced unknown. Z is tri-state or high impedance. L and H are weak "0" and weak "1". "-" is don't care. Type STD_LOGIC_VECTOR contains a one-dimensional array of STD_LOGIC bits. Using these types normally requires the inclusion of special standard logic libraries at the beginning of each VHDL module. The value of a standard logic bit can be set to '0' or '1' using single quotes. A standard logic vector constant, such as the 2-bit zero value, "00" must be enclosed in double quotes. X"F" is the four bit hexadecimal value F.

6.2 VHDL Operators

Table 6.1 lists the VHDL operators and their common function in VHDL synthesis tools.

Table 6.1 VHDL Operators.

VHDL Operator	Operation
+	Addition
-	Subtraction
*	Multiplication*
/	Division*
MOD	Modulus*
REM	Remainder*
&	Concatenation – used to combine bits
SLL**	logical shift left
SRL**	logical shift right
SLA**	arithmetic shift left
SRA**	arithmetic shift right
ROL**	rotate left
ROR**	rotate right
=	equality
/=	Inequality
<	less than
<=	less than or equal
>	greater than
>=	greater than or equal
NOT	logical NOT
AND	logical AND
OR	logical OR
NAND	logical NAND
NOR	logical NOR
XOR	logical XOR
XNOR*	logical XNOR

*Not supported in many VHDL synthesis tools. In the Quartus II tools, only multiply and divide by integers are supported. Mod and Rem are not supported in Quartus II. Efficient design of multiply or divide hardware may require the user to specify the arithmetic algorithm and design in VHDL.

** Supported only in 1076-1993 VHDL only.

Table 6.2 illustrates two useful conversion functions for type STD_LOGIC and integer.

Table 6.2 STD_LOGIC conversion functions.

Function	Example:
CONV_STD_LOGIC_VECTOR(*integer, bits*)	**CONV_STD_LOGIC_VECTOR**(7, 4)
Converts an integer to a standard logic vector. Useful to enter constants. **CONV_SIGNED** and **CONV_UNSIGNED** work in a similar way to produce signed and unsigned values.	Produces a standard logic vector of "0111".
CONV_INTEGER(*std_logic_vector*)	**CONV_INTEGER**("0111")
Converts a standard logic vector to an integer. Useful for array indexing when using a std_logic_vector signal for the array index.	Produces an integer value of 7.

6.3 VHDL Based Synthesis of Digital Hardware

VHDL can be used to construct models at a variety of levels such as structural, behavioral, register transfer level (RTL), and timing. An RTL model of a circuit described in VHDL describes the input/output relationship in terms of dataflow operations on signal and register values. If registers are required, a synchronous clocking scheme is normally used. Sometimes an RTL model is also referred to as a dataflow-style model.

VHDL simulation models often include physical device time delays. In VHDL models written for logic synthesis, timing information should not be provided.

For timing simulations, the CAD tools automatically include the actual timing delays for the synthesized logic circuit. A FPGA timing model supplied by the CAD tool vendor is used to automatically generate the physical device time delays inside the FPGA. Sometimes this timing model is also written in VHDL. For a quick overview of VHDL, several constructs that can be used to synthesize common digital hardware devices will be presented.

6.4 VHDL Synthesis Models of Gate Networks

The first example consists of a simple gate network. In this model, both a concurrent assignment statement and a sequential process are shown which generate the same gate network. X is the output on one network and Y is the output on the other gate network. The two gate networks operate in parallel.

In VHDL synthesis, inputs and outputs from the port declaration in the module will become I/O pins on the programmable logic device. Comment lines begin with "--". The Quartus II editor performs syntax coloring and is useful to quickly find major problems with VHDL syntax.

Inside a process, statements are executed in sequential order, and all processes are executed in parallel. If multiple assignments are made to a signal inside a process, the last assignment is taken as the new signal value.

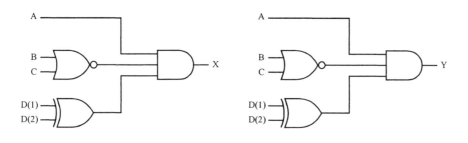

```
LIBRARY IEEE;                      -- Include Libraries for standard logic data types
USE IEEE.STD_LOGIC_1164.ALL;

                                   -- Entity name normally the same as file name
ENTITY gate_network IS             -- Ports: Declares module inputs and outputs
      PORT( A, B, C : IN     STD_LOGIC;
                                   -- Standard Logic Vector ( Array of 4 Bits )
             D     : IN     STD_LOGIC_VECTOR( 3 DOWNTO 0 );
                                   -- Output Signals
             X, Y   : OUT   STD_LOGIC );
END gate_network;


                                   -- Defines internal module architecture
ARCHITECTURE behavior OF gate_network IS
BEGIN                              -- Concurrent assignment statements operate in parallel
                                   -- D(1) selects bit 1 of standard logic vector D
      X <= A AND NOT( B OR C ) AND ( D( 1 ) XOR D( 2 ) );


                                   -- Process must declare a sensitivity list,
                                   -- In this case it is  ( A, B, C, D )
                                   -- List includes all signals that can change the outputs
   PROCESS ( A, B, C, D )
      BEGIN                        -- Statements inside process execute sequentially
             Y <= A AND NOT( B OR C) AND ( D( 1) XOR D( 2 ) );
   END PROCESS;
END behavior;
```

6.5 VHDL Synthesis Model of a Seven-segment LED Decoder

The following VHDL code implements a seven-segment decoder for seven-segment LED displays. A 7-bit standard logic vector is used to assign the value of all seven bits in a single case statement. In the logic vector, the most-significant bit is segment 'a' and the least-significant bit is segment 'g'. The logic synthesis CAD tool automatically minimizes the logic required for implementation. The signal MSD contains the 4-bit binary value to be displayed in hexadecimal. MSD is the left or most-significant digit. Another identical process with a different input variable is needed for the second display digit.

```
LED_MSD_DISPLAY:                 -- BCD to 7 Segment Decoder for LED Displays

PROCESS (MSD)
BEGIN
                                 -- Case statement implements a logic truth table
    CASE MSD IS
        WHEN "0000" =>
                MSD_7SEG <= "1111110";
        WHEN "0001" =>
                MSD_7SEG <= "0110000";
        WHEN "0010" =>
                MSD_7SEG <= "1101101";
        WHEN "0011" =>
                MSD_7SEG <= "1111001";
        WHEN "0100" =>
                MSD_7SEG <= "0110011";
        WHEN "0101" =>
                MSD_7SEG <= "1011011";
        WHEN "0110" =>
                MSD_7SEG <= "1011111";
        WHEN "0111" =>
                MSD_7SEG <= "1110000";
        WHEN "1000" =>
                MSD_7SEG <= "1111111";
        WHEN "1001" =>
                MSD_7SEG <= "1111011";
        WHEN OTHERS =>
                MSD_7SEG <= "0111110";
    END CASE;

END PROCESS LED_MSD_DISPLAY;
```

The following VHDL concurrent assignment statements provide the value to be displayed and connect the individual segments. NOT is used since a logic zero actually turns on the LED. Automatic minimization in the synthesis process will eliminate the extra inverter in the logic circuit. Pin assignments for the seven-segment display must be included in the project's *.qsf file or in the top-level schematic.

```
                                 -- Provide 4-bit value to display
MSD <= PC ( 7 DOWNTO 4 );

                                 -- Drive the seven-segments (LEDs are active low)
MSD_a <= NOT MSD_7SEG( 6 );
MSD_b <= NOT MSD_7SEG( 5 );
MSD_c <= NOT MSD_7SEG( 4 );
MSD_d <= NOT MSD_7SEG( 3 );
MSD_e <= NOT MSD_7SEG( 2 );
MSD_f <= NOT MSD_7SEG( 1 );
MSD_g <= NOT MSD_7SEG( 0 );
```

6.6 VHDL Synthesis Model of a Multiplexer

The next example shows several alternative ways to synthesize a 2-to-1 multiplexer in VHDL. Four identical multiplexers that operate in parallel are synthesized by this example. In VHDL, IF and CASE statements must be inside a process. The inputs and outputs from the multiplexers could be changed to standard logic vectors if an entire bus is multiplexed. Multiplexers with more than two inputs can also be easily constructed. Nested IF-THEN-ELSE statements generate priority-encoded logic that requires more hardware and produce a slower circuit than a CASE statement.

```
LIBRARY IEEE;
USE  IEEE.STD_LOGIC_1164.ALL;

ENTITY multiplexer IS                    -- Input Signals and Mux Control
        PORT( A, B, Mux_Control    : IN    STD_LOGIC;
              Mux_Out1, Mux_Out2,
              Mux_Out3, Mux_Out4   : OUT   STD_LOGIC );
END multiplexer;

ARCHITECTURE behavior OF multiplexer IS
BEGIN                                    -- selected signal assignment statement...

        Mux_Out1 <= A WHEN Mux_Control = '0' ELSE B;
                                -- ... with Select Statement
        WITH Mux_control SELECT

        Mux_Out2 <=   A WHEN   '0',
                      B WHEN   '1',
                      A WHEN OTHERS;     -- OTHERS case required since STD_LOGIC
                                         --    has values other than "0" or "1"
        PROCESS ( A, B, Mux_Control)
        BEGIN                            -- Statements inside a process
                IF Mux_Control = '0' THEN  --    execute sequentially.
                   Mux_Out3 <= A;
                ELSE
                   Mux_out3 <= B;
                END IF;

                CASE Mux_Control IS
                    WHEN '0' =>
                            Mux_Out4 <= A;
                    WHEN '1' =>
                            Mux_Out4 <= B;
                    WHEN OTHERS =>
                            Mux_Out4 <= A;
                END CASE;
        END PROCESS;
END behavior;
```

6.7 VHDL Synthesis Model of Tri-State Output

Tri-state gates are supported in VHDL synthesis tools and are supported in many programmable logic devices. Most programmable logic devices have tri-state output pins. Some programmable logic devices do not support internal tri-state logic. Here is a VHDL example of a tri-state output. In VHDL, the assignment of the value "Z" to a signal produces a tri-state output.

```
LIBRARY IEEE;
USE  IEEE.STD_LOGIC_1164.ALL;

ENTITY tristate IS
     PORT( A, Control      : IN      STD_LOGIC;
           Tri_out         : INOUT  STD_LOGIC); -- Use Inout for bi-directional tri-state
                                                --    signals or out for output only
END tristate;

ARCHITECTURE behavior OF tristate IS          -- defines internal module architecture
BEGIN
     Tri_out <= A WHEN Control = '0' ELSE 'Z';  -- Assignment of 'Z' value generates
END behavior;                                   --    tri-state output
```

6.8 VHDL Synthesis Models of Flip-flops and Registers

In the next example, several flip-flops will be generated. Unlike earlier combinational hardware devices, a flip-flop can only be synthesized inside a process. In VHDL, Clock'EVENT is true whenever the clock signal changes. The positive clock edge is selected by (clock'EVENT AND clock = '1') and positive edge triggered D flip-flops will be used for synthesis. The following module contains a variety of Reset and Enable options on positive edge-triggered D flip-flops. Processes with a wait statement do not need a process sensitivity list. A process can only have one clock or reset type.

The negative clock edge is selected by (clock'EVENT AND clock = '0') and negative edge-triggered D flip-flops will be used for synthesis. If (Clock = '1') is substituted for (clock'EVENT AND clock = '1') level-triggered latches will be selected for logic synthesis. Rising_edge(clock) can also be used instead of clock'EVENT AND clock = '1'. Falling_edge(clock) is also supported for negative clock edges.

```
LIBRARY IEEE;
USE  IEEE.STD_LOGIC_1164.ALL;

ENTITY DFFs IS
     PORT( D, Clock, Reset, Enable   : IN    STD_LOGIC;
           Q1, Q2, Q3, Q4            : OUT  STD_LOGIC );
END DFFs;

ARCHITECTURE behavior OF DFFs IS
```

BEGIN

```
    PROCESS                          -- Positive edge triggered D flip-flop
    BEGIN                            -- If WAIT is used no sensitivity list is used
        WAIT UNTIL ( Clock 'EVENT AND Clock = '1' );
            Q1 <= D;
    END PROCESS;
```

```
    PROCESS                          -- Positive edge triggered D flip-flop
    BEGIN                            --    with synchronous reset
        WAIT UNTIL ( Clock 'EVENT AND Clock = '1' );
                IF reset = '1'  THEN
                    Q2 <= '0';
                ELSE
                    Q2 <= D;
                END IF;
    END PROCESS;
```

```
    PROCESS (Reset,Clock)            -- Positive edge triggered D flip-flop
    BEGIN                            --    with asynchronous reset
        IF reset = '1' THEN
            Q3 <= '0';
        ELSIF ( clock 'EVENT AND clock = '1' ) THEN
            Q3 <= D;
        END IF;
    END PROCESS;
```

```
    PROCESS (Reset,Clock)            -- Positive edge triggered D flip-flop
    BEGIN                            --    with asynchronous reset and
                                     --    enable
        IF reset = '1' THEN
            Q4 <= '0';
        ELSIF ( clock 'EVENT AND clock = '1' ) THEN
            IF Enable = '1' THEN
                Q4 <= D;
            END IF;
        END IF;
    END PROCESS;
END behavior;
```

In VHDL, as in any digital logic designs, it is not good design practice to AND or gate other signals with the clock. Use a flip-flop with a clock enable instead to avoid timing and clock skew problems. In some limited cases, such as power management, a single level of clock gating can be used. This works only when a small amount of clock skew can be tolerated and the signal gated with the clock is known to be hazard or glitch free. A particular programmable logic device may not support every flip-flop or latch type and Set/Reset and Enable option.

If D and Q are replaced by standard logic vectors in these examples, registers with the correct number of bits will be generated instead of individual flip-flops.

6.9 Accidental Synthesis of Inferred Latches

Here is a very common problem to be aware of when coding VHDL for synthesis. If a non-clocked process has any path that does not assign a value to an output, VHDL assumes you want to use the previous value. A level triggered latch is automatically generated or inferred by the synthesis tool to save the previous value. In many cases, this can cause serious errors in the design. Edge-triggered flip-flops should not be mixed with level-triggered latches in a design or serious timing problems will result. Typically this can happen in CASE statements or nested IF statements. In the following example, the signal OUTPUT2 infers a latch when synthesized. Assigning a value to OUTPUT2 in the last ELSE clause will eliminate the latch.

```
LIBRARY IEEE;
USE  IEEE.STD_LOGIC_1164.ALL;
ENTITY ilatch IS
        PORT(  A, B                  : IN      STD_LOGIC;
                Output1, Output2     : OUT     STD_LOGIC );
END ilatch;

ARCHITECTURE behavior OF ilatch IS
BEGIN
        PROCESS ( A, B )
        BEGIN
            IF A = '0' THEN
                Output1 <= '0';
                Output2 <= '0';
            ELSE
              IF B = '1' THEN
                  Output1 <= '1';
                  Output2 <= '1';
              ELSE                  -- Latch inferred since no value is assigned
                  Output1 <= '0';   --    to output2 in the else clause!
              END IF;
            END IF;
        END PROCESS;
END behavior;
```

6.10 VHDL Synthesis Model of a Counter

Here is an 8-bit counter design. This design performs arithmetic operations on standard logic vectors. Since this example includes arithmetic operations, two new libraries must be included at the beginning of the module. Either signed or unsigned libraries can be selected, but not both. Since the unsigned library was used, an 8-bit magnitude comparator is automatically synthesized for the internal_count < max_count comparison.

Compare operations between standard logic and integer types are supported. The assignment internal_count <= internal_count + 1 synthesizes an 8-bit incrementer. An incrementer circuit requires less hardware than an adder that adds one. The operation, "+1", is treated as a special incrementer case by synthesis tools. VHDL does not allow reading of an "OUT" signal so an internal_count signal is used which is always the same as count. This is the first example that includes an internal signal. Note its declaration at the beginning of the architecture section.

```
LIBRARY IEEE;
USE IEEE.STD_LOGIC_1164.ALL;
USE IEEE.STD_LOGIC_ARITH.ALL;
USE IEEE.STD_LOGIC_UNSIGNED.ALL;


ENTITY Counter IS
    PORT( Clock, Reset  : IN    STD_LOGIC;
          Max_count     : IN    STD_LOGIC_VECTOR( 7 DOWNTO 0 );
          Count         : OUT STD_LOGIC_VECTOR( 7 DOWNTO 0 )  );
END Counter;


ARCHITECTURE behavior OF Counter IS         -- Declare signal(s) internal to module
    SIGNAL internal_count:      STD_LOGIC_VECTOR( 7 DOWNTO 0 );
BEGIN
    count <= internal_count;

    PROCESS ( Reset,Clock )
        BEGIN                                  -- Reset counter
            IF reset = '1' THEN
                internal_count <= "00000000";
            ELSIF ( clock 'EVENT AND clock = '1' ) THEN
                IF internal_count < Max_count THEN    -- Check for maximum count
                    internal_count <= internal_count + 1;  -- Increment Counter
                ELSE                                  -- Count >= Max_Count
                    internal_count <= "00000000";     --    reset Counter
                END IF;
            END IF;
        END PROCESS;
END behavior;
```

6.11 VHDL Synthesis Model of a State Machine

The next example shows a Moore state machine with three states, two inputs and a single output. A state diagram of the example state machine is shown in Figure 6.1. In VHDL, an enumerated data type is specified for the current state using the TYPE statement. This allows the synthesis tool to assign the actual "0" or "1" values to the states. In many cases, this will produce a smaller hardware design than direct assignment of the state values in VHDL.

Depending on the synthesis tool settings, the states may be encoded or constructed using the one-hot technique. Outputs are defined in the last WITH... SELECT statement. This statement lists the output for each state and eliminates possible problems with inferred latches. To avoid possible timing problems, unsynchronized external inputs to a state machine should be synchronized by passing them through one or two D flip-flops that are clocked by the state machine's clock.

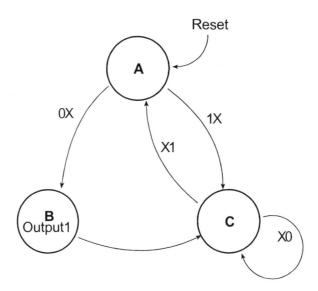

Figure 6.1 State Diagram for st_mach VHDL example

```
LIBRARY IEEE;
USE IEEE.STD_LOGIC_1164.ALL;

ENTITY st_mach IS
      PORT( clk, reset      : IN       STD_LOGIC;
            Input1, Input2   : IN       STD_LOGIC;
            Output1          : OUT      STD_LOGIC);
END st_mach;

ARCHITECTURE A OF st_mach IS
                                         -- Enumerated Data Type for State
      TYPE STATE_TYPE IS ( state_A, state_B, state_C );
      SIGNAL state: STATE_TYPE;

BEGIN
      PROCESS ( reset, clk )
      BEGIN
            IF reset = '1' THEN               -- Reset State
               state <= state_A;
            ELSIF clk 'EVENT AND clk = '1' THEN

                CASE state IS                 -- Define Next State Transitions using a Case
                                              --    Statement based on the Current State
                    WHEN state_A =>
```

```
                    IF Input1 = '0' THEN
                        state <= state_B;
                    ELSE
                        state <= state_C;
                    END IF;

                WHEN state_B =>
                    state <= state_C;

                WHEN state_C =>
                    IF Input2 = '1' THEN
                        state <= state_A;
                    END IF;

                WHEN OTHERS =>
                    state <= state_A;
            END CASE;
        END IF;
END PROCESS;

WITH state SELECT                        -- Define State Machine Outputs
    Output1   <=   '0' WHEN state_A,
                   '1' WHEN state_B,
                   '0' WHEN state_C;

END a;
```

6.12 VHDL Synthesis Model of an ALU with an Adder/Subtractor and a Shifter

Here is an 8-bit arithmetic logic unit (ALU), that adds, subtracts, bitwise ANDs, or bitwise ORs, two operands and then performs an optional shift on the output. The most-significant two bits of the Op-code select the arithmetic logical operation. If the least-significant bit of the op_code equals '1' a 1-bit left-shift operation is performed. An addition and subtraction circuit is synthesized for the "+" and "-" operator. Depending on the number of bits and the speed versus area settings in the synthesis tool, ripple carry or carry-lookahead circuits will be used. Several "+" and "-" operations in multiple assignment statements may generate multiple ALUs and increase the hardware size, depending on the VHDL CAD tool and compiler settings used. If a single ALU is desired, muxes can be placed at the inputs and the "+" operator would be used only in a single assignment statement.

```
LIBRARY IEEE;
USE IEEE.STD_LOGIC_1164.ALL;
USE IEEE.STD_LOGIC_ARITH.ALL;
USE IEEE.STD_LOGIC_UNSIGNED.ALL;


ENTITY ALU IS
    PORT( Op_code                : IN      STD_LOGIC_VECTOR( 2 DOWNTO 0 );
```

```
            A_input, B_input        : IN    STD_LOGIC_VECTOR( 7 DOWNTO 0 );
            ALU_output              : OUT  STD_LOGIC_VECTOR( 7 DOWNTO 0 ) );
END ALU;


ARCHITECTURE behavior OF ALU IS
                                            -- Declare signal(s) internal to module here
    SIGNAL temp_output                 :    STD_LOGIC_VECTOR( 7 DOWNTO 0 );
BEGIN

    PROCESS ( Op_code, A_input, B_input )
    BEGIN
        CASE Op_Code ( 2 DOWNTO 1 ) IS    -- Select Arithmetic/Logical Operation
            WHEN "00" =>
                    temp_output <= A_input  +    B_input;
            WHEN "01" =>
                    temp_output <= A_input  -    B_input;
            WHEN "10" =>
                    temp_output <= A_input  AND B_input;
            WHEN "11" =>
                    temp_output <= A_input  OR  B_input;
            WHEN OTHERS =>
                    temp_output <= "00000000";
        END CASE;
```

-- *Select Shift Operation: Shift bits left with zero fill using concatenation operator*
--   *Can also use VHDL 1076-1993 shift operator such as SLL*

```
        IF Op_Code( 0 ) = '1' THEN
            Alu_output <= temp_output( 6 DOWNTO 0 ) & '0';
        ELSE
            Alu_output <= temp_output;
        END IF;
    END PROCESS;
END behavior;
```

6.13 VHDL Synthesis of Multiply and Divide Hardware

In the Quartus II tool, integer multiply and divide is supported using VHDL's "*" and "/" operators. Mod and Rem are not supported in Quartus II. In current generation tools, efficient design of multiply or divide hardware typically requires the use of a vendor-specific library function or even the specification of the arithmetic algorithm and hardware implementation in VHDL.

A wide variety of multiply and divide algorithms that trade off time versus hardware size can be found in most computer arithmetic texts. Several such references are listed at the end of this chapter. These algorithms require a sequence of add/subtract and shift operations that can be easily synthesized in VHDL using the standard operators. The LPM_MULT function in Quartus II can be used to synthesize integer multipliers. LPM_DIVIDE, is also available. When using LPM functions, **Tools ⇨   MegaWizard Plug-in Manager** can be used to help generate VHDL code. The LPM functions also support pipeline options. Array multiply and divide hardware for more than a few bits requires extensive hardware and a large FPGA.

```
LIBRARY IEEE;
USE  IEEE.STD_LOGIC_1164.ALL;
USE  IEEE.STD_LOGIC_ARITH.ALL;
USE  IEEE.STD_LOGIC_UNSIGNED.ALL;
LIBRARY lpm;
USE lpm.lpm_components.ALL;

ENTITY mult IS
      PORT( A, B     : IN   STD_LOGIC_VECTOR(  7 DOWNTO 0 );
             Product  : OUT STD_LOGIC_VECTOR( 15 DOWNTO 0 ) );
END mult;

ARCHITECTURE a OF mult IS
BEGIN                                  -- LPM 8x8 multiply function P = A * B
      multiply: lpm_mult
      GENERIC MAP(  LPM_WIDTHA              => 8,
                     LPM_WIDTHB              => 8,
                     LPM_WIDTHS              => 16,
                     LPM_WIDTHP              => 16,
                     LPM_REPRESENTATION      => "UNSIGNED" )

      PORT MAP (    data  => A,
                     datab => B,
                     result => Product );
END a;
```

Floating-point operations can be implemented on very large FPGAs; however, performance is lower than current floating-point DSP and microprocessor chips. The floating-point algorithms must be coded by the user in VHDL using integer add, multiply, divide, and shift operations. The LPM_CLSHIFT function is useful for the barrel shifter needed in a floating-point ALU. Some floating point IP cores are starting to appear. Many FPGA vendors also have optimized arithmetic packages for DSP applications such as FIR filters.

6.14 VHDL Synthesis Models for Memory

Typically, it is more efficient to call a vendor-specific function to synthesize RAM. These functions typically use the FPGA's internal RAM blocks rather than building a RAM using FPGA logic elements. The memory function in the Altera toolset is the ALTSYNCRAM function. On the UP 2 board's older FPGA, the LPM_RAM_DQ memory function can also be used. The memory can be set to an initial value using a separate memory initialization file with the extension *.mif. A similar call, LPM_ROM, can be used to synthesize ROM.

If small blocks of multi-ported or other special-purpose RAM are needed, they can be synthesized using registers with address decoders for the write operation and multiplexers for the read operation. Additional read or write ports can be added to synthesized RAM. An example of this approach is a dual-ported register file for a computer processor core. Most RISC processors need to read two registers on each clock cycle and write to a third register.

VHDL Memory Model - Example One

The first memory example synthesizes a memory that can perform a read and a write operation every clock cycle. Memory is built using arrays of positive edge-triggered D flip-flops. Memory write, memwrite, is gated with an address decoder output and used as an enable to load each memory location during a write operation. A synchronous write operation is more reliable. Asynchronous write operations respond to any logic hazards or momentary level changes on the write signal. As in any synchronous memory, the write address must be stable before the rising edge of the clock signal. A non-clocked mux is used for the read operation. If desired, memory can be initialized by a reset signal.

```
LIBRARY IEEE;
USE IEEE.STD_LOGIC_1164.ALL;

ENTITY memory IS
    PORT( read_data      :   OUT  STD_LOGIC_VECTOR( 7 DOWNTO 0 );
          read_address   :   IN   STD_LOGIC_VECTOR( 2 DOWNTO 0 );
          write_data     :   IN   STD_LOGIC_VECTOR( 7 DOWNTO 0 );
          write_address  :   IN   STD_LOGIC_VECTOR( 2 DOWNTO 0 );
          Memwrite       :   IN   STD_LOGIC;
          clock,reset    :   IN   STD_LOGIC );
END memory;

ARCHITECTURE behavior OF memory IS
    SIGNAL mem0, mem1  :          STD_LOGIC_VECTOR( 7 DOWNTO 0 );

BEGIN
    PROCESS (read_address, mem0, mem1)  -- Process for memory read operation
    BEGIN
        CASE read_address IS
            WHEN "000" =>
                    read_data <= mem0;
            WHEN "001" =>
                    read_data <= mem1;
            WHEN OTHERS =>              -- Unimplemented memory locations
                    read_data <= X"FF" ;
        END CASE;
    END PROCESS;

    PROCESS
    BEGIN
        WAIT UNTIL clock 'EVENT AND clock = '1';
            IF ( reset = '1' ) THEN
                mem0 <= X"55" ; -- Initial values for memory (optional)
                mem1 <= X"AA" ;
            ELSE
                IF memwrite = '1' THEN          -- Write to memory?
```

```
                CASE write_address IS          -- Use a flip-flop with
                    WHEN "000" =>              --    an enable for memory
                        mem0 <= write_data;
                    WHEN "001" =>
                        mem1 <= write_data;
                    WHEN OTHERS =>             -- unimplemented memory locations
                        NULL;
                END CASE;
                END IF;
            END IF;
    END PROCESS;
END behavior;
```

VHDL Memory Model - Example Two

The second example uses an array of standard logic vectors to implement memory. This approach is easier to write in VHDL since the array index generates the address decoder and multiplexers automatically; however, it is a little more difficult to access the values of individual array elements during simulation. There are a few VHDL synthesis tools that do not support array types. Synthesizing RAM requires a vast amount of programmable logic resources. Only a few hundred bits of RAM can be synthesized, even on large devices. Each bit of RAM requires 10 to 20 logic gates and a large amount of FPGA interconnect resources. Some tools may automatically detect synthesized RAM and use the FPGA's embedded memory blocks.

```
LIBRARY IEEE;
USE IEEE.STD_LOGIC_1164.ALL;

ENTITY memory IS
    PORT(  read_data      :      OUT  STD_LOGIC_VECTOR( 7 DOWNTO 0 );
            read_address  :      IN   STD_LOGIC_VECTOR( 2 DOWNTO 0 );
            write_data    :      IN   STD_LOGIC_VECTOR( 7 DOWNTO 0 );
            write_address :      IN   STD_LOGIC_VECTOR( 2 DOWNTO 0 );
            Memwrite      :      IN   STD_LOGIC;
            Clock         :      IN   STD_LOGIC  );
END memory;
ARCHITECTURE behavior OF memory IS
                                -- define new data type for memory array
    TYPE memory_type IS ARRAY ( 0 TO 7 ) OF STD_LOGIC_VECTOR( 7 DOWNTO 0 );
    SIGNAL memory        : memory_type;
BEGIN
    -- Read Memory and  convert array index to an integer with CONV_INTEGER
    read_data <= memory( CONV_INTEGER( read_address( 2 DOWNTO 0 ) ) );

    PROCESS                    -- Write Memory?
    BEGIN
        WAIT UNTIL clock 'EVENT AND clock = '1';
        IF ( memwrite = '1' ) THEN

-- convert array index to an integer with CONV_INTEGER
            memory( CONV_INTEGER( write_address( 2 DOWNTO 0 ) ) ) <= write_data;
```

```
        END IF;
    END PROCESS;
END behavior;
```

VHDL Memory Model - Example Three

The third example shows the use of the ALTSYNCRAM megafunction to implement a block of memory. An additional library is needed for the megafunctions. For more information on the megafunctions see the online help guide in the Quartus II tool. In single port mode, the ALTSYNCRAM memory can do either a read or a write operation in a single clock cycle since there is only one address bus. In dual port mode, it can do both a read and write. If this is the only memory operation needed, the ALTSYNCRAM function produces a more efficient hardware implementation than synthesis of the memory in VHDL. In the ALTSYNCRAM megafunction, the memory address must be clocked into a dedicated address register located inside the FPGA's synchronous memory block. Asynchronous memory operations without a clock can cause timing problems and are not supported on many FPGAs including the Cyclone.

```
LIBRARY IEEE;
USE IEEE.STD_LOGIC_1164.ALL;
LIBRARY Altera_mf;
USE altera_mf.altera_mf_components.all;

ENTITY amemory IS
    PORT( read_data            :       OUT     STD_LOGIC_VECTOR( 7 DOWNTO 0 );
          memory_address       :       IN      STD_LOGIC_VECTOR( 2 DOWNTO 0 );
          write_data           :       IN      STD_LOGIC_VECTOR( 7 DOWNTO 0 );
          Memwrite             :       IN      STD_LOGIC;
          clock,reset          :       IN      STD_LOGIC );
END amemory;

ARCHITECTURE behavior OF amemory IS
BEGIN
    data_memory: altsyncram             -- Altsyncram memory function
    GENERIC MAP ( operation_mode => "SINGLE_PORT",
                  width_a            => 8,
                  widthad_a          => 3,
                  lpm_type => "altsyncram",
                  outdata_reg_a       => "UNREGISTERED",
                                  -- Reads in mif file for initial data values (optional)
                  init_file           => "memory.mif",
                  intended_device_family => "Cyclone"     )

    PORT MAP (wren_a => Memwrite,  clock0  => clock,
              address_a => memory_address( 2 DOWNTO 0 ),
              data_a => write_data,  q_a => read_data );
END behavior;
```

On the Cyclone FPGA chip, the memory can be implemented using the M4K memory blocks, which are separate from the FPGA's logic cells. In the Cyclone EP1C6 chip there are 20 M4K RAM blocks at 4Kbits each for a total of 92,160 bits. In the Cyclone EP1C12 there are 52 M4K blocks for a total of 239,616 bits. Each M4K block can be setup to be 4K by 1, 2K by 2, 1K by 4, 512 by 8, 256 by 16, 256 by 18, 128 by 32 or 128 by 36 bits wide. The **Tools** ⇨ **MegaWizard Plug-in Manager** feature is useful to configure the Altsyncram parameters.

6.15 Hierarchy in VHDL Synthesis Models

Large VHDL models should be split into a hierarchy using a top-level structural model in VHDL or by using the symbol and graphic editor in the Quartus II tool. In the graphical editor, a VHDL file can be used to define the contents of a symbol block. Synthesis tools run faster using a hierarchy on large models and it is easier to write, understand, and maintain a large design when it is broken up into smaller modules.

An example of a hierarchical design with three submodules is seen in the schematic in Figure 6.2. Following the schematic, the same design using a top-level VHDL structural model is shown. This VHDL structural model provides the same connection information as the schematic seen in Figure 6.2.

Debounce, Onepulse, and Clk_div are the names of the VHDL submodules. Each one of these submodules has a separate VHDL source file. In the Quartus II tool, compiling the top-level module will automatically compile the lower-level modules.

In the example, VHDL structural model, note the use of a component declaration for each submodule. The component statement declares the module name and the inputs and outputs of the module. Internal signal names used for interconnections of components must also be declared at the beginning of the component list.

In the final section, port mappings are used to specify the module or component interconnections. Port names and their order must be the same in the VHDL submodule file, the component instantiations, and the port mappings. Component instantiations are given unique labels so that a single component can be used several times.

Note that node names in the schematic or signals in VHDL used to interconnect modules need not always have the same names as the signals in the components they connect. Just like signal or wire names in a schematic are not always the same as the pin names on chips that they connect. As an example, pb_debounced on the debounce component connects to an internal signal with a different name, pb1_debounced.

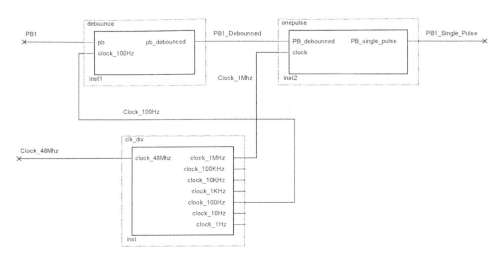

Figure 6.2 Schematic of Hierarchical Design Example

```
LIBRARY IEEE;
USE  IEEE.STD_LOGIC_1164.ALL;
USE  IEEE.STD_LOGIC_ARITH.ALL;
USE  IEEE.STD_LOGIC_UNSIGNED.ALL;
ENTITY hierarch IS
    PORT (   clock_48MHz, pb1   : IN STD_LOGIC;
             pb1_single_pulse   : OUT STD_LOGIC);
END hierarch;
ARCHITECTURE structural OF hierarch IS
-- Declare internal signals needed to connect submodules
SIGNAL clock_1MHz, clock_100Hz, pb1_debounced : STD_LOGIC;
-- Use Components to Define Submodules and Parameters
    COMPONENT debounce
            PORT(   pb, clock_100Hz      : IN       STD_LOGIC;
                    pb_debounced         : OUT      STD_LOGIC);
    END COMPONENT;

    COMPONENT onepulse
            PORT(pb_debounced, clock    : IN       STD_LOGIC;
                 pb_single_pulse        : OUT      STD_LOGIC);
    END COMPONENT;

    COMPONENT clk_div
            PORT(   clock_48MHz          : IN       STD_LOGIC;
                    clock_1MHz           : OUT      STD_LOGIC;
                    clock_100kHz         : OUT      STD_LOGIC;
                    clock_10kHz          : OUT      STD_LOGIC;
                    clock_1kHz           : OUT      STD_LOGIC;
                    clock_100Hz          : OUT      STD_LOGIC;
                    clock_10Hz           : OUT      STD_LOGIC;
                    clock_1Hz            : OUT      STD_LOGIC);
    END COMPONENT;
BEGIN
```

-- Use Port Map to connect signals between components in the hierarchy
debounce1 : debounce **PORT MAP** (pb => pb1, clock_100Hz = >clock_100Hz,
                                pb_debounced = >pb1_debounced);

prescalar : clk_div **PORT MAP** (clock_48MHz = >clock_48MHz,
                                clock_1MHz =>clock_1MHz,
                            clock_100hz = >clock_100hz);

single_pulse : onepulse **PORT MAP** (pb_debounced = >pb1_debounced,
                                clock => clock_1MHz,
                            pb_single_pulse => pb1_single_pulse);
    END structural;

6.16 Using a Testbench for Verification

Complex VHDL synthesis models are frequently verified by simulation of the model's behavior in a specially written entity called a testbench. As seen in Figure 6.3, the top–level testbench module contains a component instantiation of the hardware unit under test (UUT). The testbench also contains VHDL code used to automatically generate input stimulus to the UUT and automatically monitor the response of the UUT for correct operation.

The testbench contains test vectors and timing information used in testing the UUT. The testbench's VHDL code is used only for testing, and it is not synthesized. This keeps the test-only code portion of the VHDL model separate from the UUT's hardware synthesis model. Third party simulation tools such as ModelSIM or Active-HDL are typically required for this approach. Unfortunately, full versions of these third party simulation tools are currently very expensive for students or individuals.

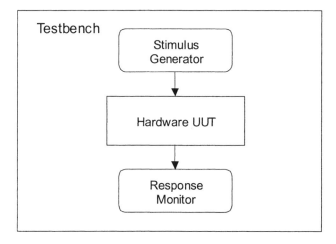

Figure 6.3 Using a testbench for automatic verification during simulation.

The testbench approach is critical in large ASIC designs where any errors are costly. Automatic Test Equipment (ATE) can also use a properly written testbench and its test vector and timing information to physically test each ASIC chip for correct operation after production. In large designs, the testbench can require as much time and effort as the UUT's synthesis model. By performing both a functional simulation and a timing simulation of the UUT with the same test vectors, it is also possible to check for any synthesis-related errors.

6.17 For additional information

The chapter has introduced the basics of using VHDL for digital synthesis. It has not explored all of the language options available. The Altera online help contains VHDL syntax and templates. A large number of VHDL reference textbooks are also available. Unfortunately, only a few of them currently examine using VHDL models that can be used for digital logic synthesis. One such text is *HDL Chip Design* by Douglas J. Smith, Doone Publications, 1996.

A number of alternative integer multiply, divide, and floating-point algorithms with different speed versus area tradeoffs can be found in computer arithmetic textbooks. Two such examples are *Digital Computer Arithmetic Design and Implementation* by Cavanagh, McGraw Hill, 1984, and *Computer Arithmetic Algorithms* by Israel Koren, Prentice Hall, 1993.

6.18 Laboratory Exercises

1. Rewrite and compile the VHDL model for the seven-segment decoder in Section 6.5 replacing the PROCESS and CASE statements with a WITH...SELECT statement.

2. Write a VHDL model for the state machine shown in the following state diagram and verify correct operation with a simulation using the Altera CAD tools. A and B are the two states, X is the output, and Y is the input. Use the timing analyzer to determine the maximum clock frequency on the Cyclone EP1C6Q240C8 device.

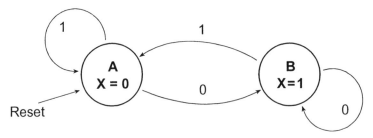

3. Write a VHDL model for a 32-bit, arithmetic logic unit (ALU). Verify correct operation with a simulation using the Altera CAD tools. A and B are 32-bit inputs to the ALU, and Y is the output. A shift operation follows the arithmetic and logical operation. The opcode controls ALU functions as follows:

Opcode	Operation	Function
000XX	ALU_OUT <= A	Pass A
001XX	ALU_OUT <= A + B	Add
010XX	ALU_OUT <= A-B	Subtract
011XX	ALU_OUT <= A AND B	Logical AND
100XX	ALU_OUT <= A OR B	Logical OR
101XX	ALU_OUT <= A + 1	Increment A
110XX	ALU_OUT <= A-1	Decrement A
111XX	ALU_OUT <= B	Pass B
XXX00	Y <= ALU_OUT	Pass ALU_OUT
XXX01	Y<= SHL(ALU_OUT)	Shift Left
XXX10	Y<=SHR(ALU_OUT)	Shift Right (unsigned-zero fill)
XXX11	Y<= 0	Pass 0's

4. Use the Cyclone chip as the target device. Determine the worst case time delay of the ALU using the timing analyzer. Examine the report file and find the device utilization. Use the logic element (LE) device utilization percentage found in the compilation report to compare the size of the designs.

5. Explore different synthesis options for the ALU from problem 3. Change the area and speed synthesis settings in the compiler under **Assignments ⇨Settings ⇨Analysis and Synthesis Settings**, rerun the timing analyzer to determine speed, and examine the report file for hardware size estimates. Include data points for the default, optimized for speed, balanced, and optimized for area settings. Build a plot showing the speed versus area trade-offs possible in the synthesis tool. Use the logic element (LE) device utilization percentage found in the compilation report to compare the size of the designs.

6. Develop a VHDL model of one of the TTL chips listed below. The model should be functionally equivalent, but there will be timing differences. Compare the timing differences between the VHDL FPGA implementation and the TTL chip. Use a data book or find a data sheet using the World Wide Web.

   A. 7400 Quad nand gate

   B. 74LS241 Octal buffer with tri-state output

   C. 74LS273 Octal D flip-flop with Clear

   D. 74163 4-bit binary counter

   E. 74LS181 4-bit ALU

7. Replace the 8count block used in the tutorial in Chapter 4, with a new counter module written in VHDL. Simulate the design and download a test program to the UP 3 board.

8. Implement a 128 by 32 RAM using VHDL and the Altsyncram function. Do not use registered output options. Target the design to the Cyclone EP1C6240C8 device. Use the timing analyzer to determine the worst-case read and write access times for the memory.

9. Study the VHDL code in the LCD Display UP3core function and draw a state diagram of the initialization and data transfer operations and explain its operation. You may find it helpful to examine the data sheet for the LCD display's microcontroller.

CHAPTER 7

Using Verilog for Synthesis of Digital Hardware

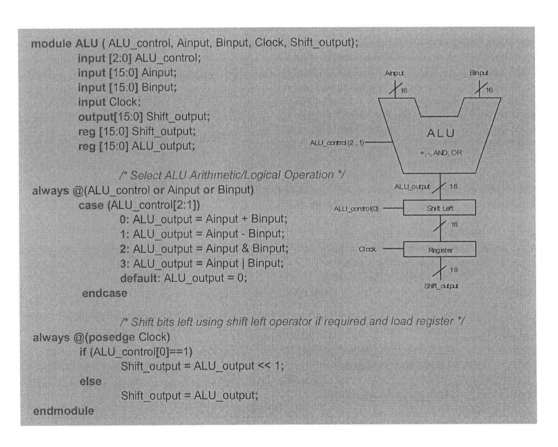

```verilog
module ALU ( ALU_control, Ainput, Binput, Clock, Shift_output);
        input [2:0] ALU_control;
        input [15:0] Ainput;
        input [15:0] Binput;
        input Clock;
        output[15:0] Shift_output;
        reg [15:0] Shift_output;
        reg [15:0] ALU_output;

                /* Select ALU Arithmetic/Logical Operation */
        always @(ALU_control or Ainput or Binput)
                case (ALU_control[2:1])
                        0: ALU_output = Ainput + Binput;
                        1: ALU_output = Ainput - Binput;
                        2: ALU_output = Ainput & Binput;
                        3: ALU_output = Ainput | Binput;
                        default: ALU_output = 0;
                endcase

                /* Shift bits left using shift left operator if required and load register */
        always @(posedge Clock)
                if (ALU_control[0]==1)
                        Shift_output = ALU_output << 1;
                else
                        Shift_output = ALU_output;
endmodule
```

7 Using Verilog for Synthesis of Digital Hardware

Verilog is another language that, like VHDL, is widely used to model and design digital hardware. In the early years, Verilog was a proprietary language developed by one CAD vendor, Gateway. Verilog was developed in the 1980's and was initially used to model high-end ASIC devices. In 1990, Verilog was released into the public domain, and Verilog now is the subject of IEEE standard 1364. Today, Verilog is supported by numerous CAD tool and programmable logic vendors. Verilog has a syntax style similar to the C programming language. Schools are more likely to cover VHDL since it was in the public domain several years earlier; however, in the FPGA industry, VHDL and Verilog have an almost equal market share for new design development.

Conventional programming languages are based on a sequential operation model. Digital hardware devices by their very nature operate in parallel. This means that conventional programming languages cannot accurately describe or model the operation of digital hardware since they are based on the sequential execution of statements. Like VHDL, Verilog is designed to model parallel operations.

> IT IS **CRUCIAL** TO REMEMBER THAT VERILOG MODULES AND CONCURRENT STATEMENTS ALL OPERATE IN PARALLEL.

In this section, a brief introduction to Verilog for logic synthesis will be presented. It is assumed that the reader is already familiar with basic digital logic devices and some basic C syntax.

Whenever you need help with Verilog syntax, Verilog templates of common statements are available in the Quartus II online help. In the text editor, just click the right mouse button and Insert ⇨Templates select Verilog.

7.1 Verilog Data Types

For logic synthesis, Verilog has simple data types. The net data type, wire, and the register data type, reg. A model with a net data type, wire, has a corresponding electrical connection or wire in the modeled device. Type reg is updated under the control of the surrounding procedural flow constructs typically inside an always statement. Type reg does not necessarily imply that the synthesized hardware for a signal contains a register, digital storage device, or flip-flop. It can also be purely combinational logic.

Table 7.1 lists the Verilog operators and their common function in Verilog synthesis tools.

7.2 Verilog Based Synthesis of Digital Hardware

Verilog can be used to construct models at a variety of abstraction levels such as structural, behavioral, register transfer level (RTL), and timing. An RTL model of a circuit described in Verilog describes the input/output relationship in terms of dataflow operations on signal and register values. If registers are required, a synchronous clocking scheme is normally used. Sometimes an RTL model is also referred to as a dataflow-style model.

Verilog simulation models often include physical device time delays. In Verilog models written for logic synthesis, timing information should not be provided.

For timing simulations, the CAD tools automatically include the actual timing delays for the synthesized logic circuit. An FPGA timing model supplied by the CAD tool vendor is used to automatically generate the physical device time delays inside the FPGA. Sometimes this timing model is also written in Verilog. For a quick overview of Verilog, several constructs that can be used to synthesize common digital hardware devices will be presented.

7.3   Verilog Operators

Table 7.1 lists the Verilog operators and their common function in Verilog synthesis tools.

Table 7.1 Verilog Operators.

Verilog Operator	Operation
+	Addition
-	Subtraction
*	Multiplication*
/	Division*
%	Modulus*
{ }	Concatenation – used to combine bits
<<	rotate left
>>	rotate right
=	equality
!=	Inequality
<	less than
<=	less than or equal
>	greater than
>=	greater than or equal
!	logical negation
&&	logical AND
||	logical OR
&	Bitwise AND
|	Bitwise OR
^	Bitwise XOR
~	Bitwise Negation

*Not supported in some Verilog synthesis tools. In the Quartus II tools, multiply , divide, and mod of integer values is supported. Efficient design of multiply or divide hardware may require the user to specify the arithmetic algorithm and design in Verilog.

7.4 Verilog Synthesis Models of Gate Networks

The first example consists of a simple gate network. In this model, both a concurrent assignment statement and a sequential always block are shown that generate the same gate network. X is the output on one network and Y is the output on the other gate network. The two gate networks operate in parallel.

In Verilog synthesis, inputs and outputs from the module will become I/O pins on the programmable logic device. For comments "//" makes the rest of a line a comment and "/*" and "*/" can be used to make a block of lines a comment.

The Quartus II editor performs syntax coloring and is useful to quickly find major problems with Verilog syntax. Verilog is case sensitive just like C.

Verilog concurrent statements are executed in parallel. Inside an always statementstatements are executed in sequential order, and all of the always statements are executed in parallel. The always statement is Verilog's equivalent of a process in VHDL. .

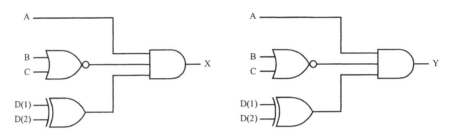

```
module gatenetwork(A, B, C, D, X, Y);
    input A;
    input B;
    input C;
    input [2:1] D;
    output X, Y;
    reg Y;
        // concurrent assignment statement
    wire X = A & ~(B|C) & (D[1] ^ D[2]);
        /* Always concurrent statement- sequential execution inside */
    always @( A or B or C or D)
        Y = A & ~(B|C) & (D[1] ^ D[2]);

endmodule
```

7.5 Verilog Synthesis Model of a Seven-segment LED Decoder

The following Verilog code implements a seven-segment decoder for seven-segment LED displays. A 7-bit vector is used to assign the value of all seven bits in a single case statement. In the 7-bit logic vector, the most-significant bit is segment 'a' and the least-significant bit is segment 'g'. The logic synthesis CAD tool automatically minimizes the logic required for implementation. The signal Hex_digit contains the 4-bit binary value to be displayed in hexadecimal.

```
module DEC_7SEG(Hex_digit, segment_a, segment_b, segment_c,
        segment_d, segment_e, segment_f, segment_g);
    input [3:0] Hex_digit;
    output segment_a, segment_b, segment_c, segment_d;
    output segment_e, segment_f, segment_g;
    reg [6:0] segment_data;

    always @(Hex_digit)
            /* Case statement implements a logic truth table using gates*/
        case (Hex_digit)
            4'b 0000:   segment_data = 7'b 1111110;
            4'b 0001:   segment_data = 7'b 0110000;
            4'b 0010:   segment_data = 7'b 1101101;
            4'b 0011:   segment_data = 7'b 1111001;
            4'b 0100:   segment_data = 7'b 0110011;
            4'b 0101:   segment_data = 7'b 1011011;
            4'b 0110:   segment_data = 7'b 1011111;
            4'b 0111:   segment_data = 7'b 1110000;
            4'b 1000:   segment_data = 7'b 1111111;
            4'b 1001:   segment_data = 7'b 1111011;
            4'b 1010:   segment_data = 7'b 1110111;
            4'b 1011:   segment_data = 7'b 0011111;
            4'b 1100:   segment_data = 7'b 1001110;
            4'b 1101:   segment_data = 7'b 0111101;
            4'b 1110:   segment_data = 7'b 1001111;
            4'b 1111:   segment_data = 7'b 1000111;
            default:    segment_data = 7'b 0111110;
        endcase
```

The following Verilog concurrent assignment statements extract the seven 1-bit values needed to connect the individual segments. The not operator (~) is used since a logic zero actually turns on most LEDs. Automatic minimization in the synthesis process will eliminate the extra inverter in the logic circuit.

```
            /* extract segment data bits and invert */
            /* LED driver circuit is inverted */
    wire segment_a = ~segment_data[6];
    wire segment_b = ~segment_data[5];
    wire segment_c = ~segment_data[4];
    wire segment_d = ~segment_data[3];
    wire segment_e = ~segment_data[2];
    wire segment_f = ~segment_data[1];
    wire segment_g = ~segment_data[0];
endmodule
```

7.6 Verilog Synthesis Model of a Multiplexer

The next example shows several alternative ways to synthesize a 2-to-1 multiplexer in Verilog. Three identical multiplexers that operate in parallel are synthesized by this example. The wire conditional continuous assignment statement can be used for a 2-to-1 mux. A concurrent assign statement can also be used instead of wire, if the output signal is already declared. In Verilog, IF and CASE statements must be inside an always statement. The inputs and outputs from the multiplexers could be changed to bit vectors if an entire bus is multiplexed. Multiplexers with more than two inputs can also be easily constructed and a case statement is preferred. Nested IF statements generate priority-encoded logic that requires more hardware and produce a slower circuit than a CASE statement.

```
                    /* Multiplexer example shows three ways to model a 2 to 1 mux */
module multiplexer(A, B, mux_control, mux_out1, mux_out2, mux_out3);
   input A;                         /* Input Signals and Mux Control */
   input B;
   input mux_control;
   output mux_out1,mux_out2, mux_out3;
   reg mux_out2, mux_out3;
                              /* Conditional Continuous Assignment Statement */
                              /*  works like an IF - ELSE */
   wire mux_out1 = (mux_control)? B:A;
                              /* If statement inside always statement */
      always @(A or B or mux_control)
         if (mux_control)
            mux_out2 = B;
         else
            mux_out2 = A;
                              /* Case statement inside always statement */
      always @(A or B or mux_control)
         case (mux_control)
            0: mux_out3 = A;
            1: mux_out3 = B;
            default: mux_out3 = A;
         endcase
endmodule
```

7.7 Verilog Synthesis Model of Tri-State Output

Tri-state gates are supported in Verilog synthesis tools and are supported in many programmable logic devices. Most programmable logic devices have tri-state output pins. Some programmable logic devices do not support internal tri-state logic. Here is a Verilog example of a tri-state output. In Verilog, the assignment of the value "Z" to a signal produces a tri-state output.

```
module tristate (a, control, tri_out);
   input a, control;
   output tri_out;
   reg tri_out;
   always @(control or a)
      if (control)
            /* Assignment of Z value generates a tri-state output */
            tri_out = 1'bZ;
      else
            tri_out = a;
endmodule
```

7.8 Verilog Synthesis Models of Flip-flops and Registers

In the next example, several flip-flops will be generated. Unlike earlier combinational hardware devices, a flip-flop can only be synthesized inside an always statement. The positive clock edge is selected by posedge clock and positive edge triggered D flip-flops will be used for synthesis. The following module contains a variety of Reset and Enable options on positive edge-triggered D flip-flops. The negative clock edge is selected by negedge clock and negative edge-triggered D flip-flops will be used for synthesis.

```verilog
module DFFs(D, clock, reset, enable, Q1, Q2, Q3, Q4);
input D;
input clock;
input reset;
input enable;
output Q1, Q2, Q3, Q4;
reg Q1, Q2, Q3, Q4;
            /* Positive edge triggered D flip-flop */
    always @(posedge clock)
        Q1 = D;
            /* Positive edge triggered D flip-flop */
            /*    with synchronous reset */
    always @(posedge clock)
        if (reset)
            Q2 = 0;
        else
            Q2 = D;
            /* Positive edge triggered D flip-flop */
            /*    with asynchronous reset */
    always @(posedge clock or posedge reset)
        if (reset)
            Q3 = 0;
        else
            Q3 = D;
            /* Positive edge triggered D flip-flop */
            /* with asynchronous reset and enable */
    always @(posedge clock or posedge reset)
        if (reset)
            Q4 = 0;
        else if (enable)
            Q4 = D;
endmodule
```

In Verilog, as in any digital logic designs, it is not good design practice to AND or gate other signals with the clock. Use a flip-flop with a clock enable instead to avoid timing and clock skew problems. In some limited cases, such as power management, a single level of clock gating can be used. This works only when a small amount of clock skew can be tolerated and the signal gated with the clock is known to be hazard or glitch free. A particular programmable logic device may not support every flip-flop or latch type and all of the Set/Reset and Enable options.

If D and Q are replaced by bit vectors in any of these examples, registers with the correct number of bits will be generated instead of individual flip-flops.

7.9 Accidental Synthesis of Inferred Latches

Here is a very common problem to be aware of when coding Verilog for synthesis. If a non-clocked process has any path that does not assign a value to an output, Verilog assumes you want to use the previous value. A level triggered latch is automatically generated or inferred by the synthesis tool to save the previous value. In many cases, this can cause serious errors in the design. Edge-triggered flip-flops should not be mixed with level-triggered latches in a design or serious timing problems will result. Typically this can happen in CASE statements or nested IF statements. In the following example, the signal Output2 infers a latch when synthesized. Assigning a value to Output2 in the last ELSE clause will eliminate the latch. Warning messages may be generated during compilation when a latch is inferred on some tools. Note the use of begin…end is somewhat different than the use of braces in C.

```
module ilatch( A, B, Output1, Output2);
input A, B;
output Output1, Output2;
reg Output1, Output2;

always@( A or B)
    if (!A)
        begin
            Output1 = 0;
            Output2 = 0;
        end
    else
        if (B)
            begin
                Output1 = 1;
                Output2 = 1;
            end
        else
            Output1 = 0;      /*latch inferred since no value */
                              /*is assigned to Output2 here */
endmodule
```

7.10 Verilog Synthesis Model of a Counter

Here is an 8-bit counter design. Compare operations such as "<" are supported and they generate a comparator logic circuit to test for the maximum count value. The assignment count = count+1; synthesizes an 8-bit incrementer. An incrementer circuit requires less hardware than an adder that adds one. The operation, "+1", is treated as a special incrementer case by synthesis tools.

```
module counter(clock, reset, max_count, count);
    input clock;
    input reset;
    input [7:0] max_count;
    output [7:0] count;
    reg [7:0] count;
            /* use positive clock edge for counter */
    always @(posedge clock or posedge reset)
        begin
            if (reset)
                count = 0;                  /* Reset  Counter */
            else if (count < max_count)     /* Check for maximum count */
                count = count + 1;          /* Increment Counter */
            else
                count = 0;                  /*  Counter set back to 0*/
        end
endmodule
```

7.11 Verilog Synthesis Model of a State Machine

The next example shows a Moore state machine with three states, two inputs and a single output. A state diagram of the example state machine is shown in Figure 7.1. Unlike VHDL, A direct assignment of the state values is required in Verilog's parameter statement. The first Always block assigns the next state using a case statement that is updated on the positive clock edge, posedge.

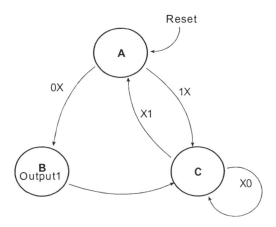

Figure 7.1 State Diagram for state_mach Verilog example

```
module state_mach (clk, reset, input1, input2 ,output1);
    input clk, reset, input1, input2;
    output output1;
    reg output1;
    reg [1:0] state;
```

```
                    /* Make State Assigments */
        parameter [1:0] state_A = 0, state_B = 1, state_C = 2;


    always@(posedge clk or posedge reset)
        begin
            if (reset)
                state = state_A;
            else
                        /* Define Next State Transitions using a Case */
                        /* Statement based on the Current State */
                case (state)
                    state_A:
                        if (input1==0)
                            state = state_B;
                        else
                            state = state_C;
                    state_B:
                        state = state_C;
                    state_C:
                        if (input2) state = state_A;
                    default:  state = state_A;
                endcase
        end
                        /* Define State Machine Outputs */
    always @(state)
        begin
            case (state)
                state_A: output1 = 0;
                state_B: output1 = 1;
                state_C: output1 = 0;
                default: output1 = 0;
            endcase
        end
    endmodule
```

7.12 Verilog Synthesis Model of an ALU with an Adder/Subtractor and a Shifter

Here is an 8-bit arithmetic logic unit (ALU), that adds, subtracts, bitwise ANDs, or bitwise ORs, two operands and then performs an optional shift on the output. The most-significant two bits of the Op-code select the arithmetic logical operation. If the least-significant bit of the op_code equals '1' a 1-bit left-shift operation is performed. An addition and subtraction circuit is synthesized for the "+" and "-" operator.

Depending on the number of bits and the speed versus area settings in the synthesis tool, ripple carry or carry-lookahead circuits will be used. Several "+" and "-" operations in multiple assignment statements may generate multiple ALUs and increase the hardware size, depending on the Verilog CAD tool and compiler settings used. If a single ALU is desired, muxes can be placed at the inputs and the "+" operator would be used only in a single assignment statement.

```
module ALU ( ALU_control, Ainput, Binput, Clock, Shift_output);
    input [2:0] ALU_control;
    input [15:0] Ainput;
     input [15:0] Binput;
    input Clock;
    output[15:0] Shift_output;
    reg [15:0] Shift_output;
    reg [15:0] ALU_output;

        /* Select ALU Arithmetic/Logical Operation */
    always @(ALU_control or Ainput or Binput)
        case (ALU_control[2:1])
            0: ALU_output = Ainput + Binput;
            1: ALU_output = Ainput - Binput;
            2: ALU_output = Ainput & Binput;
            3: ALU_output = Ainput | Binput;
            default: ALU_output = 0;
        endcase

        /* Shift bits left using shift left operator if required and load register */
    always @(posedge Clock)
        if (ALU_control[0]==1)
            Shift_output = ALU_output << 1;
        else
            Shift_output = ALU_output;
endmodule
```

7.13 Verilog Synthesis of Multiply and Divide Hardware

In the Quartus II tool, integer multiply and divide is supported using Verilog's "*" and "/" operators. In current generation tools, efficient design of multiply or divide hardware typically requires the use of a vendor-specific library function or even the specification of the arithmetic algorithm and hardware implementation in Verilog.

A wide variety of multiply and divide algorithms that trade off time versus hardware size can be found in most computer arithmetic texts. Several such references are listed at the end of this chapter. These algorithms require a sequence of add/subtract and shift operations that can be easily synthesized in Verilog using the standard operators. The LPM_MULT function in Quartus II can be used to synthesize integer multipliers. LPM_DIVIDE, is also available. When using LPM functions, **Tools ⇨ MegaWizard Plug-in Manager** can be used to help generate Verilog code. The LPM functions also support pipeline options. Array multiply and divide hardware for more than a few bits requires extensive hardware and a large FPGA. A few large FPGAs now contain multiplier blocks.

```
module mult (dataa, datab, result);
    input  [7:0] dataa;
    input  [7:0] datab;
    output [15:0]  result;

    wire [15:0] sub_wire0;
    wire [15:0] result = sub_wire0[15:0];
                /* Altera LPM 8x8 multiply function  result = dataa * datab */
    lpm_mult lpm_mult_component (
                .dataa (dataa),
                .datab (datab),
                .result (sub_wire0) );
    defparam
      lpm_mult_component.lpm_widtha = 8,
      lpm_mult_component.lpm_widthb = 8,
      lpm_mult_component.lpm_widthp = 16,
      lpm_mult_component.lpm_widths = 1,
      lpm_mult_component.lpm_type = "LPM_MULT",
      lpm_mult_component.lpm_representation = "UNSIGNED",

endmodule
```

Floating-point operations can be implemented on very large FPGAs; however, performance is lower than current floating-point DSP and microprocessor chips. The floating-point algorithms must be coded by the user in Verilog using integer add, multiply, divide, and shift operations. The LPM_CLSHIFT function is useful for the barrel shifter needed in a floating-point ALU. Some floating point IP cores are starting to appear. Many FPGA vendors also have optimized arithmetic packages for DSP applications such as FIR filters.

7.14 Verilog Synthesis Models for Memory

Typically, it is more efficient to call a vendor-specific function to synthesize RAM. These functions typically use the FPGA's internal RAM blocks rather than building a RAM using FPGA logic elements. The memory function in the Altera toolset is the ALTSYNCRAM function. On the UP 2 board's older FPGA, the LPM_RAM_DQ memory function should be used. The memory can be set to an initial value using a separate memory initialization file with the extension *.mif. A similar call, LPM_ROM, can be used to synthesize ROM.

If small blocks of multi-ported or other special-purpose RAM are needed, they can be synthesized using registers with address decoders for the write operation and multiplexers for the read operation. Additional read or write ports can be added to synthesize RAM. An example of this approach is a dual-ported register file for a computer processor core. Most RISC processors need to read two registers on each clock cycle and write to a third register.

Verilog Memory Model - Example One

The first memory example synthesizes a memory that can perform a read and a write operation every clock cycle. Memory is built using arrays of positive edge-triggered D flip-flops. Memory write, memwrite, is gated with an address decoder output and used as an enable to load each memory location during a write operation. A synchronous write operation is more reliable. Asynchronous write operations respond to any logic hazards or momentary level changes on the write signal. As in any synchronous memory, the write address must be stable before the rising edge of the clock signal. A non-clocked mux is used for the read operation. If desired, memory can be initialized by a reset signal.

```verilog
module memory(read_data, read_address, write_data, write_address,
                    memwrite, clock, reset);
output [7:0] read_data;
input [2:0] read_address;
input [7:0] write_data;
input [2:0] write_address;
input memwrite;
input clock;
input reset;
reg [7:0] read_data, mem0, mem1;

            /* Block for memory read */
always @(read_address or mem0 or mem1)
    begin
        case(read_address)
            3'b 000: read_data = mem0;
            3'b 001: read_data = mem1;
            /* Unimplemented memory */
            default: read_data = 8'h FF;
        endcase
    end


            /* Block for memory write */
always @(posedge clock or posedge reset)
    begin
    if (reset)
        begin
            /* Initial values for memory (optional) */
            mem0 = 8'h AA ;
            mem1 = 8'h 55;
        end
    else if (memwrite)
            /* write new value to memory */
        case (write_address)
            3'b 000 : mem0 = write_data;
            3'b 001 : mem1 = write_data;
        endcase
    end
endmodule
```

Verilog Memory Model - Example Two

The second example shows the use of Altera's ALTSYNCRAM megafunction to implement a block of memory. For more information on the megafunctions see the online help guide in the Quartus II tool. In single port mode, the ALTSYNCRAM memory can do either a read or a write operation in a single clock cycle since there is only one address bus. In dual port mode, it can do both a read and write. If this is the only memory operation needed, the ALTSYNCRAM function produces a more efficient hardware implementation than synthesis of the memory in Verilog. In the ALTSYNCRAM megafunction, the memory address must be clocked into a dedicated address register located inside the FPGA's synchronous memory block. Asynchronous memory operations without a clock can cause timing problems and are not supported on many FPGAs including the Cyclone.

```verilog
module amemory ( write_data, write_enable, address, clock, read_data);

input [7:0]  write_data;
input   write_enable;
input [2:0]  address;
input   clock;
output [7:0]  read_data;
wire [7:0] sub_wire0;
wire [7:0] read_data = sub_wire0[7:0];
                    /* Use Altera Altsyncram function for memory */
altsyncram  altsyncram_component (
            .wren_a (write_enable),
            .clock0 (clock),
            .address_a (address),
            .data_a (write_data),
            .q_a (sub_wire0));
defparam
    altsyncram_component.operation_mode = "SINGLE_PORT",
                    /* 8 data bits, 3 address bits, and no register on read data */
    altsyncram_component.width_a = 8,
    altsyncram_component.widthad_a = 3,
    altsyncram_component.outdata_reg_a = "UNREGISTERED",
                    /* Reads in mif file for initial memory data values (optional) */
    altsyncram_component.init_file = "memory.mif";
endmodule
```

On the Cyclone FPGA chip, the memory can be implemented using the M4K memory blocks, which are separate from the FPGA's logic cells. In the Cyclone EP1C6 chip there are 20 M4K RAM blocks at 4Kbits each for a total of 92,160 bits. In the Cyclone EP1C12 there are 52 M4K blocks for a total of 239,616 bits. Each M4K block can be setup to be 4K by 1, 2K by 2, 1K by 4, 512 by 8, 256 by 16,256 by 18, 128 by 32 or 128 by 36 bits wide. The **Tools** ⇨ **Megawizard Plug-in Manager** feature is useful to configure the Altsyncram parameters.

7.15 Hierarchy in Verilog Synthesis Models

Large Verilog models should be split into a hierarchy using a top-level structural model in Verilog or by using the symbol and graphic editor in the Quartus II tool. In the graphical editor, a Verilog file can be used to define the contents of a symbol block. Synthesis tools run faster using a hierarchy on large models and it is easier to write, understand, and maintain a large design when it is broken up into smaller modules.

An example of a hierarchical design with three submodules is seen in the schematic in Figure 7.2. Following the schematic, the same design using a top-level Verilog structural model is shown. This Verilog structural model provides the same connection information as the schematic seen in Figure 7.2.

Debounce, Onepulse, and Clk_div are the names of the Verilog submodules. Each one of these submodules has a separate Verilog source file. In the Quartus II tool, compiling the top-level module will automatically compile the lower-level modules.

In the example Verilog structural model for Figure 7.2, note the use of a component instantiation statement for each of the three submodules. The component instantiation statement declares the module name and connects inputs and outputs of the module. New internal signal names used for interconnections of modules should also be declared at the beginning of the top level module.

The order of each module's signal names must be the same as in the Verilog submodule files. Each instantiation of a module is given a unique name so that a single module can be used several times. As an example, the single instantiation of the debounce module is called debounce1 in the example code.

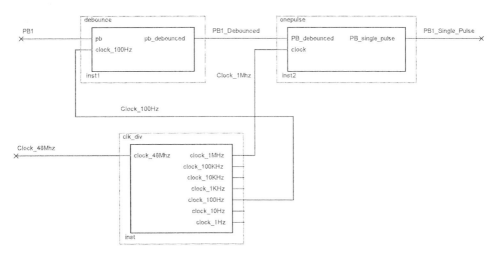

Figure 7.2 Schematic of Hierarchical Design Example

Note that node names in the schematic or signals in Verilog used to interconnect modules need not always have the same names as the signals in the components they connect. As an example, PB_debounced on the debounce component connects to an internal signal with a different name, PB1_debounced.

```
module hierarch(Clock_48MHz, PB1, PB1_Single_Pulse);
input Clock_48MHz, PB1;
output PB1_Single_Pulse;
                /* Declare internal interconnect signals */
reg Clock_100Hz, Clock_1MHz, PB1_Debounced;


            /* declare and connect all three modules in the hiearchy */
debounce debounce1( PB1, Clock_100Hz, PB1_Debounced);

clk_div clk_div1( Clock_48MHz, Clock_1MHz, Clock_100Hz);

onepulse onepulse1( PB1_Debounced, Clock_100Hz, PB1_Single_Pulse);

endmodule
```

7.16 For additional information

The chapter has introduced the basics of using Verilog for digital synthesis. It has not explored all of the language options available. The Altera online help contains Verilog syntax and templates. A number of Verilog reference textbooks are also available. Unfortunately, not all of them currently contain Verilog models that can be used for digital logic synthesis. Two recommendations are *HDL Chip Design* by Douglas J. Smith, Doone Publications, 1996 and *Modeling, Synthesis, and Rapid Prototyping with the Verilog HDL* by Michael Ciletti, 1999. An interesting free VHDL to Verilog conversion program is also available at www.ocean-logic.com/downloads.htm.

7.17 Laboratory Exercises

1. Write a Verilog model for the state machine shown in the following state diagram and verify correct operation with a simulation using the Altera CAD tools. A and B are the two states, X is the output, and Y is the input. Use the timing analyzer to determine the maximum clock frequency on the Cyclone EP1C6Q240C8 device.

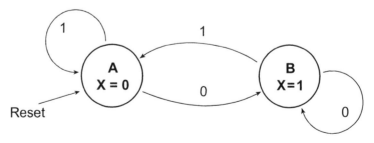

2. Write a Verilog model for a 32-bit, arithmetic logic unit (ALU). Verify correct operation with a simulation using the Altera CAD tools. A and B are 32-bit inputs to the ALU, and

Y is the output. A shift operation follows the arithmetic and logical operation. The opcode controls ALU functions as follows:

Opcode	Operation	Function
000XX	ALU_OUT <= A	Pass A
001XX	ALU_OUT <= A + B	Add
010XX	ALU_OUT <= A-B	Subtract
011XX	ALU_OUT <= A AND B	Logical AND
100XX	ALU_OUT <= A OR B	Logical OR
101XX	ALU_OUT <= A + 1	Increment A
110XX	ALU_OUT <= A-1	Decrement A
111XX	ALU_OUT <= B	Pass B
XXX00	Y <= ALU_OUT	Pass ALU_OUT
XXX01	Y<= SHL(ALU_OUT)	Shift Left
XXX10	Y<=SHR(ALU_OUT)	Shift Right (unsigned-zero fill)
XXX11	Y<= 0	Pass 0's

3. Use the Cyclone chip as the target device. Determine the worst case time delay of the ALU using the timing analyzer. Examine the report file and find the device utilization. Use the logic element (LE) device utilization percentage found in the compilation report to compare the size of the designs.

4. Explore different synthesis options for the ALU from problem 3. Change the area and speed synthesis settings in the compiler under **Assignments** ⇨**Settings** ⇨**Analysis and Synthesis Settings**, rerun the timing analyzer to determine speed, and examine the report file for hardware size estimates. Include data points for the default, optimized for speed, balanced, and optimized for area settings. Build a plot showing the speed versus area trade-offs possible in the synthesis tool. Use the logic element (LE) device utilization percentage found in the compilation report to compare the size of the designs.

5. Develop a Verilog model of one of the TTL chips listed below. The model should be functionally equivalent, but there will be timing differences. Compare the timing differences between the Verilog FPGA implementation and the TTL chip. Use a data book or find a data sheet using the World Wide Web.

    F. 7400 Quad nand gate

    G. 74LS241 Octal buffer with tri-state output

    H. 74LS273 Octal D flip-flop with Clear

    I. 74163 4-bit binary counter

    J. 74LS181 4-bit ALU

6. Replace the 8count block used in the tutorial in Chapter 4, with a new counter module written in Verilog. Simulate the design and download a test program to the UP 3 board.

7. Implement a 128 by 32 RAM using Verilog and the Altsyncram function. Do not use registered output options. Target the design to the Cyclone device. Use the timing analysis tools to determine the worst-case read and write access times for the memory.

CHAPTER 8

State Machine Design: The Electric Train Controller

8 State Machine Design: The Electric Train Controller

8.1 The Train Control Problem

The track layout of a small electric train system is shown in Figure 8.1. Two trains, we'll call A and B, run on the tracks, hopefully without colliding. To avoid collisions, the trains require a safety controller that allows trains to move in and out of intersections without mishap.

In Figure 8.1, assume for a moment that Train A is at Switch 3 and moving counterclockwise. Let's also assume that Train B is moving counterclockwise and is at Sensor 2. Since Train B is entering the common track (Track 2), Train A must be stopped when it reaches Sensor 1, and must wait until Train B has passed Sensor 3. At this point, Train A will be allowed to enter Track 2, and Train B will move toward Sensor 2.

The controller is a state machine that uses the sensors as inputs. The controller's outputs control the power to the tracks, the direction of the trains, and the position of the switches. However, the state machine does not control the speed of the train. This means that the system controller must function correctly independent of the speed of the two trains.

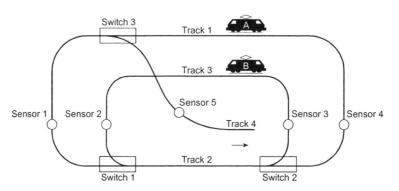

Figure 8.1 Track Layout with Input Sensors and Output Switches and Output Tracks.

The following sections describe how the state machine should control each signal to operate the trains properly. Figure 8.2 demonstrates how the actual electric train system was built with relays. A relay is an electrically controlled switch. A UP 3-based "virtual" train simulation will be used that emulates this relay setup. Since there are no actual relays on the UP 3 board, it is only intended to give you a visual diagram of how the output signals work in the real system.

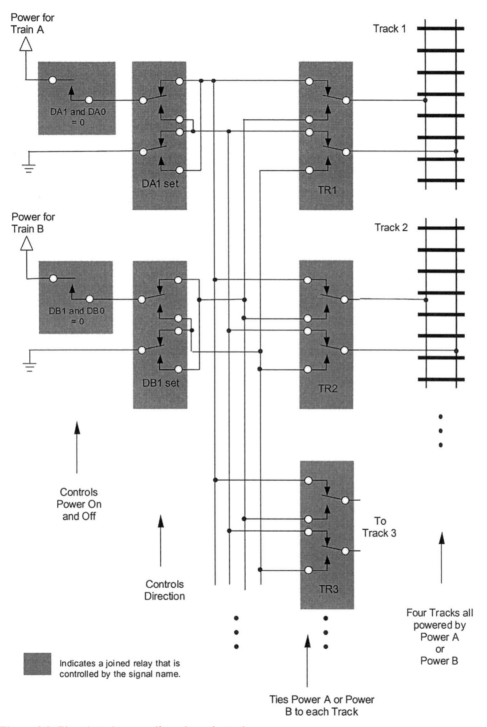

Figure 8.2 Electric train controller relay schematic.

8.2 Track Power (T1, T2, T3, and T4)

The track power signals (T1, T2, T3, T4) determine which power supply is attached to which track. Note that since these are binary signals, one of the power supplies is always connected to a track.

The track power connections are based on the actual switching relay system. There are two power supplies (Power A and Power B) that can be connected to the tracks. What these two power supplies actually allow you to do on the real system is assign different speeds to each train. This is part of the simulator and is controlled by the DIP switches on the UP 3 board. Speed will NOT be controlled by your state machine, only Stop, Forward, and Reverse.

As illustrated in Figure 8.2, if either direction switch is set, the power supply is connected to the next level of switches. Another set of relays determines the polarity (direction of trains), while a third connects either Power A or Power B to each track. Each track is either assigned power from source A (Tn = 0) or from source B (Tn = 1). In other words, if all four signals (T1, T2, T3, and T4) are asserted high, all tracks will be powered from power source B and would all be assigned the same Direction (see the next section under Track Direction for controlling train direction). See Figure 8.3 for an example.

8.3 Track Direction (DA1-DA0, and DB1-DB0)

The direction for each track is controlled by four signals (two for each power source), DA (DA1-DA0) for source A, and DB (DB1-DB0) for source B. When these signals indicate forward "01" for a particular power source, any train on a track assigned to that power source will move counterclockwise (on track 4, the train moves toward the outer track). When the signals imply reverse "10", the train(s) will move clockwise. The "11" value is illegal and should not be used. When these signals are set to "00", any train assigned to the given source will stop. (See Figures 8.2 and 8.3.)

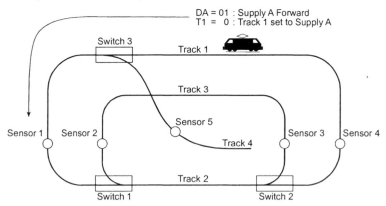

Figure 8.3 Track Power is connected to one of Two Power Sources: A and B.

8.4 Switch Direction (SW1, SW2, and SW3)

Switch directions are controlled by asserting SW1, SW2, and SW3 either high (outside connected with inside track) or low (outside tracks connected). That is, anytime all of the switches are set to 1, the tracks are setup such that the outside tracks are connected to the inside tracks. (See Figure 8.4.)

If a train moves the wrong direction through an open switch it will derail. Be careful. If a train is at the point labeled "Track 1" in Figure 8.4 and is moving to the left, it will derail at Switch 3. To keep it from derailing, SW3 would need to be set to 0.

Also, note that Tracks 3 and 4 cross at an intersection and care must be taken to avoid a crash at this point.

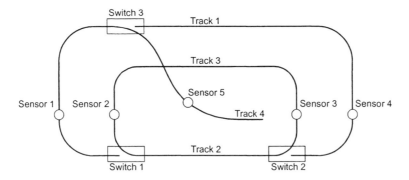

Figure 8.4 Track Direction if all Switches are Asserted (SW1 = SW2 = SW3 = 1)

8.5 Train Sensor Input Signals (S1, S2, S3, S4, and S5)

The five train sensor inputs (S1, S2, S3, S4, and S5) go high when a train is near the sensor location. It should be noted that sensors (S1, S2, S3, S4, and S5) *do not go high for only one clock cycle*. In fact, the sensors fire continuously for *many* clock cycles per passage of a train. This means that if your design is testing the same sensor from one state to another, you must wait for the signal to change from high to low.

As an example, if you wanted to count how many times that a train passes Sensor 1, you can not just have an "IF S1 GOTO count-one state" followed by "IF S1 GOTO count-two state." You would need to have a state that sees S1='1', then S1='0', then S1='1' again before you can be sure that it has passed S1 twice. If your state machine has two concurrent states that look for S1='1', the state machine will pass through both states in two consecutive clock cycles although the train will have passed S1 only once.

Another way would be to detect S1='1', then S4='1', then S1='1' if, in fact, the train was traversing the outside loop continuously. Either method will ensure that the train passed S1 twice.

The signal inputs and outputs have been summarized in the following figure:

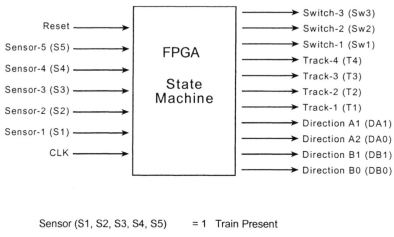

Sensor (S1, S2, S3, S4, S5)   = 1 Train Present

                                    = 0 Train not Present

Switches (SW1, SW2, SW3)   = 0 Connected to Outside Track

                                    = 1 Connected to Inside Track

Track (T1, T2, T3, T4)   = 0 A Virtual Power on Track n

                                    = 1 B Virtual Power on Track n

Direction (DA1-DA0) and (DB1-DB0)   = 00 Stop

                                    = 01 Forward (Counterclockwise)

                                    = 10 Backward (Clockwise)

Figure 8.5 Train Control State Machine I/O Configuration

8.6 An Example Controller Design

This is a working example of a train controller state machine. For this controller, two trains run counterclockwise at various speeds and avoid collisions. One Train (A) runs on the outer track and the other (B) runs on the inner track. Only one train at a time is allowed to occupy the common track.

Both an ASM chart and a classic state bubble diagram are illustrated in Figures 8.6 and 8.7 respectively. In the ASM chart, state names, ABout, Ain, Bin, Bstop, and Astop indicate the active and possible states. The rectangles contain the active (High) outputs for the given state. Outputs not listed are inactive (Low).

The diamond shapes in the ASM chart indicate where the state machine tests the condition of the inputs (S1, S2, etc.). When two signals are shown in a diamond, they are both tested at the same time for the indicated values.

A state machine classic bubble diagram is shown in Figure 8.7. Both Figures 8.6 and 8.7 contain the same information. They are simply different styles of representing a state diagram. The track diagrams in Figure 8.8 show the states visually. In the state names, "in" and "out" refer to the state of track 2, the track that is common to both loops.

Description of States in Example State Machine

All States
- T3 Asserted: The B power supply is assigned to track 3.
- All signals that are not "Asserted" are zero and imply a logical result as described.

ABout: "Trains A and B Outside"
- DA0 Asserted: Train A is on the outside track and moving counterclockwise (forward).
- DB0 Asserted: Train B is on the inner track (not the common track) and also moving forward.
- Note that by NOT Asserting DA1, it is automatically zero -- same for DB1. Hence, the outputs are DA = "01" and DB = "01".

Ain: "Train A moves to Common Track"
- Sensor 1 has fired either first or at the same time as Sensor 2.
- Either Train A is trying to move towards the common track, or
- Both trains are attempting to move towards the common track.
- Both trains are allowed to enter here; however, state Bstop will stop B if both have entered.
- DA0 Asserted: Train A is on the outside track and moving counterclockwise (forward).
- DB0 Asserted: Train B is on the inner track (not the common track) and also moving forward.

Bstop: "Train B stopped at S2 waiting for Train A to clear common track"
- DA0 Asserted: Train A is moving from the outside track to the common track.
- Train B has arrived at Sensor 2 and is stopped and waits until Sensor 4 fires.
- SW1 and SW2 are NOT Asserted to allow the outside track to connect to common track.
- Note that T2 is not asserted making Track 2 tied to the A Power Supply.

Bin: "Train B has reached Sensor 2 before Train A reaches Sensor 1"
- Train B is allowed to enter the common track. Train A is approaching Sensor 1.
- DA0 Asserted: Train A is on the outside track and moving counterclockwise (forward).
- DB0 Asserted: Train B is on the inner track moving towards the common track.
- SW1 Asserted: Switch 1 is set to let the inner track connect to the common track.
- SW2 Asserted: Switch 2 is set to let the inner track connect to the common track.
- T2 Asserted: The B Power Supply is also assigned to the common track.

Astop: "Train A stopped at S1 waiting for Train B to clear the common track"
- DB0 Asserted: Train B is on the inner track moving towards the common track.
- SW1 and SW2 Asserted: Switches 1 and 2 are set to connect the inner track to the common track.
- T2 Asserted: The B Power Supply is also assigned to the common track.

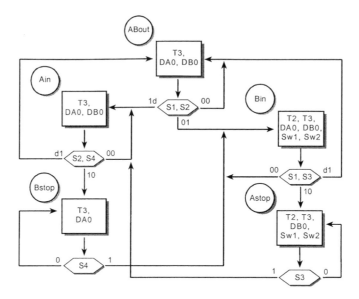

Figure 8.6 Example Train Controller ASM Chart.

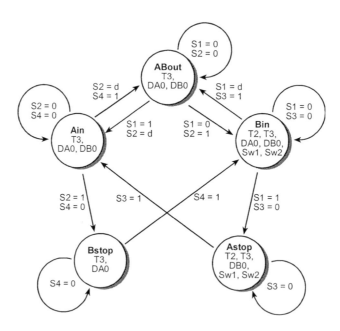

Figure 8.7 Example Train Controller State Diagram.

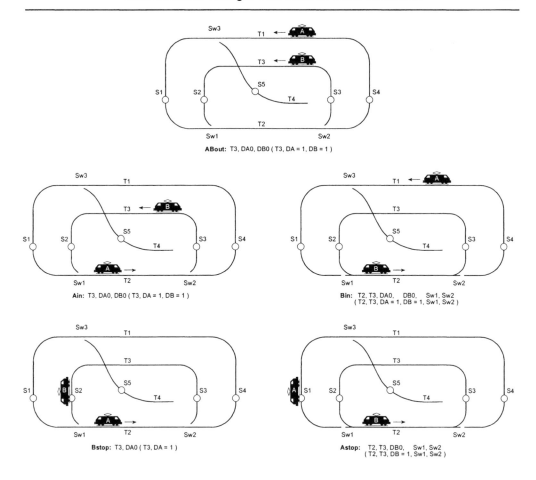

Figure 8.8 Working diagrams of train positions for each state.

Table 8.1 Outputs corresponding to states.

State	ABout	Ain	Astop	Bin	Bstop
Sw1	0	0	1	1	0
Sw2	0	0	1	1	0
Sw3	0	0	0	0	0
T1	0	0	0	0	0
T2	0	0	1	1	0
T3	1	1	1	1	1
T4	0	0	0	0	0
DA(1-0)	01	01	00	01	01
DB(1-0)	01	01	01	01	00

8.7 VHDL Based Example Controller Design

The corresponding VHDL code for the state machine in Figures 8.6 and 8.7 is shown below. A CASE statement based on the current state examines the inputs to select the next state. At each clock edge the next state becomes the current state. WITH...SELECT statements at the end of the program specify the outputs for each state. For additional VHDL help, see the help files in the Altera CAD tools or look at the VHDL examples in Chapter 6.

```vhdl
-- Example State machine to control trains-- File: Tcontrol.vhd
--
-- These libraries are required in all VHDL source files
LIBRARY IEEE;
USE IEEE.STD_LOGIC_1164.ALL;
USE IEEE.STD_LOGIC_ARITH.ALL;
USE IEEE.STD_LOGIC_UNSIGNED.ALL;

              -- This section defines state machine inputs and outputs
              -- No modifications should be needed in this section
ENTITY Tcontrol IS
PORT( reset, clock, sensor1, sensor2,
        sensor3, sensor4, sensor5    : IN      STD_LOGIC;
        switch1, switch2, switch3    : OUT     STD_LOGIC;
        track1, track2, track3, track4  : OUT  STD_LOGIC;
              -- dirA and dirB are 2-bit logic vectors(i.e. an array of 2 bits)
        dirA, dirB                   : OUT STD_LOGIC_VECTOR( 1 DOWNTO 0 ));
END Tcontrol;

              -- This code describes how the state machine operates
              -- This section will need changes for a different  state machine
ARCHITECTURE a OF Tcontrol IS

              -- Define local signals (i.e. non input or output signals) here
    TYPE STATE_TYPE IS ( ABout, Ain, Bin, Astop, Bstop );
    SIGNAL state: STATE_TYPE;
    SIGNAL sensor12, sensor13, sensor24   : STD_LOGIC_VECTOR(1 DOWNTO 0);

BEGIN
              -- This section describes how the state machine behaves
              -- this process runs once every time reset or the clock changes
    PROCESS ( reset, clock )
    BEGIN
              -- Reset to this state (i.e. asynchronous reset)
      IF reset = '1' THEN
            state <= ABout;
      ELSIF clock'EVENT AND clock = '1' THEN

              -- clock'EVENT means value of clock just changed
              --This section will execute once on each positive clock edge
              --Signal assignments in this section will generate D flip-flops
              -- Case statement to determine next state
        CASE state IS
```

```
WHEN ABout =>
    -- This Case checks both sensor1 and sensor2 bits
    CASE Sensor12 IS
        -- Note: VHDL's use of double quote for bit vector versus
        -- a single quote for only one bit!
        WHEN "00" => state <= About;
        WHEN "01" => state <= Bin;
        WHEN "10" => state <= Ain;
        WHEN "11" => state <= Ain;
        -- Default case is always required
        WHEN OTHERS => state <= ABout;
    END CASE;

WHEN Ain =>
    CASE Sensor24 IS
        WHEN "00" => state <= Ain;
        WHEN "01" => state <= ABout;
        WHEN "10" => state <= Bstop;
        WHEN "11" => state <= ABout;
        WHEN OTHERS => state <= ABout;
    END CASE;

WHEN Bin =>
    CASE Sensor13 IS
        WHEN "00" => state <= Bin;
        WHEN "01" => state <= ABout;
        WHEN "10" => state <= Astop;
        WHEN "11" => state <= About;
        WHEN OTHERS => state <= ABout;
    END CASE;

WHEN Astop =>
    IF Sensor3 = '1' THEN
        state <= Ain;
    ELSE
        state <= Astop;
    END IF;

WHEN Bstop =>
    IF Sensor4 = '1' THEN
        state <= Bin;
    ELSE
        state <= Bstop;
    END IF;
        END CASE;
    END IF;
END PROCESS;
                    -- combine sensor bits for case statements above
                    -- "&" operator combines bits
sensor12 <= sensor1 & sensor2;
sensor13 <= sensor1 & sensor3;
sensor24 <= sensor2 & sensor4;
```

```
                                    -- These outputs do not depend on the state
Track1  <= '0';
Track4  <= '0';
Switch3 <= '0';
                        -- Outputs that depend on state, use state to select value
                        -- Be sure to specify every output for every state
                        -- values will not default to zero!
WITH state SELECT
        Track3   <=     '1'     WHEN ABout,
                        '1'     WHEN Ain,
                        '1'     WHEN Bin,
                        '1'     WHEN Astop,
                        '1'     WHEN Bstop;
WITH state SELECT
        Track2   <=     '0'     WHEN ABout,
                        '0'     WHEN Ain,
                        '1'     WHEN Bin,
                        '1'     WHEN Astop,
                        '0'     WHEN Bstop;
WITH state SELECT
        Switch1  <=     '0'     WHEN ABout,
                        '0'     WHEN Ain,
                        '1'     WHEN Bin,
                        '1'     WHEN Astop,
                        '0'     WHEN Bstop;
WITH state SELECT
        Switch2  <=     '0'     WHEN ABout,
                        '0'     WHEN Ain,
                        '1'     WHEN Bin,
                        '1'     WHEN Astop,
                        '0'     WHEN Bstop;
WITH state SELECT
        DirA     <=     "01"    WHEN ABout,
                        "01"    WHEN Ain,
                        "01"    WHEN Bin,
                        "00"    WHEN Astop,
                        "01"    WHEN Bstop;
WITH state SELECT
        DirB     <=     "01"    WHEN ABout,
                        "01"    WHEN Ain,
                        "01"    WHEN Bin,
                        "01"    WHEN Astop,
                        "00"    WHEN Bstop;
END a;
```

8.8 Simulation Vector file for State Machine Simulation

The vector waveform file, *tcontrol.vwf*, controls the simulation and tests the state machine. A vector waveform file specifies the simulation stimulus and display. This file sets up a 40ns clock and specifies sensor patterns (inputs to the state machine), which will be used to test the state machine. These patterns were chosen by picking a path in the state diagram that moves to all of the different states. The sensor-input patterns will need to be changed to test a different state machine. Sensor inputs should not change faster than the clock cycle time of 40ns. As a minimum, try to test all of the states and arcs in your state machine simulation.

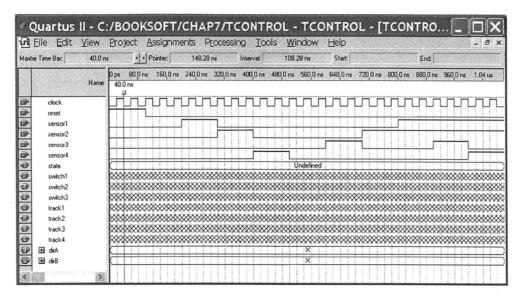

Figure 8.9 Tcontrol.vwf vector waveform file for simulation.

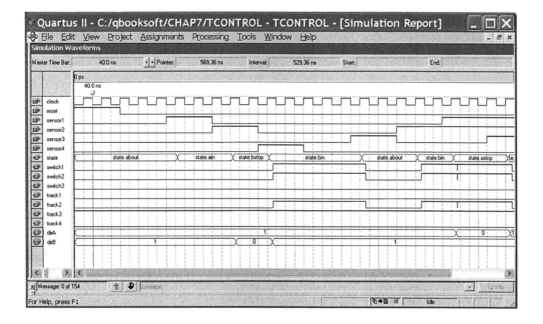

Figure 8.10 Simulation of Tcontrol.vhd using the Tcontrol.vwf vector waveform file in Figure 8.9.

8.9 Running the Train Control Simulation

Follow these steps to compile and simulate the state machine for the electric train controller.

Select Current Project

Make Tcontrol.vhd the current project with **File ⇨ Open Project ⇨ *Name***

Then find and select **Tcontrol.vhd**.

Compile and Simulate

Select **Processing ⇨ Start Compilation and Simulation**. The simulator will run automatically if there are no compile errors. Select **Processing ⇨ Simulation Report** to see the timing diagram display of your simulation as seen in Figure 8.10. Whenever you change your VHDL source you need to repeat this step. If you get compile errors, clicking on the error will move the text editor to the error location. The Altera software has extensive online help including VHDL syntax examples.

Make any text changes to Tcontrol.vhd or Tcontrol.vwf (test vector waveform file) with **File ⇨ Open**. This brings up a special editor window. Note that the menus at the top of the screen change depending on which window is currently open.

Updating new Simulation Test Vectors

To update the simulation with new test vectors from a modified Tcontrol.vwf, select **Processing ⇨ Start Simulation**. The simulation will then run with the new test vectors. If you modify Tcontrol.vhd, you will need to recompile first.

8.10 Running the Video Train System (After Successful Simulation)

A simulated or "virtual" train system is provided to test the controller without putting trains and people at risk. The simulation runs on the Cyclone chip. The output of the simulation is displayed on a VGA monitor connected directly to the UP 3 board. A typical video output display is seen in Figure 8.11. This module is also written in VHDL and it provides the sensor inputs and uses the outputs from the state machine to control the trains. The module tcontrol.vhd is automatically connected to the train simulation.

Here are the steps to run the virtual train system simulation:

Select the top-level project

Make Train.vhd the current project with **File ⇨ Open Project ⇨ *Name***

Then find and select **Train.qpf**. Train.qsf must be in the project directory since it contains the Cyclone chip pin assignment information needed for video outputs and switch inputs. Double check that your FPGA Device type is correct.

Compile the Project

Select **Processing** ⇨ **Start Compilation**. Train.vhd will link in your tcontrol.vhd file if it is in the same directory, when compiled. This is a large program, so it will take a few seconds to compile.

Download the Video Train Simulation

Select **Tools** ⇨ **Programmer**. When the programmer window opens click on the Program/Configure box if it is not already selected. In case of problems, see the UP 3 board tutorial in Chapter 1 for more details. The UP3 board must be turned on the power supply must be connected, and the Byteblaster* cable must be plugged into the PC's printer port. When everything is setup, the start button in the programming window should highlight. If the start button is not highlighted, try closing and reopening the programmer window. Under Hardware setup the Byteblaster should be selected. To download the board, click on the highlighted **start** button. Attach a VGA monitor to the UP 3 board.

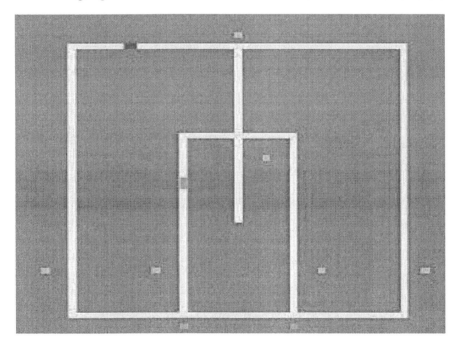

Figure 8.11 Video Image from Train System Simulation.

Viewing the Video Train Simulation

Train output should appear on the VGA monitor after downloading is complete. UP3 SW7 is run/step/stop and UP3 SW8 is the reset. Train A is displayed in black and Train B is displayed in red. Hit SW7 once to start the train simulation running. Hitting SW7 again will stop the simulation. If you SW7 twice quickly while trains are stopped, it will single step to the next track sensor state - change.

Sensor and switch values are indicated with a green or red square on the display. Switch values are the squares next to each switch location. Green indicates no train present on a sensor and it indicates switch connected to outside track for a switch. The UP 3's LCD display top line shows the values of the sensor (s), track (t) and switch (sw) signals in binary and the bottom line indicates the values of DirA and DirB in binary. The most significant bit in each field is the highest numbered bit.

If a possible train wreck is detected by two trains running on the same track segment, the simulation halts and the monitor will flash. The UP3 DIP switches control the speed of Train A (low 2 bits) and B (high 2 bits). Be sure to check operation with different train speeds. Most problems occur with a fast and a slow train.

A VERILOG VERSION OF THE TRAIN CONTROL STATE MACHINE IS ALSO AVAILABLE ON THE CD-ROM IN THE FILE, TCONTROL.V. THE /UP2 DIRECTORY CONTAINS A UP 2 VERSION OF THE TRAIN SIMULATION.

8.11 Laboratory Exercises

1. Assuming that train A now runs clockwise and B remains counterclockwise, draw a new state diagram and implement the new controller. If you use VHDL to design the new controller, you can modify the code presented in section 8.7. Simulate the controller and then run the video train simulation.

2. Design a state machine to operate the two trains avoiding collisions but minimizing their idle time. Trains must not crash by moving the wrong direction into an open switch. Develop a simulation to verify your state machine is operating correctly before running the video train system.

   The trains are assumed to be in the initial positions as shown in Figure 8.12. Train A is to move counterclockwise around the outside track until it comes to Sensor 1, then move to the inside track stopping at Sensor 5 and waiting for B to pass Sensor 3 twice. Trains can move at different speeds so no assumption should be made about the train speeds. A train hitting a sensor can be stopped before entering the switch area.

Once B has passed Sensor 3 twice, Train A moves to the outside track and continues around counterclockwise until it picks up where it left off at the starting position as shown in Figure 8.13. Train B is to move as designated only stopping at a sensor to avoid collisions with A.

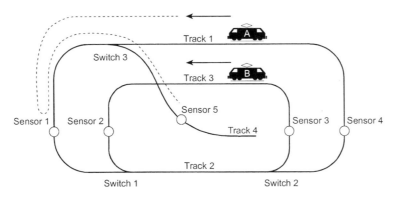

Figure 8.12 Initial Positions of Trains at State Machine Reset with Initial Paths Designated.

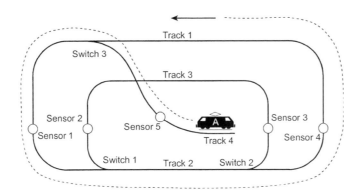

Figure 8.13 Return Path of Train A.

Train B will then continue as soon as there is no potential collision and continue as designated. Trains A and B should run continuously, stopping only to avoid a potential collision.

3. Use the single pulse UP3core functions on each raw sensor input to produce state machine sensor inputs that go High for only one clock cycle per passage of a train. Rework the state machine design with this assumption and repeat problem 1 or 2.

4. Develop another pattern of train movement and design a state machine to implement it.

CHAPTER 9

A Simple Computer Design: The μP 3

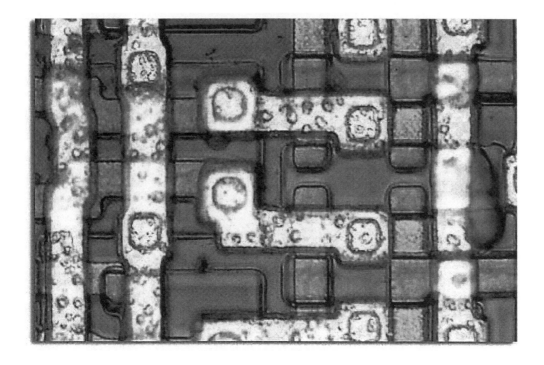

A partial die photograph of individual transistors about 10 microns tall on the Intel i4004 microprocessor is seen above. The 1971 Intel 4004 was the world's first single chip microprocessor. Prior to the 4004, Intel made memory chips. The 4004 was a 4-bit CPU with a clock rate of 108 kHz that contains 2,300 transistors. Photograph ©1995-2004 courtesy of Michael Davidson, http://micro.magnet.fsu.edu/chipshots.

9 A Simple Computer Design: The μP 3

A traditional digital computer consists of three main units, the processor or central processing unit (CPU), the memory that stores program instructions and data, and the input/output hardware that communicates to other devices. As seen in Figure 9.1, these units are connected by a collection of parallel digital signals called a bus. Typically, signals on the bus include the memory address, memory data, and bus status. Bus status signals indicate the current bus operation, memory read, memory write, or input/output operation.

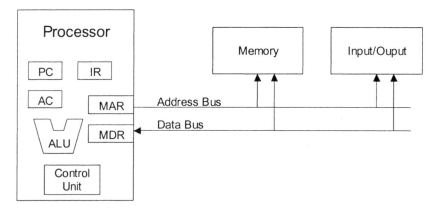

Figure 9.1 Architecture of a Simple Computer System.

Internally, the CPU contains a small number of registers that are used to store data inside the processor. Registers such as PC, IR, AC, MAR and MDR are built using D flip-flops for data storage. One or more arithmetic logic units (ALUs) are also contained inside the CPU. The ALU is used to perform arithmetic and logical operations on data values. Common ALU operations include add, subtract, and logical and/or operations. Register-to-bus connections are hard wired for simple point-to-point connections. When one of several registers can drive the bus, the connections are constructed using multiplexers, open collector outputs, or tri-state outputs. The control unit is a complex state machine that controls the internal operation of the processor.

The primary operation performed by the processor is the execution of sequences of instructions stored in main memory. The CPU or processor reads or fetches an instruction from memory, decodes the instruction to determine what operations are required, and then executes the instruction. The control unit controls this sequence of operations in the processor.

9.1 Computer Programs and Instructions

A computer program is a sequence of instructions that perform a desired operation. Instructions are stored in memory. For the following simple μP 3 computer design, an instruction consists of 16 bits. As seen in Figure 9.2 the high eight bits of the instruction contain the opcode. The instruction operation code or "opcode" specifies the operation, such as add or subtract, that will be performed by the instruction. Typically, an instruction sends one set of data values through the ALU to perform this operation. The low eight bits of each instruction contain a memory address field. Depending on the opcode, this address may point to a data location or the location of another instruction. Some example instructions are shown in Figure 9.3.

Opcode	Address
15	8 7 0

Figure 9.2 Simple μP 3 Computer Instruction Format.

Instruction Mnemonic		Operation Preformed	Opcode Value
ADD	*address*	**AC <= AC + contents of memory address**	00
STORE	*address*	**contents of memory address <= AC**	01
LOAD	*address*	**AC <= contents of memory address**	02
JUMP	*address*	**PC <= address**	03
JNEG	*address*	**If AC < 0 Then PC <= address**	04

Figure 9.3 Basic μP 3 Computer Instructions.

An example program to compute $A = B + C$ is shown in Figure 9.4. This program is a sequence of three instructions. Program variables such as A, B, and C are typically stored in dedicated memory locations. The symbolic representation of the instructions, called assembly language, is shown in the first column. The second column contains the same program in machine language (the binary pattern that is actually loaded into the computer's memory).

The machine language can be derived using the instruction format in Figure 9.2. First, find the opcode for each instruction in the first column of Figure 9.3. This provides the first two hexadecimal digits in machine language. Second, assign the data values of A, B, and C to be stored in hexadecimal addresses 10, 11, and 12 in memory. The address provides the last two hexadecimal digits of each machine instruction.

Assembly Language	Machine Language
LOAD B	0211
ADD C	0012
STORE A	0110

Figure 9.4 Example Computer Program for A = B + C.

The assignment of the data addresses must not conflict with instruction addresses. Normally, the data is stored in memory after all of the instructions in the program. In this case, if we assume the program starts at address 0, the three instructions will use memory addresses 0,1, and 2.

The instructions in this example program all perform data operations and execute in strictly sequential order. Instructions such as JUMP and JNEG are used to transfer control to a different address. Jump and Branch instructions do not execute in sequential order. Jump and Branch instructions must be used to implement control structures such as an IF...THEN statement or program loops. Details are provided in an exercise at the end of this section.

Assemblers are computer programs that automatically convert the symbolic assembly language program into the binary machine language. Compilers are programs that automatically translate higher-level languages, such as C or Pascal, into a sequence of machine instructions. Many compilers also have an option to output assembly language to aid in debugging.

The programmer's view of the computer only includes the registers (such as the program counter) and details that are required to understand the function of assembly or machine language instructions. Other registers and control hardware, such as the instruction register (IR), memory address register (MAR), and memory data register (MDR), are internal to the CPU and are not described in the assembly language level model of the computer. Computer engineers designing the processor must understand the function and operation of these internal registers and additional control hardware.

9.2 The Processor Fetch, Decode and Execute Cycle

The processor reads or fetches an instruction from memory, decodes the instruction to determine what operations are required, and then executes the instruction as seen in Figure 9.5. A simple state machine called the control unit controls this sequence of operations in the processor. The fetch, decode, and execute cycle is found in machines ranging from microprocessor-based PCs to supercomputers. Implementation of the fetch, decode, and execute cycle requires several register transfer operations and clock cycles in this example design.

The program counter contains the address of the current instruction. Normally, to fetch the next instruction from memory the processor must increment the program counter (PC). The processor must then send the address value in the PC to memory over the bus by loading the memory address register (MAR) and start a memory read operation on the bus. After a small delay, the instruction data will appear on the memory data bus lines, and it will be latched into the memory data register (MDR).

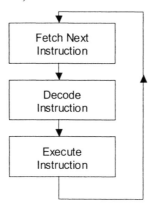

Figure 9.5 Processor Fetch, Decode and Execute Cycle.

Execution of the instruction may require an additional memory cycle so the instruction is normally saved in the CPU's instruction register (IR). Using the value in the IR, the instruction can now be decoded. Execution of the instruction will require additional operations in the CPU and perhaps additional memory operations.

The Accumulator (AC) is the primary register used to perform data calculations and to hold temporary program data in the processor. After completing execution of the instruction the processor begins the cycle again by fetching the next instruction.

The detailed operation of a computer is often modeled by describing the register transfers occurring in the computer system. A variety of register transfer level (RTL) languages such as VHDL or Verilog are designed for this application. Unlike more traditional programming languages, RTL languages can model parallel operations and map easily into hardware designs. Logic synthesis tools can also be used to implement a hardware design automatically using an RTL description.

To explain the function and operation of the CPU in detail, consider the example computer design in Figure 9.1. The CPU contains a general-purpose data register called the accumulator (AC) and the program counter (PC). The arithmetic logic unit (ALU) is used for arithmetic and logical operations.

The fetch, decode, and execute cycle can be implemented in this computer using the sequence of register transfer operations shown in Figure 9.6. The next instruction is fetched from memory with the following register transfer operations:

> **MAR = PC**
> **Read Memory, MDR = Instruction value from memory**
> **IR = MDR**
> **PC = PC + 1**

After this sequence of operations, the current instruction is in the instruction register (IR). This instruction is one of several possible machine instructions such as ADD, LOAD, or STORE. The opcode field is tested to decode the specific machine instruction. The address field of the instruction register contains the address of possible data operands. Using the address field, a memory read is started in the decode state.

The decode state transfers control to one of several possible next states based on the opcode value. Each instruction requires a short sequence of register transfer operations to implement or execute that instruction. These register transfer operations are then performed to execute the instruction. Only a few of the instruction execute states are shown in Figure 9.6. When execution of the current instruction is completed, the cycle repeats by starting a memory read operation and returning to the fetch state. A small state machine called a control unit is used to control these internal processor states and control signals.

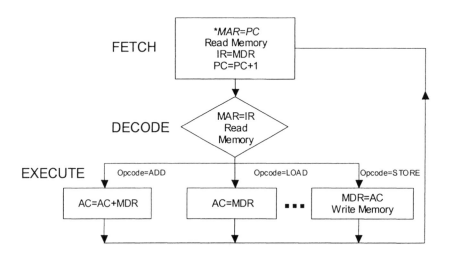

Figure 9.6 Detailed View of Fetch, Decode, and Execute for the μP 3 Computer Design.

Figure 9.7 is the datapath used for the implementation of the μP 3 Computer. A computer's datapath consists of the registers, memory interface, ALUs, and the bus structures used to connect them. The vertical lines are the three major busses used to connect the registers. On the bus lines in the datapath, a "/" with a number indicates the number of bits on the bus. Data values present on the active busses are shown in hexadecimal. MW is the memory write control line.

A reset must be used to force the processor into a known state after power is applied. The initial contents of registers and memory produced by a reset can also be seen in Figure 9.7. Since the PC and MAR are reset to 00, program execution will start at 00.

Note that memory contains the machine code for the example program presented earlier. Recall that the program consists of a LOAD, ADD, and

STORE instruction starting at address 00. Data values for this example program are stored in memory locations, 10, 11, and 12.

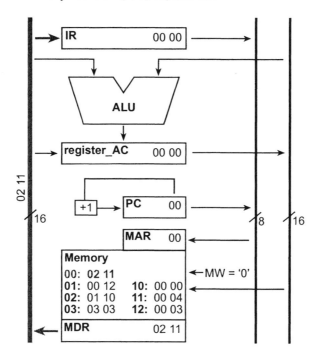

Figure 9.7 Datapath used for the μP 3 Computer Design after applying reset.

Consider the execution of the ADD machine instruction (0012) stored at program location 01 in detail. The instruction, ADD *address*, adds the contents of the memory location at address 12 to the contents of AC and stores the result in AC. The following sequence of register transfer operations will be required to fetch and execute this instruction.

FETCH: *REGISTER TRANSFER CYCLE 1*:

MAR = PC prior to fetch, **read memory, IR = MDR, PC = PC + 1**

First, the memory address register is loaded with the PC. In the example program, the ADD instruction (0012) is at location 01 in memory, so the PC and MAR will both contain 01. In this implementation of the computer, the MAR=PC operation will be moved to the end of the fetch, decode, and execute loop to the execute state in order to save a clock cycle. To fetch the instruction, a memory read operation is started. After a small delay for the memory access time, the ADD instruction is available at the input of the instruction register. To set up for the next instruction fetch, one is added to the program counter. The last two operations occur in parallel during one clock cycle using two different data busses. At the rising edge of the clock signal, the decode state is entered. A block diagram of the register transfer operations for the fetch state is seen in Figure 9.8. Inactive busses are not shown.

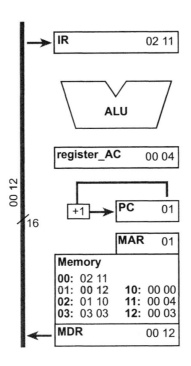

Figure 9.8 Register transfers in the ADD instruction's Fetch State.

DECODE: *REGISTER TRANSFER CYCLE 2*:

Decode Opcode to find Next State, MAR = IR, and start memory read

Using the new value in the IR, the CPU control hardware decodes the instruction's opcode of 00 and determines that this is an ADD instruction. Therefore, the next state in the following clock cycle will be the execute state for the ADD instruction.

Instructions typically are decoded in hardware using combinational circuits such as decoders, programmable logic arrays (PLAs), or perhaps even a small ROM. A memory read cycle is always started in decode, since the instruction may require a memory data operand in the execute state.

The ADD instruction requires a data operand from memory address 12. In Figure 9.9, the low 8–bit address field portion of the instruction in the IR is transferred to the MAR. At the next clock, after a small delay for the memory access time, the ADD instruction's data operand value from memory (0003) will be available in the MDR.

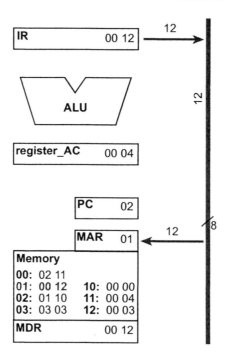

Figure 9.9 Register transfers in the ADD instruction's Decode State.

EXECUTE ADD: *REGISTER TRANSFER CYCLE 3*:
AC = AC + MDR, MAR = PC*, and GOTO FETCH

The two values can now be added. The ALU operation input is set for addition by the control unit. As shown in Figure 9.10, the MDR's value of 0003 is fed into one input of the ALU. The contents of register AC (0004) are fed into the other ALU input. After a small delay for the addition circuitry, the sum of 0007 is produced by the ALU and will be loaded into the AC at the next clock. To provide the address for the next instruction fetch, the MAR is loaded with the current value of the PC (02). Note that by moving the operation, MAR=PC, to every instruction's final execute state, the fetch state can execute in one clock cycle. The ADD instruction is now complete and the processor starts to fetch the next instruction at the next clock cycle. Since three states were required, an ADD instruction will require three clock cycles to complete the operation.

After considering this example, it should be obvious that a thorough understanding of each instruction, the hardware organization, busses, control signals, and timing is required to design a processor. Some operations can be performed in parallel, while others must be performed sequentially. A bus can only transfer one value per clock cycle and an ALU can only compute one value per clock cycle, so ALUs, bus structures, and data transfers will limit those operations that can be done in parallel during a single clock cycle. In the states examined, a maximum of three buses were used for register transfers.

Timing in critical paths, such as ALU delays and memory access times, will determine the clock speed at which these operations can be performed.

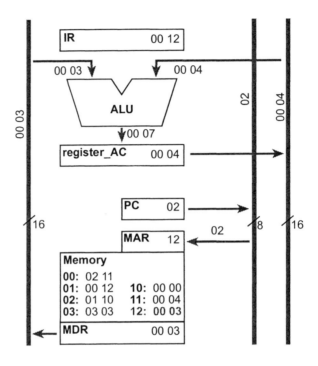

Figure 9.10 Register transfers in the ADD instruction's Execute State.

The µP 3's multiple clock cycles per instruction implementation approach was used in early generation microprocessors. These computers had limited hardware, since the VLSI technology at that time supported orders of magnitude fewer gates on a chip than is now possible in current devices. Current generation processors, such as those used in personal computers, have a hundred or more instructions, and use additional means to speedup program execution. Instruction formats are more complex with up to 32 data registers and with additional instruction bits that are used for longer address fields and more powerful addressing modes.

Pipelining converts fetch, decode, and execute into a parallel operation mode instead of sequential. As an example, with three stage pipelining, the fetch unit fetches instruction n + 2, while the decode unit decodes instruction n + 1, and the execute unit executes instruction n. With this faster pipelined approach, an instruction finishes execution every clock cycle rather than three as in the simple computer design presented here.

Superscalar machines are pipelined computers that contain multiple fetch, decode and execute units. Superscalar computers can execute several instructions in one clock cycle. Most current generation processors including those in personal computers are both pipelined and superscalar. An example of a pipelined, reduced instruction set computer (RISC) design can be found in Chapter 14.

9.3 VHDL Model of the µP 3

To demonstrate the operation of a computer, a VHDL model of the µP 3 computer is shown in Figure 9.11. The simple µP 3 computer design fits easily into a Cyclone device using less than 1% of its logic. The computer's RAM memory is implemented using the Altsyncram function which uses the FPGA's internal memory blocks.

The remainder of the computer model is basically a VHDL-based state machine that implements the fetch, decode, and execute cycle. The first few lines declare internal registers for the processor along with the states needed for the fetch, decode and execute cycle. A long CASE statement is used to implement the control unit state machine. A reset state is needed to initialize the processor. In the reset state, several of the registers are reset to zero and a memory read of the first instruction is started. This forces the processor to start executing instructions at location 00 in a predictable state after a reset.

The fetch state adds one to the PC and loads the instruction into the instruction register (IR). After the rising edge of the clock signal, the decode state starts. In decode, the low eight bits of the instruction register are used to start a memory read operation in case the instruction needs a data operand from memory. The decode state contains another CASE statement to decode the instruction using the opcode value in the high eight bits of the instruction. This means that the computer can have up to 256 different instructions, although only four are implemented in the basic model. Other instructions can be added as exercises. After the rising edge of the clock signal, control transfers to an execute state that is specific for each instruction.

Some instructions can execute in one clock cycle and some instructions may take more than one clock cycle. Instructions that write to memory will require more than one state for execute because of memory timing constraints. As seen in the STORE instruction, the memory address and data needs to be stable before and after the memory write signal is High, hence, additional states are used to avoid violating memory setup and hold times. When each instruction finishes the execute state, MAR is loaded with the PC to start the fetch of the next instruction. After the final execute state for each instruction, control returns to the fetch state.

Since the FPGA's synchronous memory block requires and contains an internal memory address and memory write register, it is necessary to make all assignments to the memory address register and memory write outside of the process to avoid having two cascaded registers. Recall that any assignment made in a clocked process synthesizes registers. Two cascaded MAR registers would require a delay of two clocks to load a new address for a memory operation.

The machine language program shown in Figure 9.12 is loaded into memory using a memory initialization file (*.mif). This produces 256 words of 16-bit

memory for instructions and data. The memory initialization file, program.mif
can be edited to change the loaded program. A write is performed only when
the memory_write signal is High. On a Cyclone FPGA device, the access time
for memory operations is in the range of 5-10ns.

```
                        -- Simple Computer Model Scomp.vhd
LIBRARY IEEE;
USE  IEEE.STD_LOGIC_1164.ALL;
USE  IEEE.STD_LOGIC_ARITH.ALL;
USE  IEEE.STD_LOGIC_UNSIGNED.ALL;
LIBRARY altera_mf;
USE  altera_mf.altera_mf_components.ALL;


ENTITY SCOMP IS
PORT(  clock, reset                        : IN STD_LOGIC;
            program_counter_out            : OUT STD_LOGIC_VECTOR( 7 DOWNTO 0 );
            register_AC_out                : OUT STD_LOGIC_VECTOR(15 DOWNTO 0 );
            memory_data_register_out       : OUT STD_LOGIC_VECTOR(15 DOWNTO 0 ));
            memory_address_register_out    : OUT STD_LOGIC_VECTOR(7 DOWNTO 0 );
            memory_write_out               : OUT STD_LOGIC);
END SCOMP;


ARCHITECTURE a OF scomp IS
TYPE STATE_TYPE IS ( reset_pc, fetch, decode, execute_add, execute_load, execute_store,
                        execute_store2, execute_jump );
SIGNAL state: STATE_TYPE;
SIGNAL instruction_register, memory_data_register    : STD_LOGIC_VECTOR(15 DOWNTO 0 );
SIGNAL register_AC                                   : STD_LOGIC_VECTOR(15 DOWNTO 0 );
SIGNAL program_counter                              :  STD_LOGIC_VECTOR( 7 DOWNTO 0 );
SIGNAL memory_address_register                      : STD_LOGIC_VECTOR( 7 DOWNTO 0 );
SIGNAL memory_write                                 : STD_LOGIC;
BEGIN
                    -- Use Altsyncram function for computer's memory (256 16-bit words)
    memory: altsyncram
            GENERIC MAP (
                    operation_mode => "SINGLE_PORT",
                    width_a => 16,
                    widthad_a => 8,
                    lpm_type => "altsyncram",
                    outdata_reg_a => "UNREGISTERED",
                            -- Reads in mif file for initial program and data values
                    init_file => "program.mif",
                    intended_device_family => "Cyclone")

            PORT MAP (wren_a => memory_write, clock0 => clock,
                    address_a =>memory_address_register,  data_a => Register_AC,
                    q_a => memory_data_register );
                            -- Output major signals for simulation
    program_counter_out            <= program_counter;
    register_AC_out                <= register_AC;
    memory_data_register_out       <= memory_data_register;
    memory_address_register_out <= memory_address_register;
```

```
PROCESS ( CLOCK, RESET )
   BEGIN
   IF reset = '1' THEN
       state <= reset_pc;
   ELSIF clock'EVENT AND clock = '1' THEN

       CASE state IS
                           -- reset the computer, need to clear some registers
       WHEN reset_pc =>
           program_counter          <= "00000000";
           register_AC              <= "0000000000000000";
           state                    <= fetch;
                           -- Fetch instruction from memory and add 1 to PC
       WHEN fetch =>
           instruction_register     <= memory_data_register;
           program_counter          <= program_counter + 1;
           state                    <= decode;
                           -- Decode instruction and send out address of any data operands
       WHEN decode =>
           CASE instruction_register( 15 DOWNTO 8 ) IS
                   WHEN "00000000" =>
                      state <= execute_add;
                   WHEN "00000001" =>
                      state <= execute_store;
                   WHEN "00000010" =>
                      state <= execute_load;
                   WHEN "00000011" =>
                      state <= execute_jump;
                   WHEN OTHERS =>
                      state <= fetch;
           END CASE;
                           -- Execute the ADD instruction
       WHEN execute_add =>
           register_ac              <= register_ac + memory_data_register;
           state                    <= fetch;
                           -- Execute the STORE instruction
                           -- (needs two clock cycles for memory write and fetch mem setup)
       WHEN execute_store =>
                           -- write register_A to memory, enable memory write
                           -- load memory address and data registers for memory write
           state                    <= execute_store2;
                           --finish memory write operation and load memory registers
                           --for next fetch memory read operation
       WHEN execute_store2 =>
           state                    <= fetch;
                           -- Execute the LOAD instruction
       WHEN execute_load =>
           register_ac              <= memory_data_register;
           state                    <= fetch;
                           -- Execute the JUMP instruction
       WHEN execute_jump =>
           program_counter          <= instruction_register( 7 DOWNTO 0 );
           state                    <= fetch;
       WHEN OTHERS =>
           state <= fetch;
```

```
        END CASE;
    END IF;
    END PROCESS;

            -- memory address register is already inside synchronous memory unit
            -- need to load its value based on current state
            -- (no second register is used - not inside a process here)
    WITH state SELECT
            memory_address_register <= "00000000"  WHEN reset_pc,
                program_counter                    WHEN fetch,
                instruction_register(7 DOWNTO 0)   WHEN decode,
                program_counter                    WHEN execute_add,
                instruction_register(7 DOWNTO 0)   WHEN execute_store,
                program_counter                    WHEN execute_store2,
                program_counter                    WHEN execute_load,
                instruction_register(7 DOWNTO 0)   WHEN execute_jump;
    WITH state SELECT
            memory_write <=    '1'                 WHEN execute_store,
                               '0'                 WHEN Others;
END a;
```

Figure 9.11 VHDL Model of μP 3 Computer.

```
DEPTH = 256;                % Memory depth and width are required    %
WIDTH = 16;                 % Enter a decimal number          %

ADDRESS_RADIX = HEX;        % Address and value radixes are optional %
DATA_RADIX = HEX;           % Enter BIN, DEC, HEX, or OCT; unless    %
                            % otherwise specified, radixes = HEX     %

                    -- Specify values for addresses, which can be single address or range
CONTENT
  BEGIN
    [00..FF]  :   0000;     % Range--Every address from 00 to FF = 0000 (Default) %
     00   :       0210;     % LOAD AC with MEM(10) %
     01   :       0011;     % ADD MEM(11) to AC %
     02   :       0112;     % STORE  AC in MEM(12) %
     03   :       0212;     % LOAD AC with MEM(12) check for new value of FFFF %
     04   :       0304;     % JUMP to 04 (loop forever) %
     10   :       AAAA;     % Data Value of B %
     11   :       5555;     % Data Value of C%
     12   :       0000;     % Data Value of A - should be FFFF after running program %
  END ;
```

Figure 9.12 Progam.mif file containg μP 3 Computer Program and DATA.

9.4 Simulation of the µP3 Computer

A simulation output from the VHDL model is seen in Figure 9.13. After a reset, the test program seen in Figure 9.12, loads, adds, and stores a data value to compute A = B + C. The final value is then loaded again to demonstrate that the memory contains the correct value for A. The program then ends with a jump instruction that jumps back to its own address producing an infinite loop. After running the program, FF is stored in location 12. Memory can be examined in the Simulator after running a program by clicking on the Logical Memories section in the left column of the Simulation Report. An example is shown in Figure 9.14. Note that the clock period is set to 20ns for simulation.

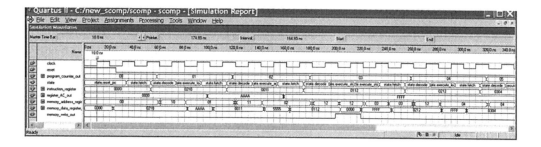

Figure 9.13 Simulation of the Simple µP 3 Computer Program.

Addr	+0	+1	+2	+3	+4	+5	+6	+7
00	0210	0011	0112	0212	0304	0000	0000	0000
08	0000	0000	0000	0000	0000	0000	0000	0000
10	AAAA	5555	FFFF	0000	0000	0000	0000	0000
18	0000	0000	0000	0000	0000	0000	0000	0000
20	0000	0000	0000	0000	0000	0000	0000	0000
28	0000	0000	0000	0000	0000	0000	0000	0000
30	0000	0000	0000	0000	0000	0000	0000	0000
38	0000	0000	0000	0000	0000	0000	0000	0000
40	0000	0000	0000	0000	0000	0000	0000	0000

Figure 9.14 Simulation display of µP 3 Computer Memory showing result stored in memory

A Verilog Version of the Simple Computer is also

available on the CD-ROM in the File, Scomp.v.

9.5 Laboratory Exercises

1. Compile and simulate the µP 3 computer VHDL or Verilog model. Rewrite the machine language program in the program.mif file to compute $A = (B + C) + D$. Store D in location 13 in memory. End the program with a Jump instruction that jumps to itself. Be sure to select the Cyclone device as the target. Find the maximum clock rate of the µP 3 computer. Examine the project's compiler report and find the logic cell (LC) percentage utilized.

2. Add the JNEG execute state to the CASE statement in the model. JNEG is Jump if AC < 0. If A >= 0 the next sequential instruction is executed. In most cases, a new instruction will just require a new execute state in the decode CASE statement. Use the opcode value of 04 for JNEG. Test the new instruction with the following test program that implements the operation, IF A >= 0 THEN B = C

	Assembly Language		Machine Language	Memory Address
	LOAD	A	0210	00
	JNEG	End_of_If	0404	01
	LOAD	C	0212	02
	STORE	B	0111	03
End_of_If:	JMP	End_of_If	0304	04

End_of_If is an example of a label; it is a symbolic representation for a location in the program. Labels are used in assembly language to mark locations in a program. The last line that starts out with End_of_If: is the address used for the End_of_If symbol in the Jump instruction address field. Assuming the program starts at address 00, the value of the End_of_If label will be 04. Test the JNEG instruction for both cases A < 0 and A >= 0. Place nonzero values in the *.mif file for B and C so that you can verify the program executes correctly.

3. Add the instructions in the table below to the VHDL model, construct a test program for each instruction, compile and simulate to verify correct operation. In JPOS and JZERO instructions, both cases must be tested.

Instruction	Function	Opcode
SUBT *address*	AC = AC - MDR	05
XOR *address*	AC = AC XOR MDR	06
OR *address*	AC = AC OR MDR	07
AND *address*	AC = AC AND MDR	08
JPOS *address*	IF AC > 0 THEN PC = address	09
JZERO *address*	IF AC = 0 THEN PC = address	0A
ADDI *address*	AC = AC + *address*	0B

In the logical XOR instruction each bit is exclusive OR'ed with the corresponding bit in each operation for a total of sixteen independent exclusive OR operations. This is called a bitwise logical operation. OR and AND are also bitwise logical operations. The add-immediate instruction, ADDI, sign extends the 8-bit address field value to 16 bits. To sign extend, copy the sign bit to all eight high bits. This allows the use of both positive and negative two's complement numbers for the 8-bit immediate value stored in the instruction.

4. Add the following two shift instructions to the simple computer model and verify with a test program and simulation.

Instruction	Function	Opcode
SHL address	AC = AC shifted left *address* bits	0C
SHR address	AC = AC shifted right *address* bits	0D

The function LPM_CLSHIFT is useful to implement multiple bit shifts. SHL and SHR can also be used if 1993 VHDL features are enabled in the compiler. Only the low four bits of the address field contain the shift amount. The other four bits are always zero.

5. Run the µP 3 computer model using the Cyclone chip on the UP 3 board. Use a debounced pushbutton for the clock and the other pushbutton for reset. Output the PC in hex to the UP 3's LCD display. Run a test program on the board and verify the correct value of the PC appears in the LCD display by stepping through the program using the pushbutton. On the UP 2, use the seven LED segment displays for the PC display.

6. Add these two input/output (I/O) instructions to the µP 3 computer model running on the UP 3 board.

Instruction	Function	Opcode
IN i/o address	AC = UP3 DIPswitch bits (low 4 bits)	0E
OUT i/o address	UP 3's LCD displays hex value of AC	0F

These instructions modify or use only the low eight bits of AC. Remove the PC display feature from the previous problem, if it was added or for more of a challenge place the AC value on the second line of the hex display by modifying the LCD display code. Test the new I/O instructions by writing a program that reads in the switches, adds one to the switch value, and outputs this value to the LED display. Repeat the input, add, and output operation in an infinite loop by jumping back to the start of the program. Add a new register, register_output, to the input of the seven-segment decoder that drives the LED display. The register is loaded with the value of AC only when an OUT instruction is executed. Compile, download, and execute the program on the UP 3 board. When several I/O devices are present, they should respond only to their own unique i/o address, just like memory.

7. Use the timing analyzer to determine the maximum clock rate for the μP 3 computer. Using this value, compute the execution time for the example program in Figure 9.4.

8. Modify the Cyclone video output display described in Chapter 9 for the MIPS computer example to display the μP 3's internal registers. While running on the UP 3 board, use the pushbuttons for clock and reset as suggested in problem 5.

9. Add video character output and keyboard input to the computer, after studying the material presented in Chapters 9 and 10.

10. Add the WAIT instruction to the simple computer model and verify with a test program and simulation. WAIT *value*, loads and starts an 8-bit ten-millisecond (10^{-2} second) timer and then waits *value*10* ms before returning to fetch for the next instruction. Use an opcode of 10 for the WAIT instruction.

11. Expand the memory address space of the μP 3 computer from eight bits to nine bits. Some registers will also need an additional bit. Use 512 locations of 16-bit memory. Expand the address field by 1-bit by reducing the size of the opcode field by 1-bit. This will limit the number of different instructions to 128 but the maximum program size can now increase from 256 locations to 512 locations.

12. Modify the μP 3 computer so that it uses two different memories. Use one memory for instructions and a new memory for data values. The new data memory should be 256 or 512 (see previous problem) locations of 16-bit data.

13. Add a subroutine CALL and RETURN instruction to the μP 3 computer design. Use a dedicated register to store the return address or use a stack with a stack pointer register. The stack should start at high addresses and as it grows move to lower addresses.

14. Implement a stack as suggested in the previous problem and add instructions to PUSH or POP register AC from the stack. At reset, set the stack pointer to the highest address of data memory.

15. Add all of the instructions and features suggested in the exercises to the μP 3 computer and use it as a microcontroller core for one of the robot projects suggested in Chapter 12. Additional instructions of your own design along with an interval timer that can be read using the IN instruction may also be useful.

16. Using the two low-bits from the opcode field, add a register address field that selects one of four different data registers A, B, C, or D for each instruction.

17. Use the implementation approach in the μP 3 computer model as a starting point to implement the basic instruction set of a different computer from your digital logic textbook or other reference manual.

CHAPTER 10

VGA Video Display Generation

```
MIPS  COMPUTER
PC    00000008
INST  00430820
REG1  00000055
REG2  000000AA
ALU   000000FF
W.B.  000000FF
BRAN  0
ZERO  0
MEMR  0
MEMW  0
CLK   ↓
RST   ↓
```

The video image above was produced by a UP 3 board design.

10 VGA Video Display Generation

To understand how it is possible to generate a video image using an FPGA board, it is first necessary to understand the various components of a video signal. A VGA video signal contains 5 active signals. Two signals compatible with TTL logic levels, horizontal sync and vertical sync, are used for synchronization of the video. Three analog signals with 0.7 to 1.0-Volt peak-to-peak levels are used to control the color. The color signals are Red, Green, and Blue. They are often collectively referred to as the RGB signals. By changing the analog levels of the three RGB signals all other colors are produced.

10.1 Video Display Technology

The first technology used to display video images dictated the nature of the video signals. Even though LCD monitors are now in common use, the major component inside early VGA computer monitors was the color CRT or Cathode Ray Tube shown in Figure 10.1. The electron beam must be scanned over the viewing screen in a sequence of horizontal lines to generate an image. The deflection yoke uses magnetic or electrostatic fields to deflect the electron beam to the appropriate position on the face of the CRT. The RGB color information in the video signal is used to control the strength of the electron beam. Light is generated when the beam is turned on by a video signal and it strikes a color phosphor dot or line on the face of the CRT. The face of a color CRT contains three different phosphors. One type of phosphor is used for each of the primary colors of red, green, and blue.

In standard VGA format, as seen in Figure 10.2, the screen contains 640 by 480 picture elements or pixels. The video signal must redraw the entire screen 60 times per second to provide for motion in the image and to reduce flicker. This period is called the refresh rate. The human eye can detect flicker at refresh rates less than 30 to 60Hz.

To reduce flicker from interference from fluorescent lighting sources, refresh rates higher than 60 Hz at around 70Hz are sometimes used in PC monitors. The color of each pixel is determined by the value of the RGB signals when the signal scans across each pixel. In 640 by 480-pixel mode, with a 60Hz refresh rate, this is approximately 40 ns per pixel. A 25MHz clock has a period of 40 ns. A slightly higher clock rate will produce a higher refresh rate.

10.2 Video Refresh

The screen refresh process seen in Figure 10.2 begins in the top left corner and paints 1 pixel at a time from left to right. At the end of the first row, the row increments and the column address is reset to the first column. Each row is painted until all pixels have been displayed. Once the entire screen has been painted, the refresh process begins again.

The video signal paints or refreshes the image using the following process. The vertical sync signal, as shown in Figure 10.3 tells the monitor to start displaying a new image or frame, and the monitor starts in the upper left corner with pixel 0,0. The horizontal sync signal, as shown in Figure 10.4, tells the monitor to refresh another row of 640 pixels.

After 480 rows of pixels are refreshed with 480 horizontal sync signals, a vertical sync signal resets the monitor to the upper left comer and the process continues. During the time when pixel data is not being displayed and the beam is returning to the left column to start another horizontal scan, the RGB signals should all be set to the color black (all zeros).

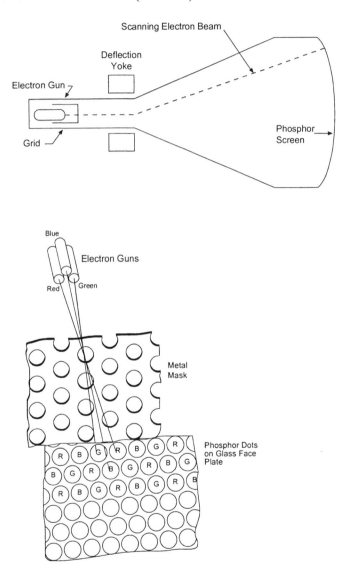

Figure 10.1 Color CRT and Phosphor Dots on Face of Display.

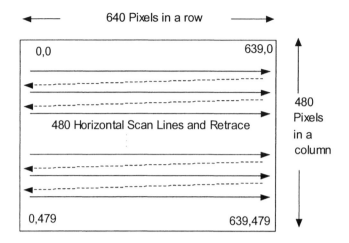

Figure 10.2 VGA Image - 640 by 480 Pixel Layout.

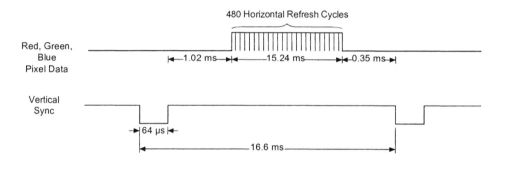

Figure 10.3 Vertical Sync Signal Timing for 640 by 480 at 60Hz.

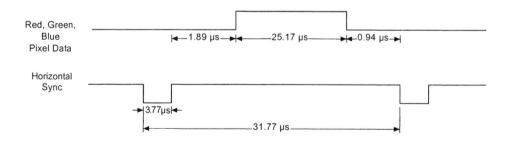

Figure 10.4 Horizontal Sync Signal Timing for 640 by 480 at 60Hz.

Many VGA monitors will shut down if the two sync signals are not the correct values. Most PC monitors have an LED that is green when it detects valid sync signals and yellow when it does not lock in with the sync signals. Modern monitors will sync up to an almost continuous range of refresh rates up to their design maximum. In a PC graphics card, a dedicated video memory location is used to store the color value of every pixel in the display. This memory is read out as the beam scans across the screen to produce the RGB signals. There is not enough memory inside current generation FPGA chips for this approach, so other techniques will be developed which require less memory.

10.3 Using an FPGA for VGA Video Signal Generation

To provide interesting output options in complex designs, video output can be developed using hardware inside the FPGA. Only five signals or pins are required, two sync signals and three RGB color signals. A simple resistor and diode circuit is used to convert TTL output pin signals from the FPGA to the low voltage analog RGB signals for the video signal. This supports two levels for each signal in the RGB data and thus produces a total of eight colors. This circuit and a VGA connector for a monitor are already installed on the Altera UP 3 board. The FPGA's Phase Locked Loop (PLL) can be used to generate clocks for a wide variety of video resolutions and refresh rates.

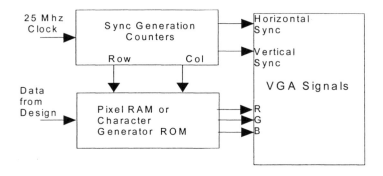

Figure 10.5 FPGA based generation of VGA Video Signals.

As seen in Figure 10.5, a 25.175 MHz clock, which is the 640 by 480 VGA pixel data rate of approximately 40ns is used to drive counters that generate the horizontal and vertical sync signals. Additional counters generate row and column addresses. In some designs, pixel resolution will be reduced from 640 by 480 to a lower resolution by using a clock divide operation on the row and column counters. The row and column addresses feed into a pixel RAM for graphics data or a character generator ROM when used to display text. The required RAM or ROM is also implemented inside the FPGA chip.

10.4 A VHDL Sync Generation Example: UP3core VGA_SYNC

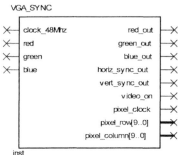

The UP3core function, VGA_SYNC is used to generate the timing signals needed for a VGA video display. Although VGA_SYNC is written in VHDL, like the other UP3core functions it can be used as a symbol in a design created with any entry method.

The following VHDL code generates the horizontal and vertical sync signals, by using 10-bit counters, H_count for the horizontal count and V_count for the vertical count. H_count and V_count generate a pixel row and column address that is output and available for use by other processes. User logic uses these signals to determine the x and y coordinates of the present video location. The pixel address is used in generating the image's RGB color data. The internal logic uses a 25 MHz clock generated by a PLL in the design file Video_PLL.vhd with counters to produce video sync timing signals like those seen in figures 10.3 and 10.4. This process is used in all of the video examples that follow.

```
LIBRARY IEEE;
USE IEEE.STD_LOGIC_1164.ALL;
USE IEEE.STD_LOGIC_ARITH.ALL;
USE IEEE.STD_LOGIC_UNSIGNED.ALL;

ENTITY VGA_SYNC IS
    PORT( clock_25MHz, red, green, blue : IN    STD_LOGIC;
          red_out, green_out, blue_out   : OUT STD_LOGIC;
          horiz_sync_out, vert_sync_out : OUT STD_LOGIC;
          pixel_row, pixel_column        : OUT STD_LOGIC_VECTOR( 9 DOWNTO 0 ));
END VGA_SYNC;

ARCHITECTURE a OF VGA_SYNC IS
    SIGNAL horiz_sync, vert_sync              : STD_LOGIC;
    SIGNAL video_on, video_on_v, video_on_h : STD_LOGIC;
    SIGNAL h_count, v_count                   : STD_LOGIC_VECTOR( 9 DOWNTO 0 );

BEGIN

                          -- video_on is High only when RGB data is displayed
video_on <= video_on_H AND video_on_V;
```

```
PROCESS
  BEGIN
  WAIT UNTIL( clock_25MHz'EVENT ) AND ( clock_25MHz = '1' );

                        --Generate Horizontal and Vertical Timing Signals for Video Signal
                        -- H_count counts pixels (640 + extra time for sync signals)
                        --
                        -- Horiz_sync -------------------------------------------      --------
                        -- H_count    0            640                659  755  799
                        --
  IF ( h_count = 799 ) THEN
      h_count <= "0000000000";
  ELSE
      h_count <= h_count + 1;
  END IF;

                        --Generate Horizontal Sync Signal using H_count
  IF ( h_count <= 755 ) AND  (h_count => 659 ) THEN
      horiz_sync <= '0';
  ELSE
      horiz_sync <= '1';
  END IF;
                        --V_count counts rows of pixels (480 + extra time for sync signals)
                        --
                        -- Vert_sync    ---------------------------------------      ------------
                        -- V_count      0               480             493-494        524
                        --
  IF ( v_count >= 524 ) AND ( h_count => 699 ) THEN
      v_count <= "0000000000";
  ELSIF ( h_count = 699 ) THEN
      v_count <= v_count + 1;
  END IF;
                        -- Generate Vertical Sync Signal using V_count
  IF ( v_count <= 494 ) AND ( v_count = >493 ) THEN
      vert_sync <= '0';
  ELSE
      vert_sync <= '1';
  END IF;
                        -- Generate Video on Screen Signals for Pixel Data
  IF ( h_count <= 639 ) THEN
      video_on_h  <= '1';
      pixel_column <= h_count;
  ELSE
      video_on_h <= '0';
  END IF;

  IF ( v_count <= 479 ) THEN
      video_on_v <= '1';
      pixel_row   <= v_count;
  ELSE
      video_on_v <= '0';
```

END IF;

```
                                        -- Put all video signals through DFFs to eliminate
                                        -- any delays that can cause a blurry image
                                        -- Turn off RGB outputs when outside video display area
red_out          <= red AND video_on;
green_out        <= green AND video_on;
blue_out         <= blue AND video_on;
horiz_sync_out <= horiz_sync;
vert_sync_out  <= vert_sync;
```

END PROCESS;
END a;

To turn off RGB data when the pixels are not being displayed the video_on signals are generated. Video_on is gated with the RGB inputs to produce the RGB outputs. Video_on is low during the time that the beam is resetting to the start of a new line or screen. They are used in the logic for the final RGB outputs to force them to the zero state. VGA_SYNC also puts the all of video outputs through a final register to eliminate any timing differences in the video outputs. VGA_SYNC outputs the pixel row and column address. See the comments at the end of VGA_SYNC.VHD for information on setting up other screen resolutions and refresh rates.

10.5 Final Output Register for Video Signals

The final video output for the RGB and sync signals in any design should be directly from a flip-flop output. Even a small time delay of a few nanoseconds from the logic that generates the RGB color signals will cause a blurry video image. Since the RGB signals must be delayed a pixel clock period to eliminate any possible timing delays, the sync signals must also be delayed by clocking them through a D flip-flop. If the outputs all come directly from a flip-flop output, the video signals will all change at the same time and a sharper video image is produced. The last few lines of VHDL code in the UP3core VGA_SYNC design generate this final output register.

10.6 Required Pin Assignments for Video Output

The UP 3 board requires the following Cyclone chip pins be defined in the project's *.qsf file, or elsewhere in your design in order to display the video signals:

```
Red          : OUTPUT_PIN = 228;      -- Red Data Signal Output Pin
Blue         : OUTPUT_PIN = 170;      -- Blue Data Signal Output Pin
Green        : OUTPUT_PIN = 122;      -- Green Data Signal Output Pin
Horiz_Sync   : OUTPUT_PIN = 227;      -- Horizontal Sync Signal Output Pin
Vert_Sync    : OUTPUT_PIN = 226;      -- Vertical Sync Signal Output Pin
```

These pins are hard wired on the UP 3 board to the VGA connector and cannot be changed. A pixel clock is also needed at the appropriate rate for the screen resolution and refresh rate. One of the Cyclone's two PLLs is an easy way to generate this clock on the UP 3. The UP 3's 48MHz reference clock is used as input for the PLL. A table of the common screen resolutions and refresh rates with the required pixel clocks and sync counter values can be found at the end of the VGA_SYNC IP core code.

10.7 Video Examples

For a simple video example with the VGA_SYNC function, the following schematic produces a video simulation of a red LED. When the PB1 pushbutton is hit, the color of the entire video screen will change from black to red.

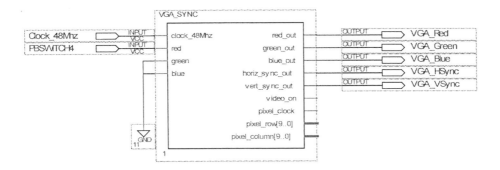

VGA_SYNC outputs the pixel row and column address. Pixel_row and Pixel_column are normally inputs to user logic that in turn generates the RGB color data. Here is a simple example that uses the pixel_column output to generate the RGB inputs. Bits 7, 6, and 5 of the pixel_column count are connected to the RGB data. Since bits 4 through 0 of pixel column are not connected, RGB color data will only change once every 32 pixels across the screen. This in turn generates a sequence of color bars in the video output. The color bars display the eight different colors that can be generated by the three digital RGB outputs.

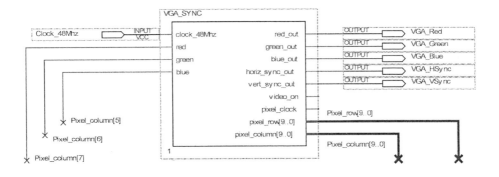

10.8 A Character Based Video Design

One option is a video display that contains mainly textual data. For this approach, a pixel pattern or font is needed to display each different character. The character font can be stored in a ROM implemented inside the Cyclone FPGA. A memory initialization file, *.mif, can be used to initialize the ROM contents during download. Given the memory limitations inside the Cyclone FPGA, one option that fits is a display of 40 characters by 30 lines.

Each letter, number, or symbol is a pixel image from the 8 by 8 character font. To make the characters larger, each dot in the font maps to a 2 by 2 pixel block so that a single character requires 16 by 16 pixels. This was done by dividing the row and column counters by 2. Recall that in binary, division by powers of two can be accomplished by truncating the lower bits, so no hardware is needed for this step. The row and column counters provide inputs to circuits that address the character font ROM and determine the color of each pixel. The clock used is the onboard 25.175MHz clock and other timing signals needed are obtained by dividing this clock down in hardware.

10.9 Character Selection and Fonts

Because the screen is constantly being refreshed and the video image is being generated on-the-fly as the beam moves across the video display, it is necessary to use other registers, ROM, or RAM inside the FPGA to hold and select the characters to be displayed on the screen. Each location in this character ROM or RAM contains only the starting address of the character font in font ROM. Using two levels of memory results in a design that is more compact and uses far less memory bits. This technique was used on early generation computers before the PC.

Here is an example implementation of a character font used in the UP3core function, char_ROM. To display an "A" the character ROM would contain only the starting address 000001 for the font table for "A". The 8 by 8 font in the character generation ROM would generate the letter "A" using the following eight memory words:

Address	Font Data
000001000 :	00011000 ;
000001001 :	00111100 ;
000001010 :	01100110 ;
000001011 :	01111110 ;
000001100 :	01100110 ;
000001101 :	01100110 ;
000001110 :	01100110 ;
000001111 :	00000000 ;

Figure 10.6 Font Memory Data for the Character "A".

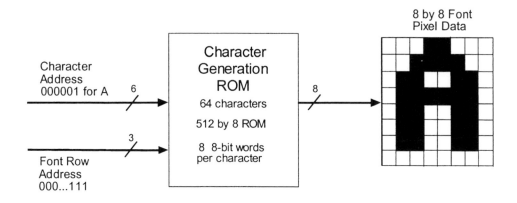

Figure 10.7 Accessing a Character Font Using a ROM.

The column counters are used to select each font bit from left to right in each word of font memory as the video signal moves across a row. This value is used to drive the logic for the RGB signals so that a "0" font bit has a different color from a "1". Using the low three character font row address bits, the row counter would select the next memory location from the character font ROM when the display moves to the next row.

A 3-bit font column address can be used with a multiplexer to select the appropriate bit from the ROM output word to drive the RGB pixel color data. Both the character font ROM and the multiplexer are contained in the UP3core char_ROM as shown below. The VHDL code declares the memory size using the LPM_ROM function and the tcgrom.mif file contains the initial values or font data for the ROM.

```
        Char_ROM
  ┌───────────────────────────────────────────┐
 ✕─┤ clock                       rom_mux_output ├─✕
 ✕─┤ character_address[5..0]                    │
 ✕─┤ font_row[2..0]                             │
 ✕─┤ font_col[2..0]                             │
  └───────────────────────────────────────────┘
        inst
```

```vhdl
LIBRARY IEEE;
USE  IEEE.STD_LOGIC_1164.ALL;
USE  IEEE.STD_LOGIC_ARITH.ALL;
USE  IEEE.STD_LOGIC_UNSIGNED.ALL;
LIBRARY lpm;
USE lpm.lpm_components.ALL;

ENTITY Char_ROM IS
   PORT(    character_address    : IN        STD_LOGIC_VECTOR( 5 DOWNTO 0 );
```

```
                        font_row, font_col    : IN      STD_LOGIC_VECTOR( 2 DOWNTO 0 );
                        rom_mux_output        : OUT     STD_LOGIC);
        END Char_ROM;

        ARCHITECTURE a OF Char_ROM IS
            SIGNAL   rom_data            : STD_LOGIC_VECTOR( 7 DOWNTO 0 );
            SIGNAL   rom_address         : STD_LOGIC_VECTOR( 8 DOWNTO 0 );
        BEGIN
                        -- Small 8 by 8 Character Generator ROM for Video Display
                        -- Each character is 8 8-bit words of pixel data
        char_gen_rom: lpm_rom
            GENERIC MAP (
                lpm_widthad             => 9,
                lpm_numwords            => 512,
                lpm_outdata             => "UNREGISTERED",
                lpm_address_control     => "UNREGISTERED",
                        -- Reads in mif file for character generator font data
                lpm_file                => "tcgrom.mif",
                lpm_width               => 8)
            PORT MAP ( address => rom_address, q = > rom_data);
                rom_address <= character_address & font_row;
                        -- Mux to pick off correct rom data bit from 8-bit word
                        -- for on screen character generation
                rom_mux_output <= rom_data (
                    (CONV_INTEGER( NOT font_col( 2 DOWNTO 0 ))) );
        END a;
```

Table 10.1 Character Address Map for 8 by 8 Font ROM.

CHAR	ADDRESS	CHAR	ADDRESS	CHAR	ADDRESS	CHAR	ADDRESS
@	00	P	20	Space	40	0	60
A	01	Q	21	!	41	1	61
B	02	R	22	"	42	2	62
C	03	S	23	#	43	3	63
D	04	T	24	$	44	4	64
E	05	U	25	%	45	5	65
F	06	V	26	&	46	6	66
G	07	W	27	'	47	7	67
H	10	X	30	(	50	8	70
I	11	Y	31	)	51	9	71
J	12	Z	32	*	52	A	72
K	13	[	33	+	53	B	73
L	14	Dn Arrow	34	,	54	C	74
M	15	]	35	-	55	D	75
N	16	Up Arrow	36	.	56	E	76
O	17	Lft Arrow	37	/	57	F	77

A 16 by 16 pixel area is used to display a single character with the character font. As the display moves to another character outside of the 16 by 16 pixel area, a different location is selected in the character RAM using the high bits of the row and column counters. This in turn selects another location in the character font ROM to display another character.

Due to limited ROM space, only the capital letters, numbers and some symbols are provided. Table 10.1 shows the alphanumeric characters followed by the high six bits of its octal character address in the font ROM. For example, a space is represented by octal code 40. The repeated letters A-F were used to simplify the conversion and display of hexadecimal values.

10.10 VHDL Character Display Design Examples

The UP3cores VGA_SYNC and CHAR_ROM are designed to be used together to generate a text display. CHAR_ROM contains an 8 by 8 pixel character font. In the following schematic, a test pattern with 40 characters across with 30 lines down is displayed. Examining the RGB inputs on the VGA_SYNC core you can see that characters will be white (111 = RGB) with a red (100 = RGB) background. Each character uses a 16 by 16 pixel area in the 640 by 480 display area. Since the low bit in the pixel row and column address is skipped in the font row and font column ROM inputs, each data bit from the font is a displayed in a 2 by 2 pixel area. Since pixel row bits 9 to 4 are used for the character address a new character will be displayed every 16th pixel row or character line. Division by 16 occurs without any logic since the low four bits are not connected.

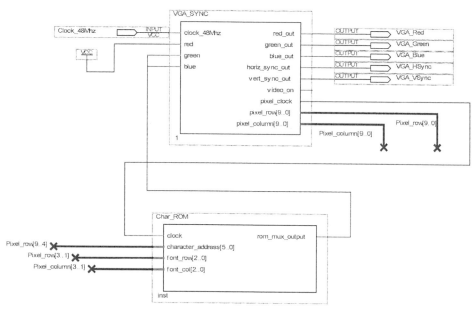

Normally, more complex user designed logic is used to generate the character address. The video example shown in Figure 10.8 is an implementation of the MIPS RISC processor core. The values of major busses are displayed in hexadecimal and it is possible to single step through instructions and watch the values on the video display. This example includes both constant and variable character display areas. The video setup is the same as the schematic, but additional logic is used to generate the character address.

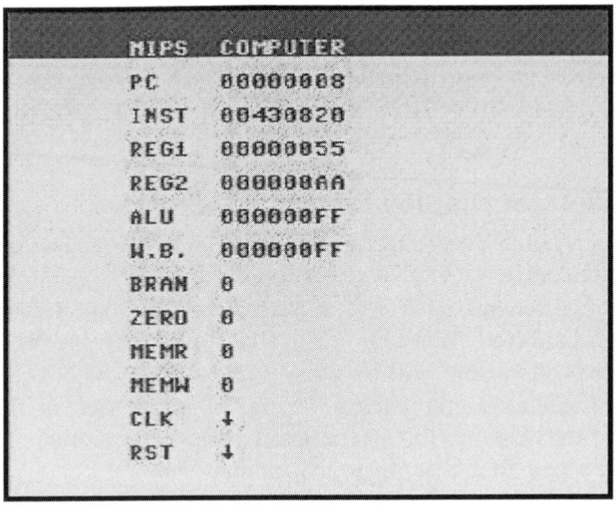

Figure 10.8 MIPS Computer Video Output.

Pixel row address and column address counters are used to determine the current character column and line position on the screen. They are generated as the image scans across the screen with the VGA_SYNC core by using the high six bits of the pixel row and pixel column outputs. Each character is a 16 by 16 block of pixels. The divide by 16 operation just requires truncation of the low four bits of the pixel row and column. The display area is 40 characters by 30 lines.

Constant character data for titles in the left column is stored in a small ROM called the character format ROM. This section of code sets up the format ROM that contains the character addresses for the constant character data in the left column of the video image for the display.

```
                              -- Character Format ROM for Video Display
                              -- Displays constant format character data
                              -- on left side of Display area
format_rom: lpm_rom
   GENERIC MAP (
      lpm_widthad          => 6,
      lpm_numwords         =>60,
      lpm_outdata          => "UNREGISTERED",
      lpm_address_control => "UNREGISTERED",
                              -- Reads in mif file for data display titles
      lpm_file             =>"format.mif",
      lpm_width            => 6)
```

Each pixel clock cycle, a process containing a series of nested CASE statements is used to select the character to display as the image scans across the screen. The CASE statements check the row and column counter outputs from the sync unit to determine the exact character column and character line that is currently being displayed. The CASE statements then output the character address for the desired character to the char_ROM UP3core.

Table 10.1 lists the address of each character in the font ROM. Alphabetic characters start at octal location 01 and numbers start at octal location 60. Octal location 40 contains a space that is used whenever no character is displayed. When the display is in the left column, data from the format_ROM is used. Any unused character display areas must select the space character that has blank or all zero font data.

Hexadecimal variables in the right column in Figure 10.8 are generated by using 4-bit data values from the design to index into the character font ROM. As an example, the value "11" & PC(7 DOWNTO 4), when used as the character address to the UP3core, char_ROM, will map into the character font for 0..9 and A..F. The actual hex character selected is based on the current value of the 4 bits in the VHDL signal, PC. As seen in the last column of Table 10.1, the letters, A..F, appear again after decimal numbers in the font ROM to simplify this hexadecimal mapping conversion.

10.11 A Graphics Memory Design Example

For another example, assume the display will be used to display only graphics data. The Cyclone EP1C6 FPGA contains 92K bits of memory. If only two colors are used in the RGB signals, one bit will be required for each pixel in the video RAM. If a 300 by 300 pixel video RAM was implemented in the Cyclone chip it would use all of the chip's 92K-bit memory. For full color RGB data of three bits per pixel, a 175 by 175 pixel RAM would use all of the 92K on-chip memory and no memory would be left for the remainder of the design.

Pixel memory must always be in read mode whenever RGB data is displayed. To avoid flicker and memory access conflicts on single port memory, designs should update pixel RAM and other signals that produce the RGB output, during the time the RGB data is not being displayed.

When the scan of each horizontal line is complete there are typically over 100 clock cycles before the next RGB value is needed, as seen in Figure 10.9. Additional clocks are available when a vertical sync signal resets the monitor to the first display line. The exact number of clocks available depends on the video resolution and refresh rate.

In most cases, calculations that change the video image should be performed during this off-screen period of time to avoid memory conflicts with the readout of video RAM or other registers which are used to produce the RGB video pixel color signals. Since on-chip pixel memory is limited, complex graphic designs with higher resolutions will require another approach.

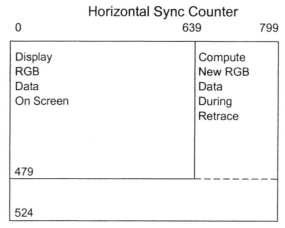

Figure 10.9 Display and Compute clock cycles available in a single Video Frame.

10.12 Video Data Compression

Here are some ideas to save memory and produce more complex graphics. Compress the video pixel data in memory and uncompress it on-the-fly as the video signal is generated. One compression technique that works well is run length encoding (RLE). The RLE compression technique only requires a simple state machine and a counter for decoding.

In RLE, the pixels in the display are encoded into a sequence of length and color fields. The length field specifies the number of sequentially scanned pixels with the same color. In simple color images, substantial data compression can be achieved with RLE and it is used in PCs to encode color bitmaps. Matlab can be used to read bitmaps into a two-dimensional array and then write the output as an RLE encoded version directly to a *.mif file. An example program is available on the CD-ROM. Bitmap file formats and some C utilities to help read bitmaps can be found on the web.

Many early video games, such as Pong, have a background color with a few moving images. In such cases, the background image can be the default color value and not stored in video RAM. Hardware comparators can check the row and column counts as the video signal is generated and detect when another image other than the background should be displayed. When the comparator signals that the row and column count matches the image location, the image's color data instead of the normal background data is switched into the RGB output using gates or a multiplexer.

The image can be made to move if its current row and column location is stored in registers and the output of these registers are used as the comparator input. Additional logic can be used to increment or decrement the image's location registers slowly over time and produce motion. Multiple hardware comparators can be used to support several fixed and moving images. These moving images are also called sprites. This approach was used in early-generation video games.

10.13 Video Color Mixing using Dithering

PC graphics cards use an analog to digital converter to drive the analog RGB color signals. Although the hardware directly supports only eight different pixel colors using digital color signals, there are some techniques that can be used to generate more colors. On analog CRTs, pixels can be overclocked at two to four times the normal rate to turn on and off the 1-bit color signal several times while scanning across a single pixel. The FPGA's PLL is handy to generate the higher frequency clocks need. Along the same lines, anding the final color signal output with the clock signal itself can further reduce the signal's on time to ½ a clock or less. Unfortunately, this technique does not work quite as well on LCD monitors due to the differences in the internal electronics.

The screen is refreshed at 60Hz, but flicker is almost undetected by the human eye at 30Hz. So, in odd refresh scans one pixel color is used and in even refresh scans another pixel color is used. This 30Hz color mixing or dithering technique works best if large areas have both colors arranged in a checkerboard pattern. Alternating scans use the inverse checkerboard colors. At 30Hz, the eye can detect color intensity flicker in large regions unless the alternating checkerboard pattern is used.

10.14 VHDL Graphics Display Design Example

This simple graphics example will generate a ball that bounces up and down on the screen. As seen in Figure 10.10, the ball is red and the background is white. This example requires the VGA_SYNC design from Section 10.4 to generate the video sync and the pixel address signals. The pixel_row signal is used to determine the current row and the pixel_column signal determines the current column. Using the current row and column addresses, the process Display_Ball generates the red ball on the white background and produces the ball_on signal which displays the red ball using the logic in the red, green, and blue equations.

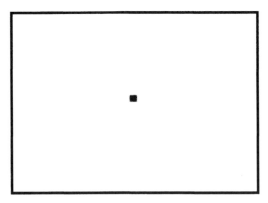

Figure 10.10 Bouncing Ball Video Output.

Ball_X_pos and Ball_y_pos are the current address of the center of the ball. Size is the size of the square ball.

The process Move_Ball moves the ball a few pixels every vertical sync and checks for bounces off of the walls. Ball_motion is the number of pixels to move the ball at each vertical sync clock. The VGA_SYNC core is also used to generate sync signals and pixel addresses, but is not shown in the code below.

```
ENTITY ball IS
    PORT(
        SIGNAL Red, Green, Blue        : OUT STD_LOGIC;
        SIGNAL vert_sync_out           : IN STD_LOGIC;
        SIGNAL pixel_row, pixel_column : IN STD_LOGIC_VECTOR( 9 DOWNTO 0 ));
END ball;
ARCHITECTURE behavior OF ball IS
                                 -- Video Display Signals
        SIGNAL reset, Ball_on, Direction  : STD_LOGIC;
        SIGNAL Size                       : STD_LOGIC_VECTOR( 9 DOWNTO 0 );
        SIGNAL Ball_Y_motion              : STD_LOGIC_VECTOR( 9 DOWNTO 0 );
        SIGNAL Ball_Y_pos, Ball_X_pos     : STD_LOGIC_VECTOR( 9 DOWNTO 0 );
BEGIN                            -- Size of Ball
    Size       <= CONV_STD_LOGIC_VECTOR (8,10 );
                                 -- Ball center  X address
    Ball_X_pos <= CONV_STD_LOGIC_VECTOR( 320,10 );
                                 -- Colors for pixel data on video signal
    Red        <= '1';           -- Turn off Green and Blue to make
                                 -- color Red when displaying ball
    Green      <= NOT Ball_on;
    Blue       <= NOT Ball_on;

Display_Ball:
    PROCESS ( Ball_X_pos, Ball_Y_pos, pixel_column, pixel_row, Size )
    BEGIN                                  -- check row & column for ball area
                                           -- Set Ball_on = '1' to display ball
        IF (  '0' & Ball_X_pos   <= pixel_column + Size ) AND
           ( Ball_X_pos + Size >= '0' & pixel_column   ) AND
           ( '0' & Ball_Y_pos    <= pixel_row + Size    ) AND
           ( Ball_Y_pos + Size >= '0' & pixel_row       ) THEN
                    Ball_on <= '1';
        ELSE
                    Ball_on <= '0';
        END IF;
    END PROCESS Display_Ball;

Move_Ball:
    PROCESS
    BEGIN
                              -- Move ball once every vertical sync
    WAIT UNTIL Vert_sync'EVENT AND Vert_sync = '1';
                              -- Bounce off top or bottom of screen
        IF ('0' & Ball_Y_pos) >= CONV_STD_LOGIC_VECTOR(480,10) - Size THEN
            Ball_Y_motion <= CONV_STD_LOGIC_VECTOR(-2,10);
```

```
        ELSIF  Ball_Y_pos <= Size   THEN
           Ball_Y_motion <= CONV_STD_LOGIC_VECTOR(2,10);
        END IF;
                          -- Compute next ball Y position
        Ball_Y_pos <= Ball_Y_pos + Ball_Y_motion;
     END PROCESS Move_Ball;
  END behavior;
```

10.15 Higher Video Resolution and Faster Refresh Rates

The UP 3's Video Sync core function is designed to support higher resolutions and refresh rates. The UP 2 can only support its Video Sync core's existing 640 by 480 60Hz video mode since it does not have an internal PLL to produce different pixel clocks. Table 10.2 shows several common screen resolutions and refresh rates. To change resolutions or refresh rates two changes are needed. First, change the UP 3 PLL's video output pixel clock to the new frequency value by editing the Video_PLL.vhd file using the MegaWizard edit feature. Second, the six counter constant values used to generate the horizontal and vertical sync signals in the Video_Sync.vhd core need to be changed to the new six values for the desired resolution and refresh rate found in the large table at the end of the Video_Sync.vhd file. Keep in mind that higher resolutions will require more pixel memory and smaller hardware delays that can support the faster clock rates needed.

Table 10.2 Pixel clock rates for some common video resolutions and refresh rates.

Mode	Refresh	Hor. Sync	Pixel clock
640x480	60Hz	31.5kHz	25.175MHz
640x480	63Hz	32.8kHz	28.322MHz
640x480	70Hz	36.5kHz	31.5MHz
640x480	72Hz	37.9kHz	31.5MHz
800x600	56Hz	35.1kHz	36.0MHz
800x600	60Hz	37.9kHz	40.0MHz
800x600	72Hz	48.0kHz	50.0MHz
1024x768	60Hz	48.4kHz	65.0MHz
1024x768	70Hz	56.5kHz	75.0MHz
1024x768	70Hz	56.25kHz	72.0MHz
1024x768	76Hz	62.5kHz	85.0MHz
1280x1024	61Hz	64.24kHz	110.0MHz
1280x1024	74Hz	78.85kHz	135.0MHz

10.16 Laboratory Exercises

1. Design a video output display that displays a large version of your initials. Hint: use the character generation ROM, the Video Sync UP3core, and some of the higher bits of the row and column pixel counters to generate larger characters.

2. Modify the bouncing ball example to bounce and move in both the X and Y directions. You will need to add code for motion in two directions and check additional walls for a bounce condition.

3. Modify the bouncing ball example to move up or down based on input from the two pushbuttons.

4. Modify the example to support different speeds. Read the speed of the ball from the UP 3's DIP switches.

5. Draw a more detailed ball in the bouncing ball example. Use a small ROM to hold a small detailed color image of a ball.

6. Make a Pong-type video game by using pushbutton input to move a paddle up and down that the ball will bounce off of.

7. Design your own video game with graphics. Some ideas include breakout, space invaders, Tetris, a slot machine, poker, craps, blackjack, pinball, and roulette. Keep the graphics simple so that the design will fit on the Cyclone chip. If the video game needs a random number generator, information on random number generation can be found in Appendix A.

8. Use the character font ROM and the ideas from the MIPS character output example to add video character output to another complex design.

9. Using Matlab or C, write a program to convert a color bitmap into a *.mif file with run-length encoding. Design a state machine to read out the memory and generate the RGB color signals to display the bitmap. Use a reduced resolution pixel size such as 160 by 120. Find a bitmap to display or create one with a paint program. It will work best if the bitmap is already 160 by 120 pixels or smaller. A school mascot or your favorite cartoon character might make an interesting choice. 92K bits of memory are available in the Cyclone EP1C6 so a 12-bit RLE format with nine bits for length and three bits for color can be used with up to 7,600 locations. This means that the bitmap can only have 7,600 color changes as the image is scanned across the display. Simple images such as cartoons have fewer color changes. A Matlab example is on the CD-ROM.

10. Add color mixing or dithering with more than 8 colors to the previous problem. The 3-bit color code in the RLE encoded memory can be used to map into a color palette. The color palette contains the RGB patterns used for color mixing or interlacing. The color palette memory selects 8 different colors. The program translating the bitmap should select the 8 closest colors for the color palette.

11. Modify the VGA Sync core to support a higher screen resolution and demonstrate it using one of the earlier example video designs.

CHAPTER 11

Interfacing to the PS/2 Keyboard and Mouse

A PS/2 mouse is shown above with the cover removed. The ball (upper right) rolls two plastic X and Y axles with a slotted wheel at one end. The slotted wheel passes through a square slotted case containing an IR emitter and detector pair. When the wheel rotates it generates pulses by interrupting the IR light beam. A microcontroller (lower left) counts the pulses and sends data packets containing mouse movement and button data to the PC.

11 Interfacing to the PS/2 Keyboard and Mouse

The PS/2 interface was originally developed for the IBM PC/AT's mouse and keyboard in 1984. The Altera UP 3 board supports the use of either a mouse or keyboard using a PS/2 connector on the board. This provides only the basic electrical connections from the PS/2 cable and the FPGA chip. It is necessary to design a hardware interface using logic in the FPGA chip to communicate with a keyboard or a mouse. Serial-to-parallel conversion using a shift register is required in the interface hardware.

11.1 PS/2 Port Connections

The PS/2 port consists of 6 pins including ground, power (VDD), keyboard data, and a keyboard clock line. The UP 3 board supplies the power to the mouse or keyboard. Two lines are not used. The keyboard data line is pin 13 on the UP 3's Cyclone chip, and the keyboard clock line is pin 12. Pins must be specified in one of the design files.

Table 11.1 PS/2 Keyboard Commands and Messages.

Commands Sent to Keyboard	Hex Value
Reset Keyboard Keyboard returns AA, 00 after self-test	FF
Resend Message	FE
Set key typematic (autorepeat) XX is scan code for key	FB, XX
Set key make and break	FC, XX
Set key make	FD, XX
Set all key typematic, make and break	FA
Set all keys make	F9
Set all keys make and break	F8
Make all keys typematic (autorepeat)	F7
Set to Default Values	F6
Clear Buffers and start scanning keys	F4

Set typematic (autorepeat) rate and delay
    Set typematic (autorepeat) rate and delay
        Bits 6 and 5 are delay (250ms to 1 sec)
        Bits 4 to 0 are rate (all 0's-30x/sec to all 1's 2x/sec)
	F3, XX

Read keyboard ID Keyboard sends FA, 83, AB	F2
Set scan code set XX is 01, 02, or 03	F0, XX
Echo	EE

Set Keyboard LEDs
    XX is 00000 Scroll, Num, and Caps Lock bits
    1 is LED on and 0 is LED off
	ED, XX

Both the clock and data lines are open collector and bi-directional. The clock line is normally controlled by the keyboard, but it can also be driven by the computer system or in this case the Cyclone chip, when it wants to stop data transmissions from the keyboard. Both the keyboard and the system can drive the data line. The data line is the sole source for the data transfer between the computer and keyboard. The keyboard and the system can exchange several commands and messages as seen in Tables 11.1 and 11.2.

Table 11.2 PS/2 Commands and messages sent by keyboard.

Messages Sent by Keyboard	Hex Value
Resend Message	FE
Two bad messages in a row	FC
Keyboard Acknowledge Command Sent by Keyboard after each command byte	FA
Response to Echo command	EE
Keyboard passed self-test	AA
Keyboard buffer overflow	00

11.2 Keyboard Scan Codes

Keyboards are normally encoded by placing the key switches in a matrix of rows and columns. All rows and columns are periodically checked by the keyboard encoder or "scanned" at a high rate to find any key state changes. Key data is passed serially to the computer from the keyboard using what is known as a scan code. Each keyboard key has a unique scan code based on the key switch matrix row and column address to identify the key pressed.

There are different varieties of scan codes available to use depending on the type of keyboard used. The PS/2 keyboard has two sets of scan codes. The default scan code set is used upon power on unless the computer system sends a command the keyboard to use an alternate set. The typical PC sends commands to the keyboard on power up and it uses an alternate scan code set. To interface the keyboard to the UP 3 board, it is simpler to use the default scan code set since no initialization commands are required.

11.3 Make and Break Codes

The keyboard scan codes consist of 'Make' and 'Break' codes. One make code is sent every time a key is pressed. When a key is released, a break code is sent. For most keys, the break code is a data stream of F0 followed by the make code for the key. Be aware that when typing, it is common to hit the next key(s) before releasing the first key hit.

Using this configuration, the system can tell whether or not the key has been pressed, and if more than one key is being held down, it can also distinguish which key has been released. One example of this is when a shift key is held down. While it is held down, the '3' key should return the value for the '#' symbol instead of the value for the '3' symbol. Also note that if a key is held down, the make code is continuously sent via the typematic rate until it is released, at which time the break code is sent.

11.4 The PS/2 Serial Data Transmission Protocol

The scan codes are sent serially using 11 bits on the bi-directional data line. When neither the keyboard nor the computer needs to send data, the data line and the clock line are High (inactive).

As seen in Figure 11.1, the transmission of a single key or command consists of the following components:

1. A start bit ('0')

2. 8 data bits containing the key scan code in low to high bit order

3. Odd parity bit such that the eight data bits plus the parity bit are an odd number of ones

4. A stop bit ('1')

The following sequence of events occur during a transmission of a command by the keyboard:

1. The keyboard checks to ensure that both the clock and keyboard lines are inactive. Inactive is indicated by a High state. If both are inactive, the keyboard prepares the 'start' bit by dropping the data line Low.

2. The keyboard then drops the clock line Low for approximately 35us.

3. The keyboard will then clock out the remaining 10 bits at an approximate rate of 70us per clock period. The keyboard drives both the data and clock line.

4. The computer is responsible for recognizing the 'start' bit and for receiving the serial data. The serial data, which is 8 bits, is followed by an odd parity bit and finally a High stop bit. If the keyboard wishes to send more data, it follows the 12th bit immediately with the next 'start' bit.

This pattern repeats until the keyboard is finished sending data at which point the clock and data lines will return to their inactive High state.

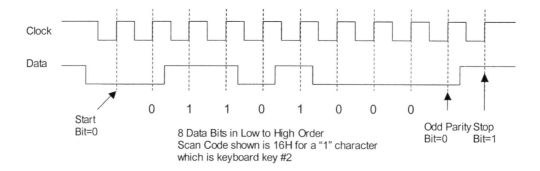

Clock

Data

Start
Bit=0   0   1   1   0   1   0   0   0   Odd Parity Stop
                                                      Bit=0   Bit=1
                 8 Data Bits in Low to High Order
                 Scan Code shown is 16H for a "1" character
                 which is keyboard key #2

Figure 11.1 Keyboard Transmission of a Scan Code.

In Figure 11.1 the keyboard is sending a scan code of 16 for the "1" key and it has a zero parity bit. When implementing the interface code, it will be necessary to filter the slow keyboard clock to ensure reliable operation with the fast logic inside the FPGA chip. Whenever an electrical pulse is transmitted on a wire, electromagnetic properties of the wire cause the pulse to be distorted and some portions of the pulse may be reflected from the end of the wire. On some PS/2 keyboards and mice there is a reflected pulse on the cable that is strong enough to cause additional phantom clocks to appear on the clock line.

Here is one approach that solves the reflected pulse problem. Feed the PS/2 clock signal into an 8-bit shift register that uses a 24MHz clock. AND the bits of the shift register together and use the output of the AND gate as the new "filtered" clock. This prevents noise and ringing on the clock line from causing occasional extra clocks during the serial-to-parallel conversion in the Cyclone chip.

A few keyboards and mice will work without the clock filter and many will not. They all will work with the clock filter, and it is relatively easy to implement. This circuit is included in the UP3cores for the keyboard and the mouse.

As seen in Figure 11.2, the computer system or FPGA chip in this case sends commands to the PS/2 keyboard as follows:

1. System drives the clock line Low for approximately 60us to inhibit any new keyboard data transmissions. The clock line is bi-directional.

2. System drives the data line Low and then releases the clock line to signal that it has data for the keyboard.

3. The keyboard will generate clock signals in order to clock out the remaining serial bits in the command.

4. The system will send its 8-bit command followed by a parity bit and a stop bit.

5. After the stop bit is driven High, the data line is released.

Upon completion of each command byte, the keyboard will send an acknowledge (ACK) signal, FA, if it received the data successfully. If the system does not release the data line, the keyboard will continue to generate the clock, and upon completion, it will send a 're-send command' signal, FE or FC, to the system. A parity error or missing stop bit will also generate a re-send command signal.

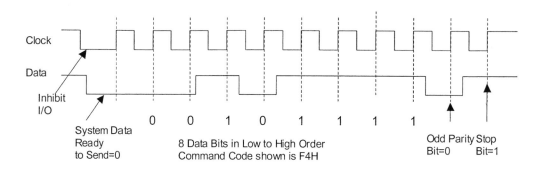

Figure 11.2 System Transmission of a Command to PS/2 Device.

11.5 Scan Code Set 2 for the PS/2 Keyboard

PS/2 keyboards are available in several languages with different characters printed on the keys. A two-step process is required to find the scan code. A key number is used to lookup the scan code. Key numbers needed for the scan code table are shown in Figure 11.3 for the English language keyboard layout.

Figure 11.3 Key Numbers for Scan Code.

Each key sends out a make code when hit and a break code when released. When several keys are hit at the same time, several make codes will be sent before a break code.

The keyboard powers up using this scan code as the default. Commands must be sent to the keyboard to use other scan code sets. The PC sends out an initialization command that forces the keyboard to use the other scan code.

The interface is much simpler if the default scan code is used. If the default scan code is used, no commands will need to be sent to the keyboard. The keys in Table 11.3 for the default scan code are typematic (i.e. they automatically repeat the make code if held down).

Table 11.3 Scan Codes for PS/2 Keyboard.

Key#	Make Code	Break Code	Key#	Make Code	Break Code	Key#	Make Code	Break Code
1	0E	F0 0E	31	1C	F0 1C	90	77	F0 77
2	16	F0 16	32	1B	F0 1B	91	6C	F0 6C
3	1E	F0 1E	33	23	F0 23	92	6B	F0 6B
4	26	F0 26	34	2B	F0 2B	93	69	F0 69
5	25	F0 25	35	34	F0 34	96	75	F0 75
6	2E	F0 2E	36	33	F0 33	97	73	F0 73
7	36	F0 36	37	3B	F0 3B	98	72	F0 72
8	3D	F0 3D	38	42	F0 42	99	70	F0 70
9	3E	F0 3E	39	4B	F0 4B	100	7C	F0 7C
10	46	F0 46	40	4C	F0 4C	101	7D	F0 7D
11	45	F0 45	41	52	F0 52	102	74	F0 74
12	4E	F0 4E	43	5A	F0 5A	103	7A	F0 7A
13	55	F0 55	44	12	F0 12	104	71	F0 71
15	66	F0 66	46	1A	F0 1A	105	7B	F0 7B
16	0D	F0 0D	47	22	F0 22	106	79	F0 79
17	15	F0 15	48	21	F0 21	110	76	F0 76
18	1D	F0 1D	49	2A	F0 2A	112	05	F0 05
19	24	F0 24	50	32	F0 32	113	06	F0 06
20	2D	F0 2P	51	31	F0 31	114	04	F0 04
21	2C	F0 2C	52	3A	F0 3A	115	0c	F0 0C
22	35	F0 35	53	41	F0 41	116	03	F0 03
23	3C	F0 3C	54	49	F0 49	117	0B	F0 0B
24	43	F0 43	55	4A	F0 4A	118	83	F0 83
25	44	F0 44	57	59	F0 59	119	0A	F0 0A
26	4D	F0 4D	58	14	F0 14	120	01	F0 01
27	54	F0 54	60	11	F0 11	121	09	F0 09
28	5B	F0 5B	61	29	F0 29	122	78	F0 78
29	5D	F0 5D	62	E0 11	E0 F0 11	123	07	F0 07

The remaining key codes are a function of the shift, control, alt, or num-lock keys.

Table 11.3 (Continued) - Scan Codes for PS/2 Keyboard.

Key	No Shift or Num Lock		Shift*		Num Lock On	
#	Make	Break	Make	Break	Make	Break
76	E0 70	E0 F0 70	E0 F0 12 E0 70	E0 F0 70 E0 12	E0 12 E0 70	E0 F0 70 E0 F0 12
76	E0 71	E0 F0 71	E0 F0 12 E0 71	E0 F0 71 E0 12	E0 12 E0 71	E0 F0 71 E0 T0 12
79	E0 6B	E0 F0 6B	E0 F0 12 E0 6B	E0 F0 6B E0 12	E0 12 E0 6B	E0 F0 6B E0 F0 12
80	E0 6C	E0 F0 6C	E0 F0 12 E0 6C	E0 F0 6C E0 12	E0 12 E0 6C	E0 F0 6C E0 F0 12
81	E0 69	E0 F0 69	E0 F0 12 E0 69	E0 F0 69 E0 12	E0 12 E0 69	E0 F0 69 E0 F0 12
83	E0 75	E0 F0 75	E0 F0 12 E0 75	E0 F0 75 E0 12	E0 12 E0 75	E0 F0 75 E0 F0 12
84	E0 72	E0 F0 72	E0 F0 12 E0 72	E0 F0 72 E0 12	E0 12 E0 72	E0 F0 72 E0 F0 12
85	E0 7D	E0 F0 7D	E0 F0 12 E0 7D	E0 F0 7D E0 12	E0 12 E0 7D	E0 F0 7D E0 F0 12
86	E0 7A	E0 F0 7A	E0 F0 12 E0 7A	E0 F0 7A E0 12	E0 12 E0 7A	E0 F0 7A E0 F0 12
89	E0 74	E0 F0 74	E0 F0 12 E0 74	E0 F0 74 E0 12	E0 12 E0 74	E0 F0 74 E0 F0 12

* When the left Shift Key is held down, the 12 - FO 12 shift make and break is sent with the other scan codes. When the right Shift Key is held down, 59 – FO 59 is sent.

Key	Scan Code		Shift Case *	
#	Make	Break	Make	Break
95	E0 4A	E0 F0 4A	E0 F0 12 E0 4A	E0 12 F0 4A

* When the left Shift Key is held down, the 12 - FO 12 shift make and break is sent with the other scan codes. When the right Shift Key is held down, 59 - FO 59 is sent. When both Shift Keys are down, both sets of codes are sent with the other scan codes.

Key	Scan Code		Control Case, Shift Case		Alt Case	
#	Make	Break	Make	Break	Make	Break
124	E0 12 E0 7C	E0 F0 7C E0 F0 l2	E0 7C	E0 F0 7C	84	F0 84

Key #	Make Code	Control Key Pressed
126 *	EI 14 77 EI F0 14 F0 77	E0 7E E0 F0 7E

* This key does not repeat

11.6 The Keyboard UP3core

The following VHDL code for the keyboard UP3core shown in Figure 11.4 reads the scan code bytes from the keyboard. In this example code, no command is ever sent to the keyboard, so clock and data are always used as inputs and the keyboard power-on defaults are used.

To send commands, a more complex bi-directional tri-state clock and data interface is required. The details of such an interface are explained in later sections on the PS/2 mouse. The keyboard powers up and sends the self-test code AA and 00 to the Cyclone chip before it is downloaded.

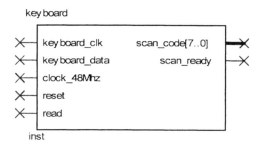

Figure 11.4 Keyboard UP3core

```
LIBRARY IEEE;
USE IEEE.STD_LOGIC_1164.ALL;
USE IEEE.STD_LOGIC_ARITH.ALL;
USE IEEE.STD_LOGIC_UNSIGNED.ALL;

ENTITY keyboard IS
    PORT(  keyboard_clk, keyboard_data, clock_48MHz ,
           reset, read          : IN    STD_LOGIC;
           scan_code            : OUT   STD_LOGIC_VECTOR( 7 DOWNTO 0 );
           scan_ready           : OUT   STD_LOGIC);
END keyboard;

ARCHITECTURE a OF keyboard IS
    SIGNAL INCNT                         : STD_LOGIC_VECTOR( 3 DOWNTO 0 );
    SIGNAL SHIFTIN                       : STD_LOGIC_VECTOR( 8 DOWNTO 0 );
    SIGNAL READ_CHAR, clock_enable       : STD_LOGIC;
    SIGNAL INFLAG, ready_set             : STD_LOGIC;
    SIGNAL keyboard_clk_filtered         : STD_LOGIC;
    SIGNAL filter                        : STD_LOGIC_VECTOR( 7 DOWNTO 0 );

BEGIN
    PROCESS ( read, ready_set )
    BEGIN
        IF read = '1' THEN
            scan_ready <= '0';
        ELSIF ready_set'EVENT AND ready_set = '1' THEN
            scan_ready <= '1';
        END IF;
    END PROCESS;
                    --This process filters the raw clock signal coming from the
                    -- keyboard using a shift register and two AND gates
Clock_filter:
    PROCESS
        BEGIN
        WAIT UNTIL clock_48MHz'EVENT AND clock_48MHz = '1';
        clock_enable <= NOT clock_enable;
        IF clock_enable = '1' THEN
```

```vhdl
                    filter ( 6 DOWNTO 0 ) <= filter( 7 DOWNTO 1 ) ;
                    filter( 7 ) <= keyboard_clk;
                    IF filter = "11111111" THEN
                    keyboard_clk_filtered <= '1';
                    ELSIF filter = "00000000" THEN
                    keyboard_clk_filtered <= '0';
                    END IF;
              END IF;
     END PROCESS Clock_filter;
                      --This process reads in serial scan code data coming from the keyboard
     PROCESS
     BEGIN
       WAIT UNTIL (KEYBOARD_CLK_filtered'EVENT AND KEYBOARD_CLK_filtered = '1');
       IF RESET = '0' THEN
           INCNT  <= "0000";
           READ_CHAR   <= '0';
       ELSE
           IF KEYBOARD_DATA = '0' AND READ_CHAR = '0' THEN
               READ_CHAR   <= '1';
               ready_set   <= '0';
           ELSE
                          -- Shift in next 8 data bits to assemble a scan code
               IF READ_CHAR = '1' THEN
                   IF INCNT < "1001" THEN
                       INCNT   <= INCNT + 1;
                       SHIFTIN( 7 DOWNTO 0 ) <= SHIFTIN( 8 DOWNTO 1 );
                       SHIFTIN( 8 )    <= KEYBOARD_DATA;
                       ready_set    <= '0';
                          -- End of scan code character, so set flags and exit loop
                   ELSE
                       scan_code  <= SHIFTIN( 7 DOWNTO 0 );
                       READ_CHAR   <='0';
                       ready_set   <= '1';
                       INCNT  <= "0000";
                   END IF;
               END IF;
           END IF;
       END IF;
     END PROCESS;
     END a;
```

The keyboard clock is filtered in the Clock_filter process using an 8-bit shift register and an AND gate to eliminate any reflected pulses, noise, or timing hazards that can be found on some keyboards. The clock signal in this process is the 48 MHz system clock divided by two to produce a 24 MHz clock rate using the clock enable signal. The output signal, keyboard_clk_filtered, will only change if the input signal, keyboard_clk, has been High or Low for eight successive 24 MHz clocks or 320ns. This filters out noise and reflected pulses on the keyboard cable that could cause an extra or false clock signal on the fast Cyclone chip. This problem has been observed to occur on some PS/2 keyboards and mice and is fixed by the filter routine.

The RECV_KBD process waits for a start bit, converts the next eight serial data bits to parallel, stores the input character in the signal, charin, and sets a flag, scan_ready, to indicate a new character was read. . The scan_ready or input ready flag is a handshake signal needed to ensure that a new scan code is read in and processed only once. Scan_ready is set whenever a new scan code is received. The input signal, read, resets the scan ready handshake signal.

The process using this code to read the key scan code would need to wait until the input ready flag, scan_ready, goes High. This process should then read in the new scan code value, scan_code. Last, read should be forced High and Low to clear the scan_ready handshake signal.

Since the set and reset conditions for scan_ready come from different processes each with different clocks, it is necessary to write a third process to generate the scan_ready handshake signal using the set and reset conditions from the other two processes. Hitting a common key will send a 1-byte make code and a 2-byte break code. This will produce at least three different scan_code values each time a key is hit and released.

A shift register is used with the filtered clock signals to perform the serial to parallel conversion. No command is ever sent the keyboard and it powers up using scan code set 2. Since commands are not sent to the keyboard, in this example clock and data lines are not bi-directional. The parity bit is not checked.

11.7 A Design Example Using the Keyboard UP3core

Here is a simple design using the Keyboard and LCD_Display UP3cores. The last six bytes of scan codes will appear in the UP3's LCD display. The block code_FIFO saves the last six scan codes for the LCD display.

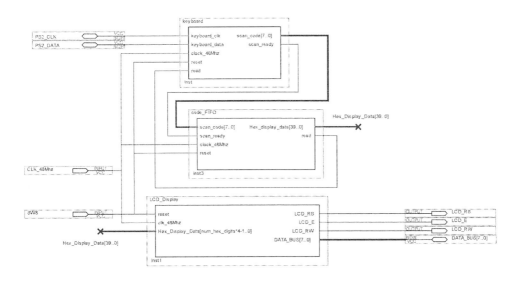

Figure 11.5 Example design using the Keyboard UP3core.

11.8 Interfacing to the PS/2 Mouse

Just like the PS/2 keyboard, the PS/2 mouse uses the PS/2 synchronous bi-directional serial communication protocol described in section 11.4 and shown in Figures 11.1 and 11.2. Internally, the mouse contains a ball that rolls two slotted wheels. The wheels are connected to two optical encoders. The two encoders sense x and y motion by counting pulses when the wheels move. It also contains two or three pushbuttons that can be read by the system and a single-chip microcontroller. The microcontroller in the mouse sends data packets to the computer reporting movement and button status.

It is necessary for the computer or in this case the FPGA chip to send the mouse an initialization command to have it start sending mouse data packets. This makes interfacing to the mouse more difficult than interfacing to the keyboard. As seen in Table 11.4, the command value needed for initialization after power up is F4, enable streaming mode.

Table 11.4 PS/2 Mouse Commands.

Commands Sent to Mouse	Hex Value
Reset Mouse	FF
Mouse returns AA, 00 after self-test	
Resend Message	FE
Set to Default Values	F6
Enable Streaming Mode	F4
Mouse starts sending data packets at default rate	
Disable Streaming Mode	F5
Set sampling rate	F3, XX
XX is number of packets per second	
Read Device Type	F2
Set Remote Mode	EE
Set Wrap Mode	EC
Mouse returns data sent by system	
Read Remote Data	EB
Mouse sends 1 data packet	
Set Stream Mode	EA
Status Request	E9
Mouse returns 3-bytes with current settings	
Set Resolution	E8, XX
XX is 0, 1, 2, 3	
Set Scaling 2 to 1	E7
Reset Scaling	E6

Table 11.5 PS/2 Mouse Messages.

Messages Sent by Mouse	Hex Value
Resend Message	FE
Two bad messages in a row	FC
Mouse Acknowledge Command Sent by Mouse after each command byte	FA
Mouse passed self-test	AA

After streaming mode is enabled, the mouse sends data to the system in three byte data packets that contain motion and pushbutton status. The format of a three-byte mouse data packet is seen in Table 11.6.

Table 11.6 PS/2 Mouse Data Packet Format.

	MSB							LSB
Bit	7	6	5	4	3	2	1	0
Byte 1	Yo	Xo	Ys	Xs	1	M	R	L
Byte 2	X7	X6	X5	X4	X3	X2	X1	X0
Byte 3	Y7	Y6	Y5	Y4	Y3	Y2	Y1	Y0

L = Left Key Status bit   (For buttons 1 = Pressed and 0 = Released)
M = Middle Key Status bit (This bit is reserved in the standard PS/2 mouse protocol, but
                             some three button mice use the bit for middle button status.)
R = Right Key Status bit
$X7 - X0$ = Moving distance of X in two's complement
                           (Moving Left = Negative; Moving Right = Positive)
$Y7 - Y0$ = Moving distance of Y in two's complement
                           (Moving Up = Positive; Moving Down = Negative)
Xo = X Data Overflow bit   (1 = Overflow)
Yo = Y Data Overflow bit   (1 = Overflow)
Xs = X Data sign bit   (1 = Negative)
Ys = Y Data sign bit   (1 = Negative)

11.9 The Mouse UP3core

The UP3core function Mouse is designed to provide a simple interface to the mouse. This function initializes the mouse and then monitors the mouse data transmissions. It outputs a mouse cursor address and button status. The internal operation of the Mouse UP3core is rather complex and the fundamentals are described in the section that follows. Like the other UP3core functions, it is written in VHDL and complete source code is provided.

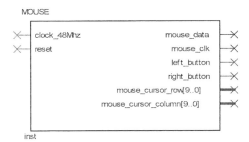

To interface to the mouse, a clock filter, serial-to-parallel conversion and parallel-to-serial conversion with two shift registers is required along with a state machine to control the various modes. See the earlier PS/2 keyboard section for an example of a clock filter design.

11.10 Mouse Initialization

Two lines are used to interface to the mouse, PS/2 clock and data. The lines must be tri-state bi-directional, since at times they are driven by the mouse and at other times by the Cyclone chip. All clock, data, and handshake signals share two tri-state, bi-directional lines, clock and data. These two lines must be declared bi-directional when pin assignments are made and they must have tri-state outputs in the interface. The mouse actually has open collector outputs that can be simulated by using a tri-state output. The mouse always drives the clock signal for any serial data exchanges. The FPGA chip can inhibit mouse transmissions by pulling the clock line Low at any time.

The FPGA chip drives the data line when sending commands to the mouse. When the mouse sends data to the FPGA chip it drives the data line. The tri-state bi-directional handshaking is described in more detail in the IBM PS/2 Technical Reference manual. A simpler version with just the basics for operation with the UP 3 is presented here. Just like the keyboard, the mouse interface is more reliable if a clock filter is used on the clock line.

At power-up, the mouse runs a self-test and sends out the codes AA and 00. The clock and data Cyclone chip outputs are tri-stated before downloading the UP 3, so they float High. High turns out to be ready to send for mouse data, so AA and 00 are sent out prior to downloading and need not be considered in the interface. This assumes that the mouse is plugged in before applying power to the UP 3 board and downloading the design.

The default power-up mode is streaming mode disabled. To get the mouse to start sending 3-byte data packets, the streaming mode must be turned on by sending the enable streaming mode command, F4, to the mouse from the FPGA chip. The clock tri-state line is driven Low by the FPGA for at least 60us to inhibit any data transmissions from the mouse. This is the only case when the FPGA chip should ever drive the clock line. The data line is then driven Low by the FPGA chip to signal that the system has a command to send the mouse.

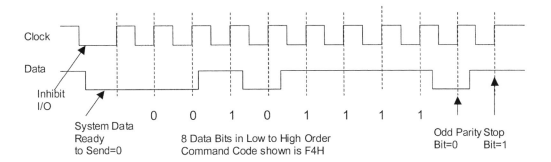

Figure 11.6 Transmission of Mouse Initialization Command.

The clock line is driven High for four clocks at 24 MHz and then tri-stated to simulate an open collector output. This reduces the rise time and reflections on the mouse cable that might be seen by the fast FPGA chip logic as the clock line returns to the High state. As an alternative, the mouse clock input to the FPGA could be briefly disabled while the clock line returns to the High state.

Next the mouse, seeing data Low and clock High, starts clocking in the serial data from the FPGA chip. The data is followed by an odd parity bit and a High stop bit. The handshake signal of the data line starting out Low takes the place of the start bit when sending commands to the mouse.

With the FPGA chip clock and data drivers both tri-stated, the mouse then responds to this message by sending an acknowledge message code, FA, back to the FPGA chip. Data from the mouse includes a Low start bit, eight data bits, an odd parity bit, and a High stop bit. The mouse, as always, drives the clock line for the serial data transmission. The mouse is now initialized.

11.11 Mouse Data Packet Processing

As long as the FPGA chip clock and data drivers remain tri-stated, the mouse then starts sending 3-byte data packets at the power-up default sampling rate of 100 per second. Bytes 2 and 3 of the data packet contain X and Y motion values as was seen in Table 11.6. These values can be positive or negative, and they are in two's complement format.

For a video mouse cursor such as is seen in the PC, the motion value will need to be added to the current value every time a new data packet is received. Assuming 640 by 480 pixel resolution, two 10-bit registers containing the current cursor row and column addresses are needed. These registers are updated every packet by adding the sign extended 8-bit X and Y motion values found in bytes 2 and 3 of the data packet. The cursor normally would be initialized to the center of the video screen at power-up.

11.12 An Example Design Using the Mouse UP3core

In this example design, the mouse drives the UP3's LCD display. The mouse cursor powers up to the center position of the 640 by 480 video screen. Note that the PS/2 mouse clock and data pins must be bi-directional. The block Mouse_LCD_interface rearranges the mouse core output signals for use by the LCD_Display core function.

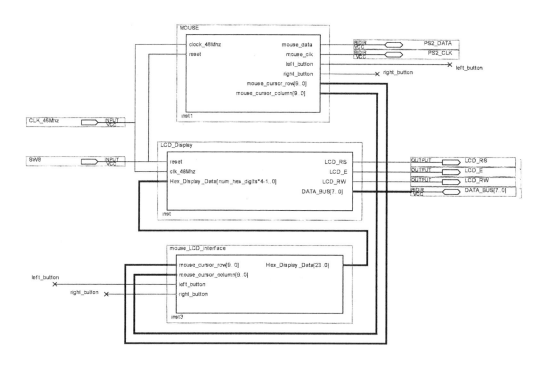

Figure 11.7 Example design using the Mouse UP3core.

11.13 For Additional Information

The IBM PS/2 Hardware Interface Technical Reference Manual, IBM Corporation, 1988 contains the original PS/2 information on the keyboard and mouse in the Keyboard and Auxiliary Device Controller Chapter. Scan codes for the alternate scan code set normally used by the PC can be found on the web and in many PC reference manuals.

11.14 Laboratory Exercises

1. Write a VHDL module to read a keyboard scan code and display the entire scan code string in hexadecimal on the VGA display using the VGA_SYNC and CHAR_ROM UP3cores. It will require the use of the read and scan ready handshake lines and a small RAM to hold the scan code bytes.

2. After reading the section on the PS/2 mouse, design an interface that can also send commands to the keyboard. Demonstrate that the design works correctly by changing the status of the keyboard LEDs after reading the new settings from the Cyclone DIP switch.

3. Develop a keyboard module that uses the alternate scan code set used by the PC.

4. Write the keyboard module in another HDL such as Verilog.

5. Use the keyboard as a new input device for a video game, the μP1 computer, or another application.

6. Generate a video display that has a moving cursor controlled by the mouse using the Mouse and VGA_Sync UP3cores. Use the mouse buttons to change the color of the cursor.

7. Use the mouse as input to a video etch-a-sketch. Use a monochrome 128 by 128 1-bit pixel RAM with the VGA_Sync core in your video design. Display a cursor. To draw a line, the left mouse button should be held down.

8. Use the mouse as an input device in another design with video output or a simple video game such as pong, breakout, or Tetris.

9. Write a mouse driver in Verilog. Use the mouse information provided in sections 11.2 and 11.3.

CHAPTER 12

Legacy Digital I/O Interfacing Standards

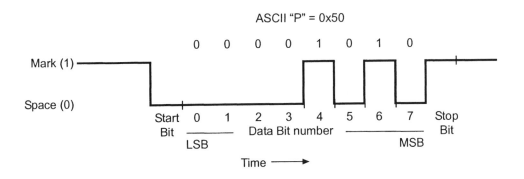

The EIA RS-232C standard is widely used in PCs on the COM ports for serial data transmission.

12 Legacy Digital I/O Interfacing Standards

Historically, several common digital interface standards have developed over the years to interface computers to their peripheral devices. This chapter will introduce several of the older standards and briefly describe how they function in a hardware design. Each standard has a unique set of hardware and performance tradeoffs. Many devices and ICs are available that use these standards. These interfaces are present in most PCs and are used on many FPGA systems such as the UP 3 board.

12.1 Parallel I/O Interface

The parallel printer interface standard was developed by Centronics in the 1970s and is a widely used standard for transferring 8-bit parallel data. Most PCs have a parallel port. Data is transferred in parallel using eight data bits and standard digital logic voltage levels. Additional status and control bits are required for the sender and receiver to exchange handshake signals that synchronize each 8-bit data transfer. Typically, the parallel printer port is interfaced to two 8-bit I/O ports on a processor. One I/O port is used for 8-bit data transfers and one I/O port for the status and control bits that are used for handshake signals.

The transfer of an 8-bit data value is shown in Figure 12.1. First, the computer waits for the printer's busy signal to go Low. Next, the computer outputs the eight data bits and the computer then sets strobe Low for at least 0.5us. The computer then waits for the printer to pulse Ack Low. The computer must wait for Ack Low before changing the data or strobe lines. The printer may go Busy after it raises Ack. The printer handshake lines are also used to force the computer to wait for events like a slow carriage return or page feed on a mechanical printer or errors like a paper out condition. Sometimes a timeout loop is used to detect conditions like paper out. The UP 3 board has a standard printer parallel port connector. With the appropriate hardware, it can be used to communicate with a standard printer.

In addition to printers, some special purpose devices also use the individual parallel port bits in a number of different ways to output digital logic bits to control external hardware. The ByteBlaster adapter you use to program the FPGA is one such example.

The original parallel interface supported only unidirectional data transfers from a computer to a printer. Recent parallel port standards such as IEEE 1284 ECP and EPP support bidirectional and faster data transfers between an external device and the computer. In these newer modes, another control bit from the computer specifies the data transfer direction and tri-state gate outputs are used in both the computer and printer to drive the data lines bidrectionally.

Parallel cables will only work for relatively short distances. The RS-232C standard in the next section supports longer cables with fewer wires, but it also has lower bandwidth and data transfer rates.

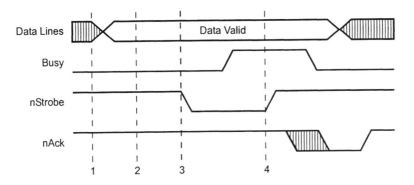

Figure 12.1 Parallel Port transfer of an 8-bit data value

12.2 RS-232C Serial I/O Interface

The Electronics Industry Association (EIA) RS-232C Serial interface is one of the oldest serial I/O standards. In Europe, is it also called V.24. 8-bit data is transmitted one bit at a time serially. Most PCs have an RS-232C serial COM port. Serial interfaces have an advantage in that they require fewer wires in the cable than a parallel interface and can support longer cables. In RS-232C's simplest implementation, only three wires are used in the cable. One wire for transmit data (TD), one for receive data (RD) and one for signal ground (GND). Individual bits are clocked in and out serially using a clock signal. The frequency of this bit clock is called the serial interface's baud rate. (Baudot was a French engineer that developed an early serial interface for the telegraph.) Since two different signal wires are used for receive and transmit, serial devices can be transferring data in both directions at the same time (full-duplex). The ASCII character code is typically used on serial devices, but they can also be used to transfer 8-bit binary values.

The baud rate clock is not synchronized by using a signal wire connected between the sending and receiving devices, rather it is asynchronous and is derived by a state machine watching the serial data bit transitions occurring at the receiver. For this to function correctly, the transmitter and receiver must be setup to operate at the same clock or baud rate. Even though they have the same clock rate, the clock phase must still be synchronized between a serial transmitter and receiver by examining the incoming serial data line. The hardware logic circuit needed for this common serial interface is called a Universal Asynchronous Receiver Transmitter (UART).

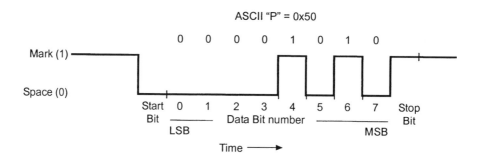

Figure 12.2 RS-232C Serial interface transmission of an 8-bit data value

Figure 12.2 shows the transmission of a single ASCII character over an RS-232C serial interface. The serial bit has two states. Mark is the high state (>3V) and Space is the low state (<-3V). Older generation serial devices will have around +12V and -12V levels for Mark and Space. Note that for the proper RS-232 voltage levels, a standard digital logic output bit will have to have its voltage levels converted for use in a serial interface. Special ICs are normally used for this RS-232C voltage conversion. To reduce the need for additional circuits, these ICs also generate the required DC supply voltages from the standard digital logic DC power supplies. This special IC chip is already present on the UP 3's serial interface. FPGA logic elements can be used to build the UART hardware function.

The idle state is High (Mark). Whenever the interface starts sending a new 8-bit data value, the line is dropped Low (Space) for one clock cycle (baud rate clock). This is called the start bit. The eight data bits are then clocked out during the next eight baud clocks in low to high bit order. The highest data bit is sometimes used as a parity bit for error detection, when only seven data bits are used instead of eight. After the data bits are clocked out, the bit goes high for one clock. This is called the Stop bit. Sometimes at low baud rates, two Stop bits are present. Note that at least 10 clocks are required to transfer an 8-bit data value.

Typically, UARTs transfer 8-bit data values in and out to other internal logic using an 8-bit parallel I/O port interfaced to a processor. Extra UART status bits can be read by the processor that indicate another 8-bit data value can be sent to the UART or another 8-bit data value is available to read in from the UART. Since serial transmission is very slow compared to a processor's clock, these status bits must be checked in software or hardware for their proper state or the processor will send/receive data faster than the UART can produce or consume it. Other status bits can also be used to detect various error conditions.

A UARTs transmitter uses a shift register clocked by the baud rate clock to convert the 8-bit parallel data to eight serial bits. Start and stop bits are automatically added by the UART's hardware.

At the other end of the serial cable, another UART's receiver uses the Stop and Start bits to reset its internal state machine that is attempting to synchronize its receive clock phase to the incoming serial bit data. This state machine synchronizes the receive clock phase whenever it sees an edge on the incoming serial line. Note that several consecutive bits could be the same value inside the eight data bits, so there is not an edge transition on every single clock.

A UART typically uses an internal clock that is eight or sixteen times the baud rate to watch for edges on the incoming serial data line. UARTs also use this faster clock and a counter to attempt to sample the data bits in the middle of each bit's time frame to minimize the possibility of reading in an incorrect value near an edge. Since long wires are allowed on an RS-232C serial interface, there will likely be noise and ringing present whenever the serial bit changes. Clocking in the bit in the middle of its time frame greatly increases the reliability of the interface. A second shift register is used for serial to parallel conversion in the UART's receiver circuit.

Some serial devices also require additional hardware handshake lines to stop and start the flow of a new 8-bit data value over the serial interface. These handshake lines require additional signal wires in the cable used to connect the serial device. Some of the more commonly used handshake lines are RTS (request to send), CTS (clear to send), DCD (data carrier detect), DSR (data set ready), and DTR (data terminal ready).

There are two types of serial devices defined in the RS-232C standard, data terminal equipment (DTE) and data carrier equipment (DCE). A standard RS-232C serial cable is designed to connect a DTE device to a DCE device. When connecting two serial devices of the same type, a special null modem cable or adapter is needed. A null modem exchanges the TD and RD signal lines at one end of the cable along with several connections on the handshake lines. If you experience problems when connecting a new serial device, the various handshake and null modem cable options can be quickly checked using a low-cost in-line RS-232C analyzer breakout box.

The UP 3 board contains a RS-232 serial connector, and it has the required voltage conversion IC needed for serial data transmission.

12.3 SPI Bus Interface

The serial peripheral interface (SPI) bus created by Motorola in the 1980s is used primarily for synchronous serial communication between a host processor and peripheral ICs. Four signal lines are used: Chip Select (CS), Serial Data Input (SDI), Serial Data Output (SDO), Serial Clock (SCKL). CS and SCKL are outputs provided by the master device. The slave devices receive their clock and chip select inputs from the master. If an SPI device is not selected, its SDO output line goes into a high impedance state (tri-state). The number of serial bits transferred to the slave device varies from device to device. Each slave device contains an internal shift register used to transfer data.

Two types on connections between master and slave devices are supported as seen in Figure 12.3. In a cascaded connection, all slaves in the chain share a single chip select line driven by the master. The master device outputs SDO and it connects as an input to a slave device's SDI input. A slave's SDO output connects to another slave's SDI input. The serial data cascades through all of the slaves and the final slave in the chain connects its SDO line to the master's SDI input to complete the chain. In this configuration, the slave devices appear as one larger slave device, the data output of one device feeds into the input of another device, thus forming one large shift register.

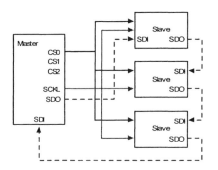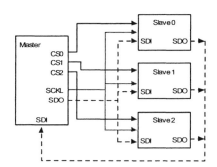

Figure 12.3 The two SPI slave device configuration options.

The second SPI configuration option supports independent slave devices, each device has its own unique chip select input line coming from the master. The master's SDO output connects to each slaves SDI input. The slave's SDO tri-state outputs are connected together and to the master's SDI input. Only the selected slave's SDO output is driven, the others are tri-stated.

Multiple masters are also supported in SPI. Several SPI modes are supported with serial data being valid on either the rising edge or the falling edge of the clock. Serial clock rates can range from 30 kHz to 3 MHz depending on the devices used. Most commonly, devices place new data on the bus during the falling clock edge and data is latched off the bus on the rising edge after it stabilizes, but you will need to check data sheets for specific master and slave devices to confirm this since some devices use the opposite clock edges.

Some Motorola literature may use different names for the SPI signals. CS may appear as SS, SDI as MOSI, and SDO as MISO. In National Semiconductor products, SPI is also known as Microwire. SPI devices are also available in several different voltage supply levels ranging from 2.3 to 5 volts. Since SPI uses a common clock, the hardware interface is simpler than RS-232C serial.

12.4 I²C Bus Interface

The Inter IC (I²C) bus is a widely used standard developed by Phillips in the 1980s for connecting ICs on the same circuit board. Many small ICs now include I²C pins to transfer data serially to other ICs. For lower bandwidth signals, a serial interface has an advantage in that it requires fewer interconnect lines. The I²C bus uses two signal wires called SCL and SDA. SCL is the clock line and SDA is the 1-bit serial data & address line. A common ground signal is also needed. The SCL and SDA lines are open drain. This means that the output is only driven Low, never High. An external pull-up resistor pulls the lines High whenever there is not a device driving the lines Low.

In an FPGA with tri-state output pins, you can simulate open drain outputs by tri-stating the output whenever the bit should go High and only driving the output signal Low. Even though there are multiple devices on the I²C bus, only one pull-up resistor is used for the entire I²C bus.

Devices on the I²C bus are masters or slaves. The slaves are the devices that respond to bus requests from the master. Each slave is assigned its own unique 7-bit I²C bus address. Since both address and data information is transferred over the bus, the protocol is a bit more involved than SPI. When the master needs to talk to a slave, it issues a start sequence on the I²C bus. In a start sequence, SDA goes from High to Low while SCL is High. To stop an I²C sequence, the master sends a stop sequence command. In a stop sequence, SDA goes from Low to High while SCL is High. Start and stop sequences are the only times a change may occur in SDA while SCL is High.

The master drives the SCL clock line to transfer each new I²C serial bit. To force a wait, a slave device can drive SCL Low. Therfore, before each new I²C SCL clock, the master checks to see if SCL is being forced Low by a slave. If it is, the master must wait. SCL clocks are typically up to 100 kHz with 400 kHz available on some new devices.

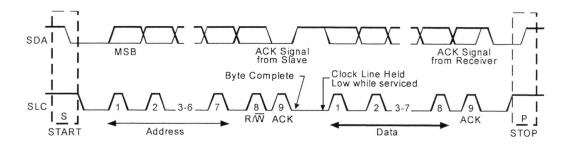

Figure 12.4 I²C interface serial transmission of an 8-bit data value

All address and data transfers contain eight bits with a final acknowledge (ACK) handshake bit for a total of nine bits. All address and data transfers send the High bits first, one per SCL clock bit High. In an address transfer, the

7-bit address is sent and the eighth bit is a R/W bit (0=read, 1=write). Some IC datasheets just append this final R/W bit to the address field and show an 8-bit address field (with even 8-bit addresses for read and odd for write).

The last bit in all data and address transfers, bit nine, is an ACK from the slave. The slave normally drives ACK Low on the last SCL cycle to indicate it is ready for another byte. If ACK is not Low, the master should send a stop sequence to terminate the transfer.

As seen in Figure 12.4, when a master wants to write data to a slave device, it issues the following bus transactions:

1. Master sends a start sequence.
2. Master sends the 7-bit I^2C address (high bits first) of the slave with the R/W bit set Low.
3. Master sends the 8-bit internal register number to write.
4. Master sends 8-bit data value(s). Highest bits first.
5. Master sends a stop sequence.

When a master wants to read data from a slave device, it issues the following bus transactions:

1. Master sends a start sequence.
2. Master sends the 7-bit I^2C address of the slave (high bits first) with the R/W bit set Low.
3. Master sends the 8-bit internal register number to read.
4. Master sends a start sequence.
5. Master sends the 7-bit I^2C address of the slave (high bits first) with the R/W bit set High.
6. Master reads the 8-bit data value(s). Highest bits first.
7. Master sends a stop sequence.

In the full I^2C standard, multiple bus masters are also supported with collision detection and bus arbitration. Collision occurs when two masters attempt to drive the bus at the same time. Arbitration schemes must decide which device can drive the bus when multiple masters are present. Some of the newest I^2C devices can support a high-speed 3.4 MHz clock rate, 10-bit addresses, programmable slave addresses, and lower supply voltages. The System Management Bus (SMB) bus developed by Intel in 1995 that is used for temperature, fan speed, and voltage measurements on many PC motherboards is based on the I^2C bus. On the UP 3 board, the real-time clock chip and the serial EEPROM chip use an I^2C bus interface. Many new TVs, automobiles, and other consumer electronics also contain I^2C interfaces between chips for control features.

SPI and I²C both offer good support for communication with low-speed devices. SPI is better suited to applications that need to transfer higher bandwidth data streams without the need for explicit address information. Some of the most common SPI examples are analog-to-digital (A/D) and digital-to-analog (D/A) converters used to continuously sample or output analog signals. Since addressing is required for I²C, it requires more hardware, but with advances in VLSI technology these additional hardware costs are minimal. In 2005, one FPGA vendor calculated that a single I/O pin on an FPGA package costs as much as 50,000 transistors inside the chip.

12.5 For Additional Information

The books *Parallel Port Complete* and *Serial Port Complete* by Jan Axelson published by Lakeview Research (www.lvr.com) contain complete details on using parallel and serial ports. The full I²C specification is available from Philips Semiconductors (www.phillipssemiconductor.com) and SMB at (www.smbus.org). The Motorola MC68HC11 data manual (www.freescale.com) and various National Semiconductor manuals (www.national.com) have more information on SPI. Analog Devices (www.analogdevices.com) makes a wide variety of A/D and D/A converters with SPI and parallel interfaces.

12.6 Laboratory Exercises

1. Interface a printer with a parallel port to the UP 3 board's parallel port. Connect the two devices using a printer cable. Design logic using a state machine or a processor core for the FPGA to transfer data and handle the handshake lines. You may want to use an older printer so that any problems with your design will not damage the printer. Be careful not to generate tri-state bus conflicts on the parallel data lines by making sure you drive the data direction bit to the proper state. Have the UP 3 print a short ASCII message on the printer ending with an ASCII form or page feed to print the message on a page. A form feed may be needed to cause the printer to print since most printers store characters in an internal page buffer.

2. Interface the UP 3 board's serial port to a PC serial port using a serial cable. Run a serial communications program on the PC. Send a short message to the PC from the UP 3 and display the data from the PC on the UP 3 board's LCD panel.

3. Design an I²C interface for the UP 3 board's real-time clock chip. Display the time from the chip on the UP 3 board's LCD display. Don't forget to check the UP 3 board's jumper settings and battery for the real-time clock chip. The data sheet for the clock chip contains address and data formats.

4. Obtain an IC chip with an SPI interface and design an interface for it on the UP 3 board. Chips with SPI interfaces include analog-to-digital converters, digital-to-analog converters and various sensor modules. Header connections are available on the UP 3 board with 5V or 3V logic levels.

CHAPTER 13

UP 3 Robotics Projects

Photo: The UP3-bot is a small robot controlled by the UP 3 board

13 UP 3 Robotics Projects

13.1 The UP3-bot Design

The UP3-bot shown in Figure 13.1 is a low-cost moving robotics platform designed for the UP 3 board. The UP3-bot is designed to be a small autonomous vehicle that is programmed to move in response to sensory input. A wide variety of sensors can be easily attached to the UP3-bot.

The round platform is cut from plastic and a readily available 7.2V R/C rechargeable battery pack is used to supply power. Two diametrically opposed drive motors move the robot. A third inactive castor wheel or skid is used to provide stability. The robot can move forward, reverse, and rotate in place. Two relatively inexpensive radio control servos are used as drive motors. The UP 3 is programmed to act as the controller. The servos are modified to act as drive motors. The servos are controlled by timing pulses produced by the UP 3 board.

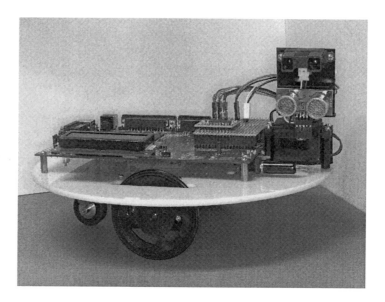

Figure 13.1 The UP3-bot uses an R/C car battery and R/C servos for drive motors.

13.2 UP3-bot Servo Drive Motors

A typical radio control servo is shown in Figure 13.2. Servos have a drive wheel that is controlled by a coded signal. The servo shown is a Futaba S3003 which is identical, internally, to the Tower TS53J servo. Radio control servomotors are mass-produced for the hobby market and are therefore relatively inexpensive and consistently available. They are ideally suited for robotics applications. Internally, the servo contains a DC drive motor (seen on the left in Figure 13.2), built-in control circuitry, and a gear reduction system.

They are small, produce a relatively large amount of torque for their size, and run at the appropriate speed for a robotics drive motor.

Figure 13.2 Left: Radio Control Servo Motor and Right: Servo with Case and Gears Removed.

The control circuitry of the servo uses a potentiometer (variable resistor) that is used to sense the angular position of the output shaft. The potentiometer is the tall component on the right in Figure 13.2. The output shaft of a servo normally travels 180-210 degrees. A single control bit is used to specify the angular position of the shaft. The timing of this bit specifies the angular position for the shaft. The potentiometer senses the angle, and if the shaft is not at the correct angle, the internal control circuit turns the motor in the correct direction until the desired angle is sensed.

The control signal bit specifies the desired angle. The desired angle is encoded using pulse width modulation (PWM). The width of the active high pulse varies from 1-2 ms. A 1ms pulse is 0 degrees, 1.5ms is 90 degrees and a 2 ms pulse is approximately 180 degrees. New timing pulses are sent to the servo every 20 ms.

13.3 Modifying the Servos to make Drive Motors

Normally, a servo has a mechanical stop that prevents it from traveling move than half a revolution. If this stop is removed along with other modifications to the potentiometer, a servo can be converted to a continuously rotating drive motor. Modifications to the servo are not reversible and they will void the warranty. Some robot kit vendors sell servos that are already modified.

To modify the servo, open the housing by removing the screws and carefully note the location of the gears, so that they can be reassembled later. The potentiometer can be replaced with two 2.2K ohm ¼ watt resistors or disconnected by cutting the potentiometer shaft shorter and setting it to the center position so that it reports the 90-degree position. A more accurate setting can be achieved by sending the servo a 1.5ms pulse and adjusting the potentiometer until the motor stops moving. The potentiometer can then be glued in place with CA glue. In the center position the potentiometer will have the same resistance from each of the outside pins to the center pin. If the potentiometer is replaced with two resistors, a resistor is connected between each of the two outside pins and the center pin.

In some servos, there will be less mechanical play if the potentiometer is disabled by cutting the center pin and modified by drilling out the stop on the potentiometer so that it can rotate freely. The two resistors are then added to replace the potentiometer in the circuit.

The largest gear in the gear train that drives the output shaft normally has a tab molded on it that serves as the mechanical stop. After removing the screw on the output shaft and removing the large gear, the mechanical stop can be carefully trimmed off with a hobby saw, knife, or small rotary-grinding tool. The servo is then carefully re-assembled.

After modifications, if a pulse shorter than 1.5 ms is sent, the motor will continuously rotate in one direction. If a pulse longer than 1.5 ms is sent the motor will continuously rotate in the other direction. The 1.5 ms or 90-degree position is sometimes called the neutral position or dead zone. The drive signal to the motor is proportional, so the farther it is from the neutral position the faster it moves. This can be used to control the speed of the motor if the neutral position is carefully adjusted. A pulse width of 0 ms or no pulse will stop the servomotor.

A servo has three wires, +4 to +6 Volt DC power, ground, and the signal wire. The assignment of the three signals on the connector varies among different servo manufacturers. For Futaba servos, the red wire is +5, black is ground, and the white or yellow wire is the pulse width signal line. For JR and Hitec servos, the orange or yellow wire is the signal line and red is +5, and black or brown is ground.

On the UP3-bot, the UP 3 board must be programmed to provide the two timing signals to control the servo drive motors.

13.4 VHDL Servo Driver Code for the UP3-bot

To drive the motors a servo signal must be sent every 20 ms with a 0, 1, or 2 ms pulse. The UP 3 board is programmed to produce the timing signals that drive the motors. If no pulse is sent, the motor stops. If a 1 ms pulse is sent, the motor moves clockwise and if a 2 ms pulse is sent the motor moves in the reverse direction, counterclockwise. To move the UP3-bot forward, one motor moves clockwise while the other motor moves counterclockwise. This is because of the way the motors are mounted to the UP3-bot base.

In the code that follows, lmotor_dir and rmotor_dir specify the direction for the left and right motor. If both signals are '1' the UP 3 bot moves forward. The VHDL code actually moves one motor in the opposite direction to move forward. If both are '0' the robot moves in reverse. If one is '1' and the other is '0', the UP 3 bot turns by rotating in place. The two speed controls are lmotor_speed and rmotor_speed. In the speed control signals, '0' is stop and '1' is run. A 1kHz clock is used for the counters in the module. The UP3core function, clk_div, can be used to provide this signal. Two more complex techniques for implementing variable speed control are discussed in problems at the end of the chapter. Acroname sells a low-cost optical encoder kit made by Nubotics that can be attached to standard R/C servo wheels and used for position feedback and more accurate motor speed control.

```
LIBRARY IEEE;
USE  IEEE.STD_LOGIC_1164.ALL;
USE  IEEE.STD_LOGIC_ARITH.ALL;
USE  IEEE.STD_LOGIC_UNSIGNED.ALL;
ENTITY motor_control IS
      PORT    (clock_1kHz                              : IN       STD_LOGIC;
               lmotor_dir, rmotor_dir        : IN STD_LOGIC;
               lmotor_speed, rmotor_speed    : IN STD_LOGIC;
               lmotor, rmotor                : OUT     STD_LOGIC);
END motor_control;

ARCHITECTURE a OF motor_control IS
    SIGNAL count_motor: STD_LOGIC_VECTOR( 4 DOWNTO 0 );
BEGIN
    PROCESS
        BEGIN
                                    -- Count_motor is a 20ms timer
                WAIT UNTIL clock_1kHz'EVENT AND clock_1kHz = '1';
                    IF count_motor /=19 THEN
                        count_motor <= count_motor + 1;
                    ELSE
                        count_motor <= "00000";
                    END IF;
                    IF count_motor >= 17 AND count_motor < 18 THEN
                            -- Don't generate any pulse for speed = 0
                        IF lmotor_speed = '0' THEN
                            lmotor <= '0';
                        ELSE
                            lmotor <= '1';
                        END IF;
                        IF rmotor_speed = '0' THEN
                            rmotor <= '0';
                        ELSE
                            rmotor <= '1';
                        END IF;
                        -- Generate a 1 or 2ms pulse for each motor
                        -- depending on direction
                        -- reverse directions between the two motors because
                        -- of servo mounting on the UP3-bot base
                    ELSIF count_motor >=18 AND count_motor <19 THEN
                        IF lmotor_speed /= '0' THEN
                            CASE lmotor_dir IS
                                -- FORWARD
                                WHEN '0' =>
```

```
                                      lmotor <= '1';
                                -- REVERSE
                                WHEN '1' =>
                                      lmotor <= '0';
                                WHEN OTHERS => NULL;
                                END CASE;
                        ELSE
                                lmotor <= '0';
                        END IF;
                        IF rmotor_speed /= '0' THEN
                           CASE rmotor_dir IS
                                -- FORWARD
                                WHEN '1' =>
                                      rmotor <= '1';
                                -- REVERSE
                                WHEN '0' =>
                                      rmotor <= '0';
                                WHEN OTHERS => NULL;
                                END CASE;
                        ELSE
                                rmotor <= '0';
                        END IF;
                ELSE
                        lmotor <= '0';
                        rmotor <= '0';
                END IF;
        END PROCESS;
END a;
```

13.5 Low-cost Sensors for a UP 3 Robot Project

A wide variety a sensors can be attached to the UP 3 board. A few of the more interesting sensors are described here. These include infrared modules to avoid objects, track lines, and support communication between UP3-bots. Other modules include sonar and IR to measure the distance to the nearest object and a digital compass to determine the orientation of the UP3-bot. Most robots will need to combine or "fuse" data from several types of sensors to provide more reliable operation.

Signal conditioning circuits are required in many cases to convert the signals to digital logic levels for interfacing to the digital inputs and outputs on the UP 3 board. Analog sensors will require an analog-to-digital converter IC to interface to the UP 3 board, so these devices pose a more challenging problem. Small low-cost A/D ICs are available with SPI interfaces that require a minimal number of FPGA pins.

Sensor module kits are available and are the easiest to use since they come with a small printed circuit board to connect the parts. Sensors can also be built using component parts and assembled on a small protoboard attached to the UP3-bot. Sensor modules are interfaced by connecting jumper wires to digital inputs and outputs on the UP 3 board's J3 and J2 expansion header connector.

Sensors with a single output bit can utilize a simple control scheme, and for basic tasks the robot can be controlled using hardware as simple as a state machine. More advanced sensors that report actual distance, location, or heading measurements will likely require a processor core on the UP 3 running a program that interprets sensor readings and implements the robot's control algorithm.

- **Line Tracker Sensor**

  A line tracker module from Lynxmotion is shown in Figure 13.3. This device uses three pairs of red LEDs and infrared (IR) phototransistor sensors that indicate the presence or absence of a black line below each sensor. When the correct voltages are applied in a circuit, an IR phototransistor operates as a switch. When IR is present the switch turns on and when no IR is present the switch turns off. The LED transmits red light that contains enough IR to trigger the phototransistor.

  Each LED and phototransistor in a pair are mounted so that the light from the LED bounces off the floor and back to the IR phototransistor. The LED and IR sensor must be mounted very close to the floor for reliable operation. Black tape or a black marker is used to draw a line on the floor. The black line does not reflect light so no IR signal is returned. Three pairs of LEDs and IR phototransistor sensors produce the three digital signals, left, center, and right. The UP3-bot can be programmed to follow a line on the floor by using these three signals to steer the robot. The mail delivery robots used in large office buildings use a similar technique to follow lines or signal cables in the floor.

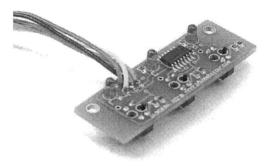

Figure 13.3 – Three LEDs and phototransistors are mounted on bottom of the Line Tracker board.

- **Infrared Proximity Detector**

  An IR proximity sensor module from Lynxmotion is seen in Figure 13.4. The UP3-bot can be outfitted with an infrared proximity detector that is activated by two off-angle infrared transmitting LEDs. The circuit utilizes a center-placed infrared sensor (Sharp GP1U5) to detect the infrared LED return as seen in Figure 13.5. The Sharp GP1U5 was originally designed to be used as the IR receiver in TV and VCR remote control units. From the diagram, one can see that the sensitivity of the sensor is based on the angle of the LEDs. The LEDs can be outfitted with short heat-shrink tubes to better direct the infrared light forward. This prevents a significant number of false reflections coming from the floor. The IR sensor will still occasionally detect a few false returns and it will function more reliably with some hardware or software filtering.

Figure 13.4 IR Proximity Sensor Module – Two IR LEDs on sides and one IR sensor in middle.

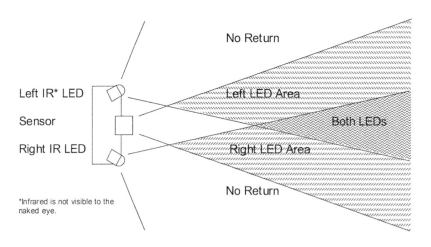

Figure 13.5 Proximity detector active sensor area.

As seen in Figure 13.6, the circuit on the IR proximity module utilizes a small feedback oscillator to set up a transmit frequency that can be easily detected by the detector module. This module utilizes a band-pass filter that essentially filters out ambient light. Some older first generation electronic ballasts used in commercial fluorescent lights can interfere with the IR sensors since they operate at the same frequency as the filter. Newer ballasts now operate at a higher frequency since they also caused problems with IR TV remote control signals.

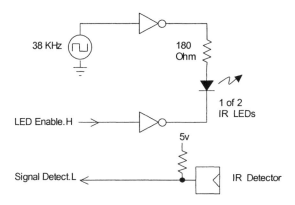

Figure 13.6 Circuit layout of one LED and the receiver module on the infrared detector.

In Figure 13.6, when the Left_LED Enable signal is High, the Low side of the IR LED is pulled to ground. This forces a voltage drop across the LED at the frequency of the 5v to ground oscillating signal. In other words, the LED produces IR light pulses at 38 kHz. Using a 38 kHz signal helps reduce noise from other ambient light sources.

Since the IR detector has an internal band-pass filter centered at 38 kHz, the detector is most sensitive to the transmitted oscillating light. The 5v pull-up resistor allows the IR Detector's open collector output to pull up the SOUT signal to High when no IR output is sensed. To detect right and left differences, the right and left LEDs are alternately switched so that the detected signals are not ambiguous. If both the left and right LEDs detect an object at the same time, the object is in front of the sensor.

If the IR sensor was built from component parts, a hardware timer implemented on the UP 3 board could be used to supply the 38-40 kHz signal. Similar IR LEDs and IR detector modules are available from Radio Shack, #276-137B, and Digikey, #160-1060. Assuming two UP3-bots are equipped with IR sensor modules, it is also possible to use this module as a serial communication link between the robots. One UP3-bot transmits using its IR LED and the other UP3-bot receives it using its IR sensor. To prevent interference, the IR LEDs are turned off on the UP3-bot acting as a receiver. Just like an IR TV remote, the IR LED and sensor must be facing each other. Bandwidth is limited by the 38kHz modulation on the IR signal and the filters inside the IR detector. (An IR sensor strip that converts IR to visible light is available from Radio Shack. This sensor can be used to confirm the operation of IR LEDs.

- **Wheel Encoder**

The Nubotics WW01 WheelWatcher incremental quadrature encoder system from Acroname is shown in Figure 13.7. This low-cost electronics board bolts onto the top of a standard-size R/C servo. The adhesive-backed codewheel attaches to a wheel mounted on the servo's output shaft. Two pairs of optical emitters and receivers bounce light beams off of the codewheel.

Note that there are 32 black stripes on the reflective codewheel. When the wheel is rotating, the encoder produces two series of digital pulses that are 90 degrees out of phase. When one of the pulses changes twice before the other pulse changes, the direction has been reversed. 128 clock pulses per revolution are produced and a separate direction signal indicates the current direction of rotation. By counting pulses with a counter or by accurately measuring the time between individual pulses using a fast hardware counter on the UP 3, it is possible to more accurately control the position and velocity of the servo motor. When used on robot drive motors, this optical encoder feedback provides more accurate position and speed control for the robot.

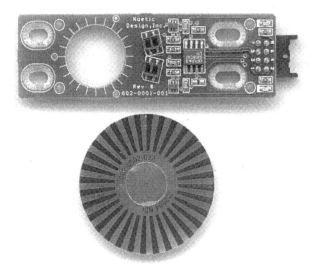

Figure 13.7 Nubotics WW-01 Wheel Watcher Incremental Encoder System.

- **Sonar Ranging Units**

  The Devantech SRF10 Sonar Module is shown in Figure 13.8. This device uses ultrasonic sound waves to measure distances from a few inches to around 35 feet. They are widely used in robotics. The timing of the sound echo indicates distance to the nearest object. The transducer first functions as a transmitter by emitting several cycles of a ultrasonic signal, and then functions as a receiver to detect sound waves returned by bouncing off nearby objects. Even though ultrasound is inaudible, the transducer also generates a slight audible click each time the device transmits. The beamwidth is rather wide, and several sonar modules facing in different directions are commonly used.

  The time it takes for the ultrasonic echo signal to return is measured using an IC mounted on the back side of the board. This time is converted to distance since sound travels out and back at 0.9 ms per foot. Only around 10-20 samples per second are possible with the device since it takes time to wait for echoes to return. Some sonar modules require external hardware to measure the pulse timing to produce the distance to target. The SRF10 device operates off +5V DC, and it sends distance measurements back to the host using an I^2C bus.

Figure 13.8 Devantech SRF10 Ultrasonic Range Finder.

- **IR Distance Sensors**

  The Sharp GPD2D02 seen in Figure 13.9 is an IR device that can provide distance measurements similar to the slightly more expensive sonar sensor. This sensor has a shorter range of 10 to 80 cm (~ 4 to 32 inches). The distance is output by the sensor on a single pin as a digital 8-bit serial stream.

Figure 13.9 Sharp IR Ranging Module.

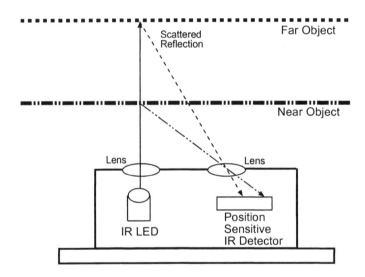

Figure 13.10 Operation of Sharp IR Ranging Module.

As shown in Figure 13.10, internally the GPD2D02 contains an IR LED and a position-sensitive IR detector. The IR LED transmits a modulated beam of infrared light. When the light strikes an object, most of the light will be reflected back to the LED. Since no surface is a perfect optical reflector, scattering of the IR beam occurs at the surface of the object and some of the light is reflected back to the position sensitive detector. By comparing the near and far object beams shown in Figure 13.10, it is apparent that the position at which the scattered reflected IR beam hits the detector is a function of the reflection angle.

The 8-bit integer value reported by the sensor in cm is approximately

$$1000 * \tan^{-1}\left(\frac{1.9}{DISTANCE}\right) + offset .$$

The constant 1.9 is the distance between the lenses in cm. The offset is the no-object present value returned by the sensor. This offset constant can vary by as much as 17 between different sensors and has a typical value of 25. Note that a close object reports a larger value and a distant object reports a smaller value. Objects closer than 10cm will report an incorrect value and should be avoided by placing the sensor away from the edge of the robot. Large objects beyond 80 cm can sometimes report an incorrect value that makes them appear closer.

A special connector (Japan Solderless Terminal #S4B-ZR) is required to connect to the GPD2D02. If desoldering equipment is available, the small connector can also be desoldered from the sensor and wires attached directly to the sensor.

In addition to +5V and ground pins, the sensor has an input, Vin, and a serial output, Vout. Vin is an input to the sensor that clocks out the serial data on Vout. When Vin is Low for around 70 ms, the sensor takes a reading. When a reading is available, Vout goes High.

On each of the next eight falling clock edges of Vin, the sensor will output a new data bit. The eight data bits should be clocked into the FPGA on the rising edges of Vin (when they are stable). When clocking out the data, the clock period on Vin should be 0.4 ms or less. The eight data bits are clocked out in high to low order. If Vin is not dropped Low within 1.5 ms after clocking out the final data bit, the sensor shuts down to save power. A shift register can be used to assemble the data bits. The demo program ir_dist.bdf on the CD-ROM contains a VHDL-based IP core for use with the GP2D02 sensor.

The sensor's Vin pin is an open-drain input. Open-drain or open-collector inputs should never be driven High. An FPGA's tri-state output pin can be connected directly to an open-drain input, if the tri-state output is never driven High. When Vin should be High, tri-state the FPGA's output pin and when the output should be Low, drive the output pin Low with the tri-state gate turned on with a low output.

Open-drain or open-collector inputs contain an internal pull-up resistor to +5V. Multiple open-drain (open-collector) outputs can be tied together to a single open drain (open-collector) input to perform a wired-AND operation. Any one of the outputs can pull the input Low. If no output pulls the signal Low, a single pull-up resistor forces the input High. This wired-AND operation occurs just by tying the open-drain (open-collector) outputs together and no physical AND gate is needed. In negative logic, a wired-OR operation occurs.

Normal gate outputs cannot be connected. This wired-AND logic only works because these gates have special output circuits that do not contain a transistor that forces the input High. This transistor is present in normal gate outputs. If a normal gate output is connected to other open-drain (open-collector) outputs, its transistor could turn on to force the input High at the same time another gate's output transistor turns on to force it Low. This would short the power supply to ground drawing excessive current that might damage the devices. An analog and longer range version of this IR distance sensor are also available.

• Magnetic Compass Sensors

Various electronic components are available that detect the magnetic field of the earth to indicate direction. A low-cost digital compass sensor is shown in Figure 13.11. The Dinsmore model 1490, often used in electronic automobile compasses, is a combination of a miniature rotor jewel suspended with four Hall-effect (magnetic) switches. Four active-low outputs are provided for the four compass directions. When the module is facing North, the North output is Low and the other three outputs will be High. Eight directions are detected by the device, since two outputs can become active simultaneously. In this way, the device can indicate the four intermediate directions, NE, SE, SW, and NW. NE for example activates the active-low North and East outputs. The device can operate off +5V.

Mount any compass device as far away from motors as possible to avoid magnetic interference from the magnets inside the motor. Four 2.2K ohm pull-up resistors to +5V are required to interface to the UP 3 board, since the four digital output pins, N, S, E, and W, all have open-collector outputs. Just like a real compass, a time delay is needed after a quick rotation to allow the outputs to stabilize. If the compass module leads are carefully bent, the compass module and the four required pull-up resistors can be mounted on a standard 20-pin DIP, machined-pin, wire-wrap socket and connected to the UP 3 header socket.

An analog version of the device is available with 1-degree accuracy, but it requires an analog-to-digital conversion chip or signal phase timing for interfacing.

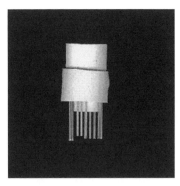

Figure 13.11 Dinsmore 1490 Digital Compass Sensor.

- **Electronic Compass Sensors**

   Low-cost electronic compass modules are also available that detect the magnetic field of the earth to indicate direction. The cost is two to three times that of the mechanical compass described in the previous section. New generation electronic compass modules offer more accuracy and faster settling times than mechanical compass sensors. An electronic compass module from PNI is shown in Fig. 13.12. This module contains a 2-axis magneto-inductive sensor and an ASIC. Heading information and magnetic field measurement data is available using a digital SPI serial interface.

Figure 13.12 PNI Electronic Compass Module.

- **Low-cost Gyros and Accelerometers**

   Gyros and accelerometers are useful sensors for robots that need a balance sense. This can include robots that balance on two wheels like the Segway Human Transporter, robots that walk on two legs, and even robots that fly. Gyros and accelerometers have traditionally been used in aircraft autopilots and inertial measurement units (IMUs). Helicopters use a gyro to stabilize and control the tail rotor. Recently, Microelectromechanical Systems (MEMS) technology has produced small low-cost piezo-gyroscope and accelerometer ICs. These devices were originally used in automobile airbags. The gyros output a voltage level that is proportional to the speed or rate of the tilt angle changes. An analog-to-digital converter will be needed to input the gyro signal. The MEMS accelerometers output a voltage level or a pulse that changes its duty cycle proportionally (e.g., PWM) to the tilt angle by sensing the change in acceleration due to gravity. Gyros will drift slowly over time and an accelerometer is needed to correct the gyro's drift. Without an accelerometer to correct for gyro drift, the tilt error slowly grows to the point where the robot would lose its balance. Accelerometers will respond more slowly to tilt than the gyro, so both a gyro and accelerometer is typically needed for each axis that needs a balance sense.

A complementary filter is used to combine or fuse sensor data from both the gyro and accelerometer to generate a more accurate tilt angle. Kalman filtering techniques can be used to improve the accuracy of noisy measurements. Noise levels are still somewhat high at very low G forces on these low-cost gyros and accelerometer IC sensors, so currently they are not useful for navigation since they cannot accurately determine the exact location of a slow moving robot by integrating the sensor measurements over time.

Analog Devices makes a variety of these sensors and sells small evaluation boards for them. It is likely that small low-cost sensor modules containing both a MEMS gyro and an accelerometer with a microcontroller will be available commercially in the near term.

Figure 13.13 Small sensor board for an aircraft autopilot system. Photograph ©2004 courtesy of Henrik Christophersen , Georgia Institute of Technology Unmanned Aerial Research Facility.

Figure 13.13 shows a sensor board for an autopilot system that is used for unmanned aircraft. In the top corner, three MEMS gyros and accelerometers are mounted at right angles to provide data on all three axes. The white square flat module between the two vertical assemblies is a GPS receiver. The black square ICs at each end of the vertical assemblies are airspeed and altitude sensors. An A/D chip with an SPI interface is used to read sensors that have analog voltage outputs. The three square modules near the bottom edge of the board are DC to DC voltage converters. The lower board contains an FPGA and a DSP processor.

- **GPS and DGPS receivers**

  The Global Positioning System was built by the US Department of Defense to provide highly precise worldwide positioning. Triangulation using radio signals from several satellites provides a position accurate to 25 meters. With an additional land-based correction signal, Differential GPS (DGPS) improves the accuracy to 3 meters. DGPS receivers provide ideal position data for robot navigation. Unfortunately, with current systems you are not likely to receive the GPS radio signals indoors in most buildings, so their use is typically limited to larger more rugged outdoor robots. Low-cost single chip GPS modules such as the Motorola FS Oncore seen in Figure 13.14 or the Ublox in Figure 13.13 are currently available. An SPI serial interface is supported. A new generation of highly sensitive GPS systems is being developed that may function indoors in some buildings.

Figure 13.14 Motorola Single Chip GPS module.

- **Thermal Image Sensors**

  Low-cost thermal image sensors can provide thermal imaging data for robots. Most thermal sensors such as those used in motion detectors and burglar alarms detect only movement. Thermopile sensors measure the temperature of a heat source. One such sensor, the Devantech TPA81 Thermopile Array is shown in Figure 13.15. It contains 8 Pyro-electric sensors arranged in a column that detect infra-red in the radiant heat range of 2um to 22um range. It contains an on-board PIC microcontroller.

When the sensor is mounted on a servo, it can be used to horizontally scan an area and generate a thermal image. Candle flames and human body heat can be detected several feet away at room temperature. It uses an I^{2}C bus for interfacing to the host controller.

Figure 13.15 Devantech TPA81 Eight Pixel Thermal Array Sensor.

- **Solid State Cameras**

  Low-cost solid state cameras can provide visual sensors for robots. Keep in mind that advanced image processing and visual pattern recognition requires complex algorithms that need a lot of processing power. The CMUCAM2 developed at Carnegie Mellon University seen in Figure 13.16 contains a PIC microcontroller and can transfer image data using a serial connection. It can track color blobs and report their location and size in an image at 26 to 50 frames per second.

  Low-cost USB cameras are another option, but they will require a USB core interface and additional image processing. The low-cost CMOS color camera assembly OV6620 or OV7620 used in the CMUCAM2 module from Omnivision (www.ovt.com) can also be directly interfaced to an FPGA. It uses an I^{2}C interface for camera control signals and a separate parallel bus is used to transfer image data.

Figure 13.16 The CMUCAM2 contains a color video camera on a chip and a microcontroller.

13.6 Assembly of the UP3-bot Body

Assembly of the UP3-bot can be accomplished in about an hour. A drill or drill press, screwdriver, scissors, a soldering iron, and a wire stripper are the only tools required. First, obtain the parts in the parts list. Next, drill out the holes in the round Plexiglas base (part #15) as shown in Figure 13.17. To prevent scratches, leave the paper covering on the Plexiglas until all of the holes are marked and drilled out. The front of the base is on the right side in Figure 13.17. The wheel slots are symmetric with respect to the center of the circle.

Proper alignment of the four screw mounting holes for the UP 3 board is critical. Unscrew the four standoffs from the bottom of the UP 3 board. Carefully place it towards the rear of the plastic base as shown in Figure 13.17, and mark the location of the screw holes using a pen or pencil. A UP 2 board can also be used, but the mounting holes will be in different locations.

Locate the cable and switch holes as shown in Figure 13.17. Exact positioning on these holes is not critical. If one is available, use an automatic center punch to help align the drill holes. The J1..J4 header pins and power switch should face towards the front of the base. The extra space in front of the UP 3 board on the plastic base is used for sensor modules. Re-attach the standoffs to the UP 3 board and set it aside. After all holes are drilled, remove the paper covering the Plexiglas.

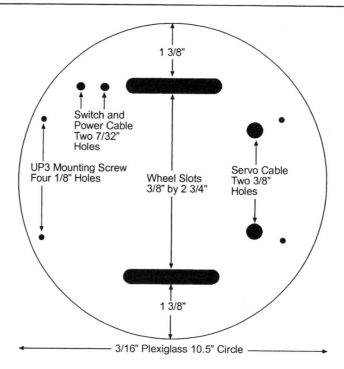

Switch and
Power Cable
Two 7/32"
Holes

1 3/8"

UP3 Mounting Screw
Four 1/8" Holes

Wheel Slots
3/8" by 2 3/4"

Servo Cable
Two 3/8"
Holes

1 3/8"

3/16" Plexiglass 10.5" Circle

Figure 13.17 UP3-bot Plexiglas Base with wheel slots and drill hole locations.

Mount the toggle switch (part #8) in the hole provided in the base. If available, Loctite or CA glue can be used on the switch mounting threads to prevent the switch nut from working lose. Solder the red wire (+7.2V) from the battery connector to one of the switch contacts. This is the connector with wires that plugs into the battery pack connector (part #4). Solder one of the twin lead wires (part #9) to the other switch terminal. Solder the other twin lead wire to the black (GND) battery connector wire and insulate the splice with heat shrink tubing or electrical tape (part #10).

Route the twin lead wire through the hole provided in the base. A small knot in the twin lead on the bottom side of the base can be used for strain relief. Solder the power connector (part #11) to the other end of the twin lead wire on the top of the base. The center conductor is +7.2V and the outer conductor is ground on the power connector.

Check the power connections with an ohmmeter for shorts and proper polarity before connecting the battery. For strain relief and extra insulation, consider sealing up the power connector with Silicone RTV or insulating one of the wire connections with heat shrink tubing. Be careful, NiCAD and NiMH batteries have been known to explode or catch on fire, if there is a short. A fuse on the battery power wire might be a good idea, if you are prone to shorting out circuits.

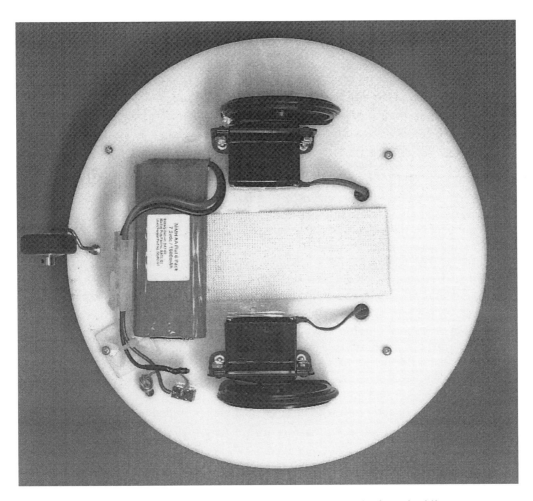

Figure 13.18 Bottom view of UP3-bot base showing battery, servos, wheels, and cabling.

Attach the battery pack (part #2) to the bottom of the Plexiglas base with sticky-back Velcro (part #16). Figure 13.18 is a close-up photo of the bottom side of the UP3-bot. A NiMH battery pack is shown in Figure 13.18. If you use a larger NiCAD battery pack, it can be mounted in the middle of the base about one inch off center towards the rear wheel, with the battery pack connector facing the rear.

The battery is moved towards the rear for balance to place the weight on the rear skid. The Velcro on the base should be around 2 inches longer than the battery pack towards the rear of the robot to allow for positioning of the battery later on to balance the robot. The wire and connector on the battery pack should also be attached to the base to prevent it from dragging on the floor. Attach a small piece of Velcro on the rear of the connector so that the battery wires can be attached to the base. Attach the battery pack to the base.

Solder a 60-pin female header socket (part #7) to the Cyclone expansion B location. Attach the UP 3 board to the top of the base with 4-40 screws (part #18), using the hex spacers provided on the UP 3 board (part #14). Figure 13.19 is a close-up photo of the top of the UP3-bot. Double check power connections and polarity with an ohmmeter. The inner contact on the power connector should be +7.2V, the outer contact is ground, and the toggle switch should turn it off. Then plug the power connector into the UP 3 board. Plug in the battery connector and flip the power switch. A green LED should light up on the UP 3 board indicating power on. The Cyclone expansion B header socket faces the front of the robot.

Mount the wheels (part #5) on two modified servos (part #3). If you are not using servo wheels, you may need to enlarge the hole in the center of each wheel by drilling it out partially with a drill bit that is the same size as the servo output shaft. The depth of the hole should be slightly shorter than the servo output shaft and not all the way through the wheel, so that the wheel does not contact the servo body. The servo output shaft screw is inserted on the side of the wheel with the smaller hole. A washer may be required on the servo screw. The wheel should not contact the servo case and must be mounted so that it is straight on the servo. CA glue or Blue Loctite can also be used to attach the wheels and screws more securely to the servo output shaft.

Attach the servos to the bottom of the base using double sided foam tape (part #17) or a more durable servo mounting bracket. The servo body faces toward the center of the base. Be sure to carefully center the wheels in the plastic-base wheel slot. If you are using foam tape, make sure all surfaces are clean and free of grease, so that the foam tape adhesive will work properly. Lightly sanding the servo case and adding a drop of CA glue helps with tape adhesion. Route the servo connector and wire through the holes provided in the base.

Attach a tail wheel to the base or a skid (part #19) at the rear of the battery pack using layers of foam tape as needed. Move the battery as needed so that the robot has proper balance and rests on the two wheels and the rear skid.

Attach another skid to the front of the battery pack using several layers of foam tape. The front skid should not contact the floor and at least ¼ inch of clearance is recommended. The front skid only serves to prevent the robot from tipping forward during abrupt stops.

Attach a 3-pin .1 inch header (part #9) to the small wire wrap protoboard in an open area. One is required for each servo on the robot. Solder wires from the appropriate pins J2 and J3 connections on the protoboard to the new header pins. The three wires on the servo are Vcc (4.8 to 6 volts), ground, and the PCM control signal wire. Some manufacturers' servos have different power connections, but they all have three pins. Wrap extra servo wire around the hex spacers underneath the UP 3 board.

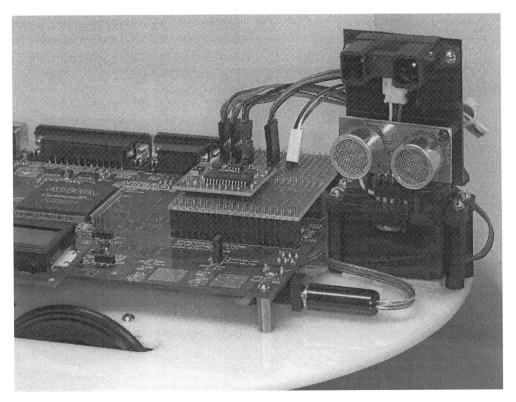

Figure 13.19 Top View of UP3-bot Base with Compass, IR, and Sonar Sensor Modules.

Optional sensor modules such as the IR proximity detector or line tracker can be attached to the base unit with foam tape. Run wires from the sensors to the Santa Cruz expansion connectors, J2 and J3. A small .1 inch wire wrap protoboard with 40-pin female header connectors soldered to the protoboard as shown in Figure 13.19 is handy for making servo and sensor connections to the UP 3 board. In Figure 13.19, a third servo is used to make a sensor turret for IR and Sonar distance sensors.

- **Parts List for the UP3-bot**

  1. **An Altera UP 3 Board**. The UP 3 board serves as the controller for the UP3-bot. It is attached to the UP3-bot body with screws. No modifications are required to the board. A UP 2 board can also be used, but mounting holes will be in different locations.

- **Parts Available from a Hobby Store**

  2. **A 7.2V 1300-1700mAh Rechargeable NiCAD battery pack with the standard Kyosho battery connector.** This is a standard R/C car part, and it is used to power the UP3-bot. For a small additional cost, new NiMH batteries are also available that store almost twice the energy per weight. The battery will need to be charged prior to first use.

3. **Two modified R/C Servomotors**. Two identical model servos are required so that the motors run at the same speed. Servo modifications are described in section 13.3. Any servo should work. The following servos have been tested: Tower Hobbies TS53J, Futaba S148 and S3003, and HS 300. Some manufacturers' servos appear to run in the reverse direction. This is easily fixed in the hardware design since the motor controller is implemented on the UP 3 board. Several robot parts vendor sell modified servos for a slightly higher cost. Ball bearing servos are worth the extra cost, if you intend to run the robot constantly for several months. A third unmodified servo will be needed if you want a rotating sensor turret as shown on the example UP3-bot photo.

4. **Kyosho Female Battery connector with wire leads, Duratrax or Tower Hobbies #DTXC2280**. This is used to connect to battery. A connector is needed so that the battery can be disconnected from the UP3-bot and connected to a charger.

5. **Two Acroname or Lynxmotion servo wheels**. These wheels are 2 ¾ plastic wheels that are designed to attach to the servo's output shaft spline. Prather Products 2¼-inch aluminum racing wheels with rubber O ring tires, Tower Hobbies #PRAQ1810 or Hayes Products #114, 2 ¼-inch hard plastic racing wheels (also available from Tower Hobbies) can be used as a substitute. These somewhat smaller two alternative wheels will work, but they do not have the spline to match the servo output shaft and are a bit more difficult to connect reliably than the Acroname or Lynxmotion servo wheels.

6. **A Castering Wheel or Two small Teflon or Nylon Furniture Slides.** There is a bit too much mechanical play in common furniture casters for a small robot and they tend to randomly deflect the robots direction after sharp turns. Lynxmotion's #TWA-01 is a mini castering robot tail wheel built using an R/C airplane tail wheel that works well. The mounting wire needs to be bent a little off center so that the wheel quickly rotates to the direction of travel. Other robot parts vendors such as Acroname also have robot tail wheels, but a spacer may be required to adjust the height. The battery will need to be moved a bit and perhaps rotated ninety degrees to accommodate them and still maintain proper balance on the robot base. Magic Sliders 7/8-inch diameter circular discs also work well on flat surfaces. The slides are used as a skid instead of a third wheel on the UP3-bot. Metal or hard plastic will also work. Attached to the bottom of the battery with several layers of foam tape, an optional front skid can be used for stability during abrupt stops. On flat surfaces, a Teflon skid actually works better than a common small furniture caster from a hardware store.

7. **A charger for the 7.2V battery pack**. An adjustable DC power supply can be used to charge the battery if it is properly adjusted and timed so that the battery is not overcharged. Overcharged batteries will get hot and will have a shorter life. Automatic peak-detection quick chargers are the easiest and most foolproof to use. These chargers shut off automatically when the battery is charged. One quick charger can be used for several robots as a full charge is achieved in less than 30 minutes with around 5 Amps

maximum charge current. Inexpensive trickle battery chargers deliver only around 75 mA of charge current, and they will require several hours charge the battery.

- ## Parts Available from an Electronics Parts Store

8. **Three 40-pin .1-inch double row PC board mount female header sockets, DigiKey #S4310 or equivalent.** These sockets are soldered into a small 0.1" center wire wrap protoboard that fits into the Santa Cruz Expansion connector on the UP3. This is used to connect servos and sensors to the UP 3 board.

9. **A 2 to 3 inch strip of .1" single row breakaway headers. DigiKey #S1021-36 or equivalent** These headers are used to make custom servo and sensor connectors on the protoboard. They can be soldered to the protoboard.

10. **A small wire wrap protoboard with holes on .1" centers cut down to 2" by 2.8". A** This is used to make a protoboard for use with the UP3 board. The protoboard contains connectors for servos and sensor. A protoboard with solder pads makes it easier to mount the connectors.

11. **A miniature toggle switch with solder lug connections.** The switch should have a contact rating of more than two amps (Radio Shack #275-635B or equivalent). Only two contacts or single pole single throw (SPST) is needed on the switch to turn power on and off. If all of your servos and sensors connect to the UP 3 and do not use the Vunregulated supply, you could eliminate the switch by using the UP 3's power switch.

12. **Approximately 9 inches of small-gauge twin-lead speaker wire.** This part is used to connect power to the UP 3 board. The wire must fit into the DC power plug (part# 11). Typically, 20-22 gauge wire is required. Two individual wires can also be used, but twin lead is preferred.

13. **A 1-inch piece of small heat shrink tubing or electrical tape.** This part is used to insulate a splice in the twin-lead power wire.

14. **A Coaxial DC Power Plug with 5mm O.D. and 2.1mm I.D., Radio Shack Number 274-1567** or equivalent. This power plug fits the power socket on the UP 3 board. A different size plug is needed for the UP2 board, use #274-1568 that has a 2.5mm I.D.

15. **An assortment of small wire jumpers and connectors to attach wires to the male headers on the UP 3.** These are the jumper wires commonly used for protoboards. Two short jumpers are used to connect the two servo signal wires, and other jumpers are used to any connect sensor boards.

16. **Four, 1-inch hex spacers with 4-40 threads or use the shorter spacers that come with the board.** These are used to mount the UP 3 board to the Plexiglas base using the holes in the UP 3 board.

- **Parts Available from a Hardware Store**

17. **3/16-inch thick Plexiglas cut into a 10.5-inch diameter circle.** This part is the base of the robot. Colored Plexiglas such as opaque white, will not show scratches as easy as clear. Holes to cutout and drill are shown in Figure 13.11. If a band saw, jig saw, or other machine tool is not available, a local plastics fabricator can cut this out. When using a number of very large sensors, it may be necessary to increase the size slightly or add another circular deck for sensor mounting. A larger robot requires more space for maneuvering. To prevent scratches on the Plexiglas, keep the paper backing on the plastic until all of the holes have been marked and drilled out. The size of the wheel slots may need to change depending on the wheels you select.

18. **One 8-inch long strip of 2-inch wide sticky-back Velcro.** Two 8-inch long strips, 1 inch wide can also be used. The Velcro is used to attach the battery to the bottom of the Plexiglas base. Since the battery is attached with Velcro and a connector, it can be quickly replaced and removed for charging.

19. **Approximately 8 inches of 1-inch wide double-sided 3M foam tape.** This is used to attach servos, skids, and optional sensor boards to the base. Be sure to clean surfaces to remove any grease or oil prior to application of the tape for better adhesion. For a more durable servo mount, Lynxmotion has aluminum servo mounting brackets that can be used instead of the double sided tape.

20. **Four 4-40 Screws 5/16-inch or slightly longer.** The screws are used to attach the UP 3 board to Plexiglas. The screws thread into the hex spacers attached to the UP 3 board.

21. **Blue Loctite, Cyanoacrylate (CA) Glue, and Clear Silicone RTV.** These adhesives and glues are useful to secure screws, servos, and wheels. The mechanical vibration on moving robots tends to shake parts loose over time. These items can also be found at most hobby shops. Only a few drops are needed for a single robot. A single tube or container will build several robots.

13.7 I/O Connections to the UP 3's Expansion Headers

Most servos and sensor I/O signals will need to be attached to the UP 3's J3 expansion header. The FPGA I/O Pins on J3 feed through voltage level converters to support 5V operation. R/C Servos and most sensors use 5V, but be sure to check the device's datasheet. Don't forget to connect a ground signal between the device and the UP 3 board, even if the device has it's own power supply or a direct connection to a battery. Several ground pins are available on J3. A 5V 1A power supply pin is available on the UP 3's J2 expansion connector. J4 has a 3.3V supply connection pin and JP6 can be used as another 5V supply connection.

A small protoboard can be built to connect servos and sensors to the UP 3. All of the connectors and pins on the UP 3 line up on tenth inch centers. A 0.1" perfboard or wire wrap protoboard can be cut down to 2" by 2 7/8" so that it fits over J1, J2, J3, and J4. 0.1" 40 pin connectors to attached the protoboard to connect to J1..4 can be mounted on the protoboard. A wire warp protoboard with holes every .1" has solder pads that can be used to attach connectors using solder. Point to point wiring and soldering can be used to make connections on the protoboard from the J1..J4 connectors to the .1" connectors used to attach servos and sensors. Small single row strips of .1" header pins can be snapped apart to make male connectors on the board for the servos and most sensors.

You may want to consider isolating your robot's servo or motor power supply from the supply used for the UP3's logic to control the noise generated on the supply lines by the DC motors. On larger robots, two batteries are sometimes used. A Vunregulated connection that does not go through the 5V regulator and is connected directly to the UP 3's power input jack is available on JP8 and J4. The 9V supply is connected after the input power switch on the UP 3 and to JP5. This also can be used to power servos and motors, assuming the battery voltage level is not too high. If the battery voltage is too high, another regulator can be used for the motors. At a minimum, decoupling capacitors connected across the servo's power supply connections are a good idea. If you plan on having several sensors on your robot, you may want to consider building a small PCB with header pins for the sensor power and data connections as seen in Figure 13.19. Suggestions for common UP3-bot servo motor and sensor connections are shown in parenthesis. Most R/C servos can run on 4.8 to 6V.

Table 13.1 UP3-bot J3 Expansion Header Pins

Header Pin	Cyclone Pin	Header Pin	Cyclone Pin
1	Reset	2	Gnd
3	217 (Left Servo Signal)	4	220(1C12:169)
5	216 (Right Servo Signal)	6	219
7	215 (Sensor Turret Servo)	8	218
9	206	10	221(1C12:168)
11	207 (IR distance input)	12	222
13	208 (IR distance output)	14	223
15	213(Sonar input)	16	224
17	214(Sonar output)	18	225
19	Gnd	20	NC
21	199(1C12:180)	22	Gnd
23	200	24	Gnd
25	201(Elect. Compass SDA)	26	Gnd
27	202 (Elect. Compass SDL)	28	196
29	205(1C12:NA)	30	Gnd
31	204(1C12:NA)	32	197
33	203	34	179
35	176 (IR proximity input)	36	178
37	174 (left IR proximity output)	38	124
39	173 (right IR proximity output)	40	Gnd
Note: Pin numbers in parenthesis are used on the larger 1C12 UP 3 board			

13.8 Robot Projects Based on R/C Toys, Models, and Robot Kits

A second option for building an FPGA driven robot involves modifying a low-cost radio-controlled (R/C) car or truck. Fundamentally, almost any large R/C car or truck can be modified to work with the Altera board, although some are clearly better choices than others. In our robot, we used a Radio Shack (www.radioshack.com) R/C 4WD SUV shown in Figure 13.20. The R/C platform affords a more robust drive train and control; however, turning radius and noise levels are sacrificed over the smaller UP3bot. The R/C SUV has a spring suspension and large soft tires that make it operable outdoors on rougher surfaces. Following are some R/C car selection considerations that will affect available modifications and control of the new platform.

Figure 13.20 FPGA Controlled Toy R/C Truck with IR Distance Sensors.

- **Seven-Function Controls**

    When choosing an R/C car, select one that has a remote control with at least seven remote functions (forward, backward, forward-right, forward-left, backward-right, backward-left, and stop). Note that these low-cost R/C cars do not have variable speed or variable turning controls; however, once they are interfaced to the FPGA board, variable speed and turning can be accomplished by changing the duty cycle of the command signals. (More on this later.)

    A control module built using the FPGAs logic allows a relatively inexpensive R/C car to perform with the capabilities of the more expensive cars with "digital proportional steering" and "digital proportional speed controls." Once interfaced to the FPGA board, an IP core (Robot_CTL) is used to handle control of all direction and speed control outputs. As illustrated in Figure 13.21, the IP core control module affords a higher degree of control than the original radio control. The outputs connect to the R/C cars internal control circuits that drive the DC Motors.

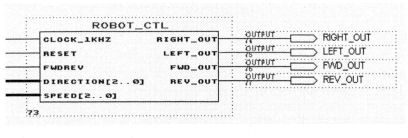

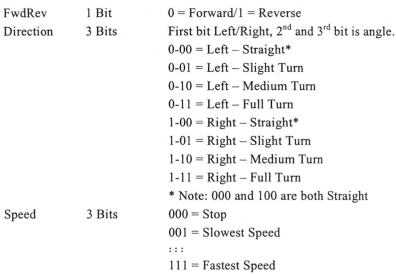

FwdRev	1 Bit	0 = Forward/1 = Reverse
Direction	3 Bits	First bit Left/Right, 2nd and 3rd bit is angle.
		0-00 = Left – Straight*
		0-01 = Left – Slight Turn
		0-10 = Left – Medium Turn
		0-11 = Left – Full Turn
		1-00 = Right – Straight*
		1-01 = Right – Slight Turn
		1-10 = Right – Medium Turn
		1-11 = Right – Full Turn
		* Note: 000 and 100 are both Straight
Speed	3 Bits	000 = Stop
		001 = Slowest Speed
		: : :
		111 = Fastest Speed

Figure 13.21 Robot Control IP Core with Pulsed Speed & Steering Control.

- **Speed**

  When considering the speed of the vehicle, a modest speed is more desirable than the faster speeds. At 800 feet per minute, our prototype FPGA controlled robot car moves fast enough to be difficult to catch. In almost all cases, the robot is operated at half the maximum speed or less. The limiting factor is generally the delay inherent in the sensor's input sampling rate and range. A fast moving car typically will hit the wall before a collision sensor can take the data samples needed to initiate avoidance.

  To control the speed (or the degree of turn) a repeated pulse train is sent to the forward or reverse signals (left or right signals for direction). Instead of a steady high signal causing the car to move forward at full speed, the pulse train varies the duty cycle to change speed (or degree of turn). By modulating the duty cycle of the pulse train, "digital proportional control" can be implemented on each control signal. In other words, changing the duty cycle can control the speed and the degree of turn. The more the duty cycle approaches 100%, the harder the turn and/or the faster the speed.

The frequency of the pulsed control signal used must be higher than the natural mechanical frequency response of the system. A very slow changing pulse will cause the motor and gears to vibrate and make additional noise. Pulse frequencies of a few kHz are typically used to avoid this problem.

When reversing direction on a moving DC motor, it is common practice to include a small time delay with the motor turned off to reduce the inductive voltage spikes produced by the motor windings. Recall that changing current flow through an inductor produces voltage. Without the delay in some circuits, these high voltage spikes can damage or reduce the life of the transistors controlling the motor. This delay can be incorporated in the IP control core, if needed.

Figure 13.22 illustrates the relationship between turn angle, speed, and duty cycle on the control signals. The figure implies that the duty cycle is linear, i.e., a 50% duty cycle produces half speed. Actually, the duty cycle is very non-linear and highly dependent on the type of car, size of the DC motors, and power. (An R/C car with a dying battery performs as if the duty cycle is considerably less.) By experimenting with patterns of the 16-bit speed and direction vectors used in the IP core controller, a more linear relationship can be established between the command bit patterns and the actual performance of the vehicle.

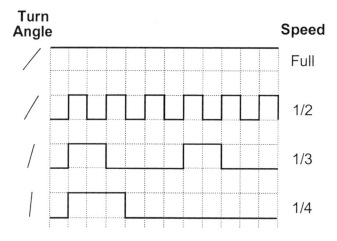

Figure 13.22 Affect of Duty Cycle on Turning Angle and Speed.

- **Battery Choice**

  The choice of car will also dictate the type of batteries and charger that will be needed. Note that some cars come with 9.6V packs and others come with 7.2V packs. Both should work well with the UP 3 board as a controller. The prototype used the 7.2V pack that discharged quickly and required a second pack placed in parallel with the first to support longer run times. The cars with a 9.6V pack should give the UP 3 board's 5V onboard regulator a better regulator margin and a longer life between recharges.

- **Mounting the UP 3 Board**

    Before you select an R/C car, make sure that there is a good place to mount the FPGA board. If the car is large enough, there is usually a large flat area under the car body cover molding to secure the FPGA board.

- **Interfacing the UP 3 Board to the R/C Car**

    Remove appropriate body cover screws and expose the PC board receiver and control module. Most current low-cost R/C toy cars have a single electronic PC board that contains both control circuits. Generally, there is one 16 or 18-pin DIP radio command demodulator chip in the center of the board that converts the radio signals into simple digital control signals. These digital control signals then activate the H-bridge circuit that controls the DC motors that drive the wheels of the R/C car.

    An H-bridge is a standard electronic circuit used to control DC motors. It allows for both forward and reverse operation of the same DC motor. H-bridge circuits contain four large power transistors that are needed to turn on and reverse a DC motor. Discrete transistors may be used to build the H-bridge or it may be in an IC or packaged module that connects directly to the motors.

> NEVER ATTEMPT TO DRIVE A MOTOR OR RELAY WITH AN FPGA PIN DIRECTLY, IT CANNOT SUPPLY THE HIGH CURRENT LEVELS THAT THESE DEVICES NEED. THE FPGA'S OUTPUT PIN DRIVER CIRCUIT MAY BE DESTROYED.

If the car supports seven functions, it will have at least four pins coming off of the DIP chip package that break down into Left, Right, Forward, and Reverse. Using a voltmeter or an oscilloscope, test which pins change when the remote control is set to each of the four directions. From each of the designated command pins on the chip, the trace on the PC board will run to separate H-bridge circuits for each motor.

By clipping or desoldering and pulling out the four control pins on the chip going to the board and soldering wires from each chip pin hole pad trace to the FPGA (Figures 13.23 and 13.24), the FPGA board can control the four directions and speed of the R/C car using the car's existing H-bridge circuits. In our modification, we desoldered the entire chip and put a socket on the board. To have the original control signals coming from the radio control module also sent to the FPGA board, run four more wires from the clipped chip pins to the one of the FPGA's headers. If the four clipped chip pins (RCx) are connected to the FPGA board, logic can be designed to include the original radio control functions supplied by the handheld remote control unit. You can also make this connection at the H-bridge circuit if necessary. On the UP 3, you will want to use the UP 3s 5V I/O pins for this interface.

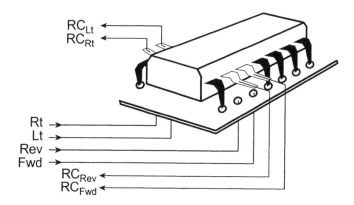

Figure 13.23 Interfacing to the R/C Car's Internal Control Signals at the Demodulator IC.

Figure 13.24 Photo Showing Control Modifications to R/C Car Control Board.

- **Hobbyist R/C Models, Robot Kits, and Commercial Robot Bases**

  For those with a larger budget, higher quality R/C hobbyist cars are available with built-in proportional steering and pulsed electronic speed controls. The control interface to these cars uses standard R/C PWM signals that are identical to the PWM servo control bit described at the end of Section 13.2. An example R/C Hummer can be seen in Figure 13.25. This robot is controlled using a C program running on the FPGA's Nios processor core. Various robot kits without control electronics or a computer are also available. Almost any of these robot kits can be controlled by the UP 2 or UP 3 board provided they are large enough to carry it and can power it from their battery or carry a second battery for the UP 3. An interesting walking robot kit containing 12 R/C servos is seen in Figure 13.26.

Figure 13.25 Hobbyist R/C model with a CMU camera and R/C PWM servos controlled by an FPGA

Figure 13.26 Lynxmotion Hexpod Walking Robot Kit with 12 R/C servos

Figure 13.27 ActiveMedia's Amigobot robot base controlled by an FPGA with a Nios Processor

Some commercial robot bases are also available such as the Amigobot as seen in Figure 13.27, the ER1 from Evolution Robotics and the mobile robot platform from Drrobot. The Amigobot uses an RS-232C serial interface and the ER1 uses USB for motor control. A robot base contains a motor, drive electronics, sensors, and a battery, but it has no high-level controller.

Once you develop a working robot and want to run existing demos, you may want to program the UP 3 flash memory configuration device so that your design automatically runs whenever the board is turned on. Instructions are provided in Appendix E.

13.9 For Additional Information

Radio-controlled cars and parts such as batteries, battery chargers, and servos can be obtained at a local hobby shop or via mail order at a lower cost from:

Tower Hobbies   http://www.towerhobbies.com
P.O. Box 9078
Champaign, IL 61826-9078
800-637-6050

IR Proximity, Line Tracker, Sonar sensors, Servos, Wheels, and Robot kits can be obtained via mail order from:

Lynxmotion, Inc.   http://www.lynxmotion.com
104 Partridge Road
Pekin, IL 61554-1403
309-382-1254

Mondotronics   http://www.robotstore.com
4286 Redwood Highway #226
San Raphael, CA 94903
800–374–5764

Sensors & Robot kits, Servo wheels for robots, and Servo wheel encoder kits can also be obtained via mail order from:

Acroname   http://www.acroname.com
P.O. Box 1894
Nederland, CO 80466
303-258-3161

Low-cost digital and analog compass sensors are available via mail order from:

Dinsmore Instrument Co.   http://www.dinsmoresensors.com
P.O. Box 345
Flint, Michigan 48501
810-744-1790

Electronic Compass Modules are available from:

PNI Corp   http://www.pnicorp.com
5464 Skylane Blvd. Suite A
Santa Rosa, CA 95403

GPS and DGPS ICs and modules are available from:

Motorola TCG   http://www.motorola.com/gps
GPS Products
2900 South Diablo Way
Tempe, AZ 85282

u-blox AG   http://www.u-blox.com
Zürcherstrasse 68
8800 Thalwil
Schweiz

A wide array of robot sensor modules is available from:

Devantech Ltd (Robot Electronics)   http://www.robot-electronics.co.uk
Unit 2B Gilray Road
Diss
Norfolk
IP22 4EU
England

A longer list of robot parts vendors and sites can be found at:

http://users.ece.gatech.edu/~hamblen/4006/robot_links.htm

13.10 Laboratory Exercises

1. Develop a counter design to find the dead zone of a converted R/C servo motor. The dead or null zone is the time near 1.5ms that actually makes the servo motor stop moving. As in the example motor driver code, send a width adjusted pulse every 20ms. You will need a resolution of at least .01ms to find the dead zone, so a clock faster than the example code is required. For example, the motor might actually stop at 1.54ms instead of 1.50ms. Use the clk_div UP3core function to provide the clock. The design should increase the width of the timing pulse if one pushbutton is hit and decrease the width if the other pushbutton is hit. Display the width of the timing pulse in the seven-segment LEDs. Use a Cyclone DIP-switch input to select the motor to examine. By hitting the pushbuttons, you should be able to stop and reverse the motor. The dead zone will be between the settings where the drive wheel reverses direction. At the dead zone, the drive wheel should stop. Settings near the dead zone will make the motor run slower. Record the dead zone for both the left and right motor.

2. Using the dead zone settings from problem 1, design a motor speed controller. Settings within around .2ms of the dead zone will make the motor run slower. The closer to the dead zone the slower the motor will run. Include at least four speed settings for each motor. See if you can get the robot to move in a straight line at a slow speed.

3. Develop a speed controller for the robot drive motors by pulsing the drive motors on and off. The motors are sent a pulse of 1ms for reverse and 2ms for forward at full speed. If no pulse is sent for 20ms, the motor stops. If a motor is sent a 1 or 2ms pulse followed by no pulse in a repeating pattern, it will move slower. To move even slower use pulse, no pulse, no pulse in a repeating pattern. To move faster use pulse, no pulse, pulse in a repeating pattern. Using this approach, develop a speed controller for the robot with at least five speeds and direction. Send no pulse for the stop speed. Some additional mechanical noise will result from pulsing the motors at slow speeds. See if the robot will move in a straight line at a slow speed.

4. Use an IR LED and IR sensor to add position feedback to the motors. You can build it yourself or a similar servo wheel encoder kit built by Nubotics is available from Acroname. Some sensor modules are available that have both the IR LED and IR sensor mounted in a single plastic case. For reflective sensors, mark the wheels with radial black paint stripes or black drafting tape and count the pulses from the IR sensor to determine movement of the wheel. Another option would be to draw the radial stripes using a PC drawing program and print it on clear adhesive labels made for laser printers. The labels could then be placed on the flat side of the wheel. If a transmissive sensor arrangement is used, holes can be drilled in the main wheel or a second smaller slotted wheel could be attached to the servo output shaft that periodically interrupts the IR light beam from the LED to the sensor. In this case, the LED and sensor are mounted on opposite sides of the wheel. This same optical sensing technique is used in many mice to detect movement of the mouse ball. Use the position feedback to implement more accurate variable speed and position control for the motors.

5. Design a state machine using a counter/timer that will move the robot in the following fixed pattern:

    *   Move forward for 6 seconds.

    *   Turn right and go forward for 4 seconds (do not count the time it takes to turn).

    *   Turn left and go forward for 2 seconds.

    *   Stop, pause for 2 seconds, turn 180 degrees, and start over.

Determine the amount of time required for 90- and 180-degree turns by trial and error. A 10Hz or 100Hz clock should be used for the timer. Use the clk_div UP3core to divide the UP 3 on-board clock. The state machine should check the timer to see if the correct amount of time has elapsed before moving to the next state in the path. The timer is reset when moving to a new portion of the path. Use an initial state that turns off the motors until a pushbutton is hit, so that it is easier to control the robot during download. Since there is no motor position feedback, all turns and the actual distance traveled by the UP3-bot will vary slightly.

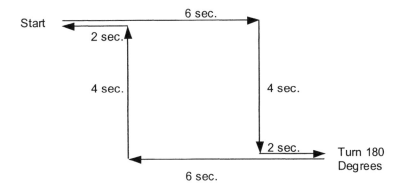

Figure 13.28 Simple path for state machine without sensor response.

6. Using a ROM, develop a ROM-based state machine that reads a motor direction and time from the ROM. Put a complex pattern such as a dance step in the ROM using a MIF file. For looping, another field in the ROM can be used to specify a jump to a different next address.

7. Using the keyboard UP3core, design an interface to the keyboard that allows the keyboard to be used as a remote control device to move the robot. Pick at least five different keys to command to robot to move, turn left, turn right, or stop.

8. Interface an IR proximity sensor module to the UP3-bot using jumpers connected to the Cyclone male header socket. Attach the module in front of the header socket using foam tape. Alternate driving the left and right IR LEDs at 100Hz. Check for an IR sensor return and develop two signals, LEFT and RIGHT to indicate if the IR sensor return is from the left or right IR LED. The IR LEDs may need to be adjusted or shielded with some heat shrink tubing so that the floor does not reflect IR to the sensor. Use the LEFT and RIGHT signals to drive the decimal points on two LEDs to help adjust the sensor. It may be necessary to filter the IR returns using a counter with a return/no return threshold for reliable operation. Using a clock faster than 100Hz, for example 10kHz, only set LEFT or RIGHT if the return was present for several clock cycles.

9. Using IR sensor input, develop a design for the UP3-bot that follows a person. The person must be within a foot or so of the UP3-bot. When a left signal is present turn left, when a right signal is present turn right, and when both signals are present, move

forward a few inches and stop. When all signals are lost, the UP3-bot should rotate until an IR return is acquired.

10. Use motor speed control and a state machine with a timer to perform a small figure eight with the UP3-bot.

11. Once the IR proximity sensor module from problem 8 is interfaced, design a state machine for the robot that moves forward and avoids obstacles. If it sees an obstacle to the left, turn right, and if there is an obstacle to the right, turn left. If both left and right obstacles are present, the robot should go backwards by reversing both motors.

12. With two UP3-bots facing each other, develop a serial communications protocol using the IR LEDs and sensors. Assume the serial data is fixed in length and always starts with a known pattern at a fixed clock rate. The IR LEDs are pulsed at around 40kHz and the sensor has a 40kHz filter, so this will limit the bandwidth to a few kHz. Transmit the 8-bit value from the Cyclone DIP switches and display the value in the receiving UP3-bot seven-segment LED displays. Display the raw IR sensor input in the decimal point LED to aid in debugging and alignment.

13. Interface the line-following module to the UP3-bot, and design a state machine that follows a line. The line-following module has three sensor signals, left, center, and right. If the line drifts to the left, turn right, and if the line drifts to the right, turn left. Adjust turn constants so that the UP3-bot moves along the line as fast as possible. If speed control was developed for the UP3-bot as suggested in earlier problems, try using speed control for smaller less abrupt turns.

14. Using a standard IR remote control unit from a television or VCR and an IR sensor interfaced to the UP3-bot, implement a remote control for the UP3-bot. Different buttons on the remote control unit generate a different sequence of timing pulses. A digital oscilloscope or logic analyzer can be used to examine the timing pulses.

15. Interface the a magnetic or electronic compass module to the UP3-bot, and design a state machine that performs the following operation:

   • Turn North.
   • Move forward 4 seconds.
   • Turn East.
   • Move forward 4 seconds.
   • Turn Southwest.
   • Move forward 6.6 seconds.
   • Stop and repeat when the pushbutton is hit.

The mechanical compass has a small time delay due to the inertia of the magnetized rotor. Just like a real compass, it will swing back and forth for awhile before stopping. With care, the leads on the compass module can be plugged into a DIP socket with wire wrapped power supply and pull-up resistor connections on a small protoboard or make a printed circuit board for the compass with jumper wires to plug into the Cyclone female header socket. Make sure the compass module is mounted so that it is level and as far away from the motors magnets as possible.

16. Interface a Sonar-ranging module to the UP3-bot and perform the following operation:

    • Scan the immediate area 360 degrees by rotating the robot
      and locate the nearest object.

    • Move close to the object and stop.

17. Attach the Sonar transducer to an unmodified servo's output shaft. Use the new servo to scan the area and locate the closest object. To sweep the unmodified servo back and forth, a timing pulse that slowly increases from 1ms to 2ms and back to 1ms is required. Move close to the nearest object and stop.

18. Attach several IR ranging sensors to the UP3-bot and use the sensor data to develop a wall following robot.

19. Interface additional sensors, switches, etc., to the UP3-bot so that it can navigate a maze. If several robots are being developed, consider a contest such as best time through the maze or best time after learning the maze.

20. Use the μP 3 computer from Chapter 8 to implement a microcontroller to control the robot instead of a custom state machine. Write a μP 3 assembly language program to solve one of the previous problems. Interface a time-delay timer, the sensors, and the motor speed control unit to the μP 3 computer using I/O ports as suggested in problem 8.6. The additional machine instructions suggested in the exercises in Chapter 8 would also be useful.

21. Use a Nios processor to control the robot with C code using the UP 3 Nios II reference design in Chapters 16 & 17.

22. Develop and hold a UP3-bot design contest. Information on previous and current robotics contests can be found online at various web sites. Here are some ideas that have been used for other robot design contests:

    • Robot Maze Solving

    • Robot Dance Contest

    • Sumo Wrestling

    • Robot Soccer Teams

    • Robot Laser Tag

    • Fire Fighting Robots

    • Robots that collect objects

    • Robots that detect mines

CHAPTER 14

A RISC Design: Synthesis of the MIPS Processor Core

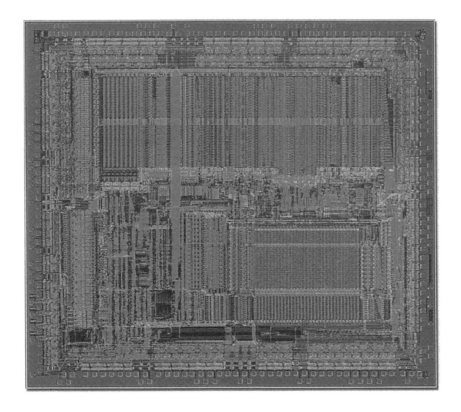

A full die photograph of the MIPS R2000 RISC Microprocessor is shown above. The 1986 MIPS R2000 with five pipeline stages and 450,000 transistors was the world's first commercial RISC microprocessor. Photograph ©1995-2004 courtesy of Michael Davidson, Florida State University, http://micro.magnet.fsu.edu/chipshots.

14 A RISC Design: Synthesis of the MIPS Processor Core

14.1 The MIPS Instruction Set and Processor

The MIPS is an example of a modern reduced instruction set computer (RISC) developed in the 1980s. The MIPS instruction set is used by NEC, Nintendo, Motorola, Sony, and licensed for use by numerous other semiconductor manufacturers. It has fixed-length 32-bit instructions and thirty-two 32-bit general-purpose registers. Register 0 always contains the value 0. A memory word is 32 bits wide.

As seen in Table 14.1, the MIPS has only three instruction formats. Only I-format LOAD and STORE instructions reference memory operands. R-format instructions such as ADD, AND, and OR perform operations only on data in the registers. They require two register operands, Rs and Rt. The result of the operation is stored in a third register, Rd. R-format shift and function fields are used as an extended opcode field. J-format instructions include the jump instructions.

Table 14.1 MIPS 32-bit Instruction Formats.

Field Size	6-bits	5-bits	5-bits	5-bits	5-bits	6-bits
R- Format	Opcode	Rs	Rt	Rd	Shift	Function
I - Format	Opcode	Rs	Rt	Address/immediate value		
J - Format	Opcode	Branch target address				

LW is the mnemonic for the Load Word instruction and SW is the mnemonic for Store Word. The following MIPS assembly language program computes A = B + C.

```
LW  $2, B          ;Register 2 = value of memory at address B
LW  $3, C          ;Register 3 = value of memory at address C
ADD $4, $2, $3     ;Register 4 = B + C
SW  $4, A          ;Value of memory at address A = Register 4
```

The MIPS I-format instruction, BEQ, branches if two registers have the same value. As an example, the instruction BEQ $1, $2, LABEL jumps to LABEL if register 1 equals register 2. A branch instruction's address field contains the offset from the current address. The PC must be added to the address field to compute the branch address. This is called PC-relative addressing.

LW and SW instructions contain an offset and a base register that are used for array addressing. As an example, LW $1, 100($2) adds an offset of 100 to the contents of register 2 and uses the sum as the memory address to read data from. The value from memory is then loaded into register 1. Using register 0, which always contains a 0, as the base register disables this addressing feature.

Table 14.2 MIPS Processor Core Instructions.

Mnemonic	Format	Opcode Field	Function Field	Instruction
Add	R	0	32	Add
Addi	I	8	-	Add Immediate
Addu	R	0	33	Add Unsigned
Sub	R	0	34	Subtract
Subu	R	0	35	Subtract Unsigned
And	R	0	36	Bitwise And
Or	R	0	37	Bitwise OR
Sll	R	0	0	Shift Left Logical
Srl	R	0	2	Shift Right Logical
Slt	R	0	42	Set if Less Than
Lui	I	15	-	Load Upper Immediate
Lw	I	35	-	Load Word
Sw	I	43	-	Store Word
Beq	I	4	-	Branch on Equal
Bne	I	5	-	Branch on Not Equal
J	J	2	-	Jump
Jal	J	3	-	Jump and Link (used for Call)
Jr	R	0	8	Jump Register (used for Return)

A summary of the basic MIPS instructions is shown in Table 14.2. In depth explanations of all MIPS instructions and assembly language programming examples can be found in the references listed in section 14.11.

A hardware implementation of the MIPS processor core based on the example in the widely used textbook, *Computer Organization and Design The Hardware/Software Interface* by Patterson and Hennessy, is shown in Figure 14.1. This implementation of the MIPS performs fetch, decode, and execute in one clock cycle. Starting at the left in Figure 14.1, the program counter (PC) is used to fetch the next address in instruction memory. Since memory is byte addressable, four is added to address the next 32-bit (or 4-byte) word in memory. At the same time as the instruction fetch, the adder above instruction memory is used to add four to the PC to generate the next address. The output of instruction memory is the next 32-bit instruction.

The instruction's opcode is then sent to the control unit and the function code is sent to the ALU control unit. The instruction's register address fields are used to address the two-port register file. The two-port register file can perform two independent reads and one write in one clock cycle. This implements the decode operation.

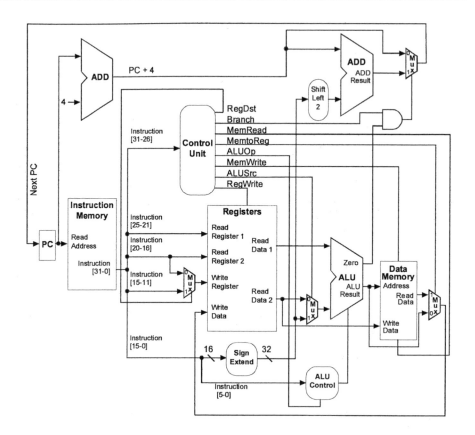

Figure 14.1 MIPS Single Clock Cycle Implementation.

The two outputs of the register file then feed into the data ALU inputs. The control units setup the ALU operation required to execute the instruction. Next, Load and Store instructions read or write to data memory. R-format instructions bypass data memory using a multiplexer. Last, R-format and Load instructions write back a new value into the register file.

PC-relative branch instructions use the adder and multiplexer shown above the data ALU in Figure 14.1 to compute the branch address. The multiplexer is required for conditional branch operations. After all outputs have stabilized, the next clock loads in the new value of the PC and the process repeats for the next instruction.

RISC instruction sets are easier to pipeline. With pipelining, the fetch, decode, execute, data memory, and register file write operations all work in parallel. In a single clock cycle, five different instructions are present in the pipeline. The basis for a pipelined hardware implementation of the MIPS is shown in Figure 14.2.

Additional complications arise because of data dependencies between instructions in the pipeline and branch operations. These problems can be resolved using hazard detection, data forwarding techniques, and branch flushing. With pipelining, most RISC instructions execute in one clock cycle. Branch instructions will still require flushing of the pipeline. Exercises that add pipelining to the processor core are included at the end of the chapter.

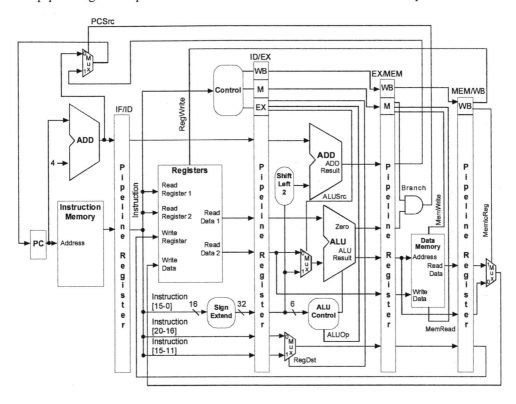

Figure 14.2 MIPS Pipelined Implementation.

14.2 Using VHDL to Synthesize the MIPS Processor Core

A VHDL-synthesis model of the MIPS single clock cycle model from Figure 14.1 will be developed in this section. This model can be used for simulation and implemented using the UP 3 board.

The full 32-bit model requires a couple minutes to synthesize. When testing new changes you might want to use the faster functional (i.e. no timing delays) simulation approach before using a full timing delay model. This approach is commonly used on larger models with long synthesis and simulation times.

A two-level hierarchy is used in the model. MIPS.VHD is the top-level of the hierarchy. It consists of a structural VHDL model that connects the five behavioral modules. The five behavioral modules are already setup so that they correspond to the different stages for the MIPS. This makes it much easier to modify when the model is pipelined in later laboratory exercises. For many synthesis tools, hierarchy is also required to synthesize large logic designs. IFETCH.VHD is the VHDL submodule that contains instruction memory and the program counter. CONTROL.VHD contains the logic for the control unit.

IDECODE.VHD contains the multi-ported register file. EXECUTE.VHD contains the data and branch address ALUs. DMEMORY.VHD contains the data memory.

14.3 The Top-Level Module

The MIPS.VHD file contains the top-level design file. MIPS.VHD is a VHDL structural model that connects the five component parts of the MIPS. This module could also be created using the schematic editor and connecting the symbols for each VHDL submodule. The inputs are the clock and reset signals. The values of major busses and important control signals are copied and output from the top level so that they are available for easy display in simulations. Signals that are not outputs at the top level will occasionally not exist due to the compilers logic optimizations during synthesis.

```
                              -- Top Level Structural Model for MIPS Processor Core
LIBRARY IEEE;
USE IEEE.STD_LOGIC_1164.ALL;
USE IEEE.STD_LOGIC_ARITH.ALL;

ENTITY MIPS IS

    PORT( reset, clock                    : IN    STD_LOGIC;
        -- Output important signals to pins for easy display in Simulator
        PC                                : OUT  STD_LOGIC_VECTOR( 7 DOWNTO 0 );
        ALU_result_out, read_data_1_out, read_data_2_out,
        write_data_out, Instruction_out   : OUT  STD_LOGIC_VECTOR( 31 DOWNTO 0 );
        Branch_out, Zero_out, Memwrite_out,
        Regwrite_out                      : OUT  STD_LOGIC );
END   TOP_SPIM;

ARCHITECTURE structure OF TOP_SPIM IS

    COMPONENT Ifetch
        PORT(   Instruction       : OUT  STD_LOGIC_VECTOR( 31 DOWNTO 0 );
                PC_plus_4_out     : OUT  STD_LOGIC_VECTOR( 9 DOWNTO 0 );
                Add_result        : IN   STD_LOGIC_VECTOR( 7 DOWNTO 0 );
                Branch            : IN   STD_LOGIC;
                Zero              : IN   STD_LOGIC;
                PC_out            : OUT  STD_LOGIC_VECTOR( 9 DOWNTO 0 );
                clock,reset       : IN   STD_LOGIC );
    END COMPONENT;

    COMPONENT Idecode
        PORT(   read_data_1       : OUT  STD_LOGIC_VECTOR( 31 DOWNTO 0 );
                read_data_2       : OUT  STD_LOGIC_VECTOR( 31 DOWNTO 0 );
                Instruction       : IN   STD_LOGIC_VECTOR( 31 DOWNTO 0 );
                read_data         : IN   STD_LOGIC_VECTOR( 31 DOWNTO 0 );
                ALU_result        : IN   STD_LOGIC_VECTOR( 31 DOWNTO 0 );
                RegWrite, MemtoReg : IN  STD_LOGIC;
```

```
                RegDst                  : IN      STD_LOGIC;
                Sign_extend             : OUT     STD_LOGIC_VECTOR( 31 DOWNTO 0 );
                clock, reset            : IN      STD_LOGIC );
END COMPONENT;


COMPONENT control
    PORT(   Opcode                      : IN      STD_LOGIC_VECTOR( 5 DOWNTO 0 );
            RegDst                      : OUT     STD_LOGIC;
            ALUSrc                      : OUT     STD_LOGIC;
            MemtoReg                    : OUT     STD_LOGIC;
            RegWrite                    : OUT     STD_LOGIC;
            MemRead                     : OUT     STD_LOGIC;
            MemWrite                    : OUT     STD_LOGIC;
            Branch                      : OUT     STD_LOGIC;
            ALUop                       : OUT     STD_LOGIC_VECTOR( 1 DOWNTO 0 );
            clock, reset                : IN      STD_LOGIC );
END COMPONENT;


COMPONENT  Execute
    PORT(   Read_data_1                 : IN      STD_LOGIC_VECTOR( 31 DOWNTO 0 );
            Read_data_2                 : IN      STD_LOGIC_VECTOR( 31 DOWNTO 0 );
            Sign_Extend                 : IN      STD_LOGIC_VECTOR( 31 DOWNTO 0 );
            Function_opcode             : IN      STD_LOGIC_VECTOR( 5 DOWNTO 0 );
            ALUOp                       : IN      STD_LOGIC_VECTOR( 1 DOWNTO 0 );
            ALUSrc                      : IN      STD_LOGIC;
            Zero                        : OUT     STD_LOGIC;
            ALU_Result                  : OUT     STD_LOGIC_VECTOR( 31 DOWNTO 0 );
            Add_Result                  : OUT     STD_LOGIC_VECTOR( 7 DOWNTO 0 );
            PC_plus_4                   : IN      STD_LOGIC_VECTOR( 9 DOWNTO 0 );
            clock, reset                : IN      STD_LOGIC );
END COMPONENT;



COMPONENT dmemory
    PORT(   read_data                   : OUT     STD_LOGIC_VECTOR( 31 DOWNTO 0 );
            address                     : IN      STD_LOGIC_VECTOR( 7 DOWNTO 0 );
            write_data                  : IN      STD_LOGIC_VECTOR( 31 DOWNTO 0 );
            MemRead, Memwrite           : IN      STD_LOGIC;
            Clock,reset                 : IN      STD_LOGIC );
END COMPONENT;

                            -- declare signals used to connect VHDL components
SIGNAL PC_plus_4        : STD_LOGIC_VECTOR( 9 DOWNTO 0 );
SIGNAL read_data_1      : STD_LOGIC_VECTOR( 31 DOWNTO 0 );
SIGNAL read_data_2      : STD_LOGIC_VECTOR( 31 DOWNTO 0 );
SIGNAL Sign_Extend      : STD_LOGIC_VECTOR( 31 DOWNTO 0 );
SIGNAL Add_result       : STD_LOGIC_VECTOR( 7 DOWNTO 0 );
SIGNAL ALU_result       : STD_LOGIC_VECTOR( 31 DOWNTO 0 );
SIGNAL read_data        : STD_LOGIC_VECTOR( 31 DOWNTO 0 );
SIGNAL ALUSrc           : STD_LOGIC;
SIGNAL Branch           : STD_LOGIC;
SIGNAL RegDst           : STD_LOGIC;
```

```
        SIGNAL Regwrite          : STD_LOGIC;
        SIGNAL Zero              : STD_LOGIC;
        SIGNAL MemWrite          : STD_LOGIC;
        SIGNAL MemtoReg          : STD_LOGIC;
        SIGNAL MemRead           : STD_LOGIC;
        SIGNAL ALUop             : STD_LOGIC_VECTOR( 1 DOWNTO 0 );
        SIGNAL Instruction       : STD_LOGIC_VECTOR( 31 DOWNTO 0 );

    BEGIN
                                 -- copy important signals to output pins for easy
                                 -- display in Simulator
     Instruction_out    <= Instruction;
     ALU_result_out     <= ALU_result;
     read_data_1_out    <= read_data_1;
     read_data_2_out    <= read_data_2;
     write_data_out     <= read_data WHEN MemtoReg = '1' ELSE ALU_result;
     Branch_out         <= Branch;
     Zero_out           <= Zero;
     RegWrite_out       <= RegWrite;
     MemWrite_out       <= MemWrite;
                                 -- connect the 5 MIPS components
     IFE : Ifetch
        PORT MAP ( Instruction       => Instruction,
                   PC_plus_4_out     => PC_plus_4,
                   Add_result        => Add_result,
                   Branch            => Branch,
                   Zero              => Zero,
                   PC_out            => PC,
                   clock             => clock,
                   reset             => reset );


     ID : Idecode
        PORT MAP ( read_data_1       => read_data_1,
                   read_data_2       => read_data_2,
                   Instruction       => Instruction,
                   read_data         => read_data,
                   ALU_result        => ALU_result,
                   RegWrite          => RegWrite,
                   MemtoReg          => MemtoReg,
                   RegDst            => RegDst,
                   Sign_extend       => Sign_extend,
                   clock             => clock,
                   reset             => reset );



     CTL:  control
        PORT MAP ( Opcode            => Instruction( 31 DOWNTO 26 ),
                   RegDst            => RegDst,
                   ALUSrc            => ALUSrc,
                   MemtoReg          => MemtoReg,
                   RegWrite          => RegWrite,
                   MemRead           => MemRead,
```

```
              MemWrite        => MemWrite,
              Branch          => Branch,
              ALUop           => ALUop,
              clock           => clock,
              reset           => reset );

EXE: Execute
   PORT MAP ( Read_data_1     => read_data_1,
              Read_data_2     => read_data_2,
              Sign_extend     => Sign_extend,
              Function_opcode => Instruction( 5 DOWNTO 0 ),
              ALUOp           => ALUop,
              ALUSrc          => ALUSrc,
              Zero            => Zero,
              ALU_Result      => ALU_Result,
              Add_Result      => Add_Result,
              PC_plus_4       => PC_plus_4,
              Clock           => clock,
              Reset           => reset );

MEM: dmemory
   PORT MAP ( read_data       => read_data,
              address         => ALU_Result,
              write_data      => read_data_2,
              MemRead         => MemRead,
              Memwrite        => MemWrite,
              clock           => clock,
              reset           => reset );
END structure;
```

14.4 The Control Unit

The control unit of the MIPS shown in Figure 14.3 examines the instruction opcode bits and generates eight control signals used by the other stages of the processor. Recall that the high six bits of a MIPS instruction contain the opcode. The opcode value is used to determine the instruction type.

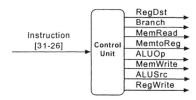

Figure 14.3 Block Diagram of MIPS Control Unit.

```
                    -- control module (implements MIPS control unit)
LIBRARY IEEE;
USE IEEE.STD_LOGIC_1164.ALL;
USE IEEE.STD_LOGIC_ARITH.ALL;
USE IEEE.STD_LOGIC_SIGNED.ALL;

ENTITY control IS
  PORT( Opcode          : IN    STD_LOGIC_VECTOR( 5 DOWNTO 0 );
         RegDst          : OUT   STD_LOGIC;
         ALUSrc          : OUT   STD_LOGIC;
         MemtoReg        : OUT   STD_LOGIC;
         RegWrite        : OUT   STD_LOGIC;
         MemRead         : OUT   STD_LOGIC;
         MemWrite        : OUT   STD_LOGIC;
         Branch          : OUT   STD_LOGIC;
         ALUop           : OUT   STD_LOGIC_VECTOR( 1 DOWNTO 0 );
         clock, reset    : IN    STD_LOGIC );
END control;


ARCHITECTURE behavior OF control IS


    SIGNAL  R_format, Lw, Sw, Beq    : STD_LOGIC;


BEGIN
                            -- Code to generate control signals using opcode bits
        R_format    <= '1' WHEN  Opcode = "000000" ELSE '0';
        Lw          <= '1' WHEN  Opcode = "100011" ELSE '0';
        Sw          <= '1' WHEN  Opcode = "101011" ELSE '0';
        Beq         <= '1' WHEN  Opcode = "000100" ELSE '0';

        RegDst      <= R_format;
        ALUSrc      <= Lw OR Sw;
        MemtoReg    <= Lw;
        RegWrite    <= R_format OR Lw;
        MemRead     <= Lw;
        MemWrite    <= Sw;
        Branch      <= Beq;
        ALUOp( 1 )  <= R_format;
        ALUOp( 0 )  <= Beq;

    END behavior;
```

14.5 The Instruction Fetch Stage

The instruction fetch stage of the MIPS shown in Figure 14.4 contains the instruction memory, the program counter, and the hardware to increment the program counter to compute the next instruction address.

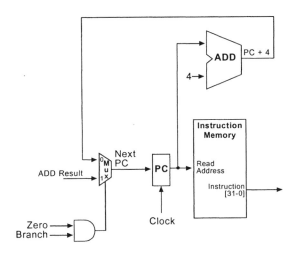

Figure 14.4 Block Diagram of MIPS Fetch Unit.

Instruction memory is implemented using the Altsyncram megafunction. 256 by 32 bits of instruction memory is available. This requires two of the Cyclone chip's M4K RAM memory blocks. Since the Altsyncram memory requires an address register, the PC register is actually implemented inside the memory block. A copy of the PC external to the memory block is also saved for use in simulation displays.

```
                         -- Ifetch module (provides the PC and instruction
                         --memory for the MIPS computer)
LIBRARY IEEE;
USE IEEE.STD_LOGIC_1164.ALL;
USE IEEE.STD_LOGIC_ARITH.ALL;
USE IEEE.STD_LOGIC_UNSIGNED.ALL;
LIBRARY altera_mf;
USE altera_mf.altera_mf_components.ALL;

ENTITY Ifetch IS
        PORT( SIGNAL Instruction    : OUT STD_LOGIC_VECTOR( 31 DOWNTO 0 );
              SIGNAL PC_plus_4_out  : OUT STD_LOGIC_VECTOR( 7 DOWNTO 0 );
              SIGNAL Add_result     : IN   STD_LOGIC_VECTOR( 7 DOWNTO 0 );
              SIGNAL Branch         : IN   STD_LOGIC;
              SIGNAL Zero           : IN   STD_LOGIC;
              SIGNAL PC_out         : OUT STD_LOGIC_VECTOR( 9 DOWNTO 0 );
              SIGNAL clock, reset   : IN   STD_LOGIC);
END Ifetch;
```

```
ARCHITECTURE behavior OF Ifetch IS
        SIGNAL PC, PC_plus_4          : STD_LOGIC_VECTOR( 9 DOWNTO 0 );
        SIGNAL next_PC                : STD_LOGIC_VECTOR( 7 DOWNTO 0 );
BEGIN
                                      --ROM for Instruction Memory
data_memory: altsyncram

    GENERIC MAP (
      operation_mode => "ROM",
      width_a => 32,
      widthad_a => 8,
      lpm_type => "altsyncram",
      outdata_reg_a => "UNREGISTERED",
                            -- Reads in mif file for initial data memory values
      init_file => "program.mif",
      intended_device_family => "Cyclone")

                            -- Fetch next instruction from memory using PC
    PORT MAP (
              clock0   => clock,
              address_a => Mem_Addr,
              q_a => Instruction
              );
                       -- Instructions always start on a word address - not byte
    PC(1 DOWNTO 0) <= "00";
                       -- copy output signals - allows read inside module
    PC_out          <= PC;
    PC_plus_4_out   <= PC_plus_4;
                       -- send word address to inst. memory address register
    Mem_Addr <= Next_PC;
                       -- Adder to increment PC by 4
    PC_plus_4( 9 DOWNTO 2 )  <= PC( 9 DOWNTO 2 ) + 1;
    PC_plus_4( 1 DOWNTO 0 )  <= "00";
                       -- Mux to select Branch Address or PC + 4
    Next_PC  <= X"00" WHEN Reset = '1'  ELSE
          Add_result  WHEN ( ( Branch = '1' ) AND ( Zero = '1' ) )
          ELSE   PC_plus_4( 9 DOWNTO 2 );

                       -- Store PC in register and load next PC on clock edge
    PROCESS
      BEGIN
        WAIT UNTIL ( clock'EVENT ) AND ( clock = '1' );
        IF reset = '1' THEN
              PC <= "0000000000" ;
        ELSE
              PC( 9 DOWNTO 2 ) <= Next_PC;
        END IF;
    END PROCESS;
END behavior;
```

The MIPS program is contained in instruction memory. Instruction memory is automatically initialized using the program.mif file shown in Figure 14.5. This initialization only occurs once during download and not at a reset.

For different test programs, the appropriate machine code must be entered in this file in hex. Note that the memory addresses displayed in the program.mif file are word addresses while addresses in registers such as the PC are byte addresses. The byte address is four times the word address since a 32-bit word contains four bytes. Only word addresses can be used in the *.mif files.

```
-- MIPS Instruction Memory Initialization File


Depth = 256;
Width = 32;
Address_radix = HEX;
Data_radix = HEX;
Content
Begin

            -- Use NOPS for default instruction memory values
[00..FF]: 00000000;        -- nop (sll r0,r0,0)
            -- Place MIPS Instructions here
            -- Note: memory addresses are in words and not bytes
            -- i.e. next location is +1 and not +4


00: 8C020000;      -- lw $2,0 ;memory(00)=55
01: 8C030001;      -- lw $3,1 ;memory(01)=AA
02: 00430820;      -- add $1,$2,$3
03: AC010003;      -- sw $1,3 ;memory(03)=FF
04: 1022FFFF;      -- beq $1,$2,-4
05: 1021FFFA;      -- beq $1,$1,-24

End;
```

Figure 14.5 MIPS Program Memory Initialization File, program.mif.

14.6 The Decode Stage

The decode stage of the MIPS contains the register file as shown in Figure 14.6. The MIPS contains thirty-two 32-bit registers. The register file requires a major portion of the hardware required to implement the MIPS. Registers are initialized to the register number during a reset. This is done to enable the use of shorter test programs that do not have to load all of the registers. A VHDL FOR...LOOP structure is used to generate the initial register values at reset.

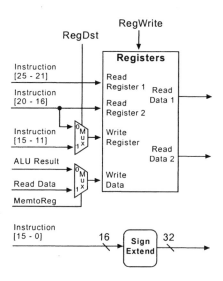

Figure 14.6 Block Diagram of MIPS Decode Unit.

```
                              -- Idecode module (implements the register file for
LIBRARY IEEE;                 -- the MIPS computer)
USE IEEE.STD_LOGIC_1164.ALL;
USE IEEE.STD_LOGIC_ARITH.ALL;
USE IEEE.STD_LOGIC_UNSIGNED.ALL;

ENTITY Idecode IS
      PORT(   read_data_1    : OUT   STD_LOGIC_VECTOR( 31 DOWNTO 0 );
              read_data_2    : OUT   STD_LOGIC_VECTOR( 31 DOWNTO 0 );
              Instruction    : IN    STD_LOGIC_VECTOR( 31 DOWNTO 0 );
              read_data      : IN    STD_LOGIC_VECTOR( 31 DOWNTO 0 );
              ALU_result     : IN    STD_LOGIC_VECTOR( 31 DOWNTO 0 );
              RegWrite       : IN    STD_LOGIC;
              MemtoReg       : IN    STD_LOGIC;
              RegDst         : IN    STD_LOGIC;
              Sign_extend    : OUT   STD_LOGIC_VECTOR( 31 DOWNTO 0 );
              clock,reset    : IN    STD_LOGIC );
END Idecode;
```

```vhdl
ARCHITECTURE behavior OF Idecode IS
TYPE register_file IS ARRAY ( 0 TO 31 ) OF STD_LOGIC_VECTOR( 31 DOWNTO 0 );

    SIGNAL register_array: register_file;
    SIGNAL write_register_address      : STD_LOGIC_VECTOR( 4 DOWNTO 0 );
    SIGNAL write_data                  : STD_LOGIC_VECTOR( 31 DOWNTO 0 );
    SIGNAL read_register_1_address     : STD_LOGIC_VECTOR( 4 DOWNTO 0 );
    SIGNAL read_register_2_address     : STD_LOGIC_VECTOR( 4 DOWNTO 0 );
    SIGNAL write_register_address_1    : STD_LOGIC_VECTOR( 4 DOWNTO 0 );
    SIGNAL write_register_address_0    : STD_LOGIC_VECTOR( 4 DOWNTO 0 );
    SIGNAL Instruction_immediate_value : STD_LOGIC_VECTOR( 15 DOWNTO 0 );

BEGIN
    read_register_1_address     <= Instruction( 25 DOWNTO 21 );
    read_register_2_address     <= Instruction( 20 DOWNTO 16 );
    write_register_address_1    <= Instruction( 15 DOWNTO 11 );
    write_register_address_0    <= Instruction( 20 DOWNTO 16 );
    Instruction_immediate_value <= Instruction( 15 DOWNTO 0 );
                                -- Read Register 1 Operation
    read_data_1 <= register_array( CONV_INTEGER( read_register_1_address) );
                                -- Read Register 2 Operation
    read_data_2 <= register_array( CONV_INTEGER( read_register_2_address) );
                                -- Mux for Register Write Address
    write_register_address <= write_register_address_1
            WHEN RegDst = '1' ELSE write_register_address_0;
                            -- Mux to bypass data memory for Rformat instructions
    write_data <= ALU_result( 31 DOWNTO 0 )
            WHEN ( MemtoReg = '0' ) ELSE read_data;
                            -- Sign Extend 16-bits to 32-bits
    Sign_extend <= X"0000" & Instruction_immediate_value
        WHEN Instruction_immediate_value(15) = '0'
        ELSE    X"FFFF" & Instruction_immediate_value;

PROCESS
    BEGIN
        WAIT UNTIL clock'EVENT AND clock = '1';
        IF reset = '1' THEN
                            -- Initial register values on reset are register = reg#
                            -- use loop to automatically generate reset logic
                            -- for all registers
            FOR i IN 0 TO 31 LOOP
                    register_array(i) <= CONV_STD_LOGIC_VECTOR( i, 32 );
            END LOOP;
                            -- Write back to register - don't write to register 0
        ELSIF RegWrite = '1' AND write_register_address /= 0 THEN
            register_array( CONV_INTEGER( write_register_address)) <= write_data;
        END IF;
    END PROCESS;
END behavior;
```

14.7 The Execute Stage

The execute stage of the MIPS shown in Figure 14.7 contains the data ALU and a branch address adder used for PC-relative branch instructions. Multiplexers that select different data for the ALU input are also in this stage.

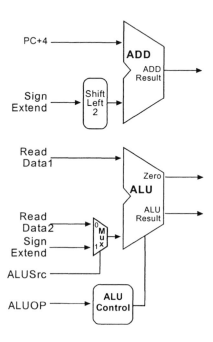

Figure 14.7 Block Diagram of MIPS Execute Unit.

```
-- Execute module (implements the data ALU and Branch Address Adder
-- for the MIPS computer)
LIBRARY IEEE;
USE IEEE.STD_LOGIC_1164.ALL;
USE IEEE.STD_LOGIC_ARITH.ALL;
USE IEEE.STD_LOGIC_SIGNED.ALL;

ENTITY  Execute IS
     PORT( Read_data_1      : IN     STD_LOGIC_VECTOR( 31 DOWNTO 0 );
           Read_data_2      : IN     STD_LOGIC_VECTOR( 31 DOWNTO 0 );
           Sign_extend      : IN     STD_LOGIC_VECTOR( 31 DOWNTO 0 );
           Function_opcode  : IN     STD_LOGIC_VECTOR( 5 DOWNTO 0 );
           ALUOp            : IN     STD_LOGIC_VECTOR( 1 DOWNTO 0 );
           ALUSrc           : IN     STD_LOGIC;
           Zero             : OUT    STD_LOGIC;
           ALU_Result       : OUT    STD_LOGIC_VECTOR( 31 DOWNTO 0 );
           Add_Result       : OUT    STD_LOGIC_VECTOR( 7 DOWNTO 0 );
           PC_plus_4        : IN     STD_LOGIC_VECTOR( 7 DOWNTO 0 );
           clock, reset     : IN     STD_LOGIC );
END Execute;
```

```
ARCHITECTURE behavior OF Execute IS
SIGNAL Ainput, Binput          : STD_LOGIC_VECTOR( 31 DOWNTO 0 );
SIGNAL ALU_output_mux          : STD_LOGIC_VECTOR( 31 DOWNTO 0 );
SIGNAL Branch_Add              : STD_LOGIC_VECTOR( 8 DOWNTO 0 );
SIGNAL ALU_ctl                 : STD_LOGIC_VECTOR( 2 DOWNTO 0 );
BEGIN
    Ainput <= Read_data_1;
                                    -- ALU input mux

    Binput <= Read_data_2
        WHEN ( ALUSrc = '0' )
        ELSE  Sign_extend( 31 DOWNTO 0 );
                                    -- Generate ALU control bits
    ALU_ctl( 0 ) <= ( Function_opcode( 0 ) OR Function_opcode( 3 ) ) AND ALUOp(1 );
    ALU_ctl( 1 ) <= ( NOT Function_opcode( 2 ) ) OR (NOT ALUOp( 1 ) );
    ALU_ctl( 2 ) <= ( Function_opcode( 1 ) AND ALUOp( 1 )) OR ALUOp( 0 );
                                    -- Generate Zero Flag
    Zero <= '1'
        WHEN ( ALU_output_mux( 31 DOWNTO 0 ) = X"00000000"  )
        ELSE '0';
                                    -- Select ALU output for SLT
    ALU_result <=  X"0000000" & B"000" & ALU_output_mux( 31 )
        WHEN   ALU_ctl = "111"
        ELSE   ALU_output_mux( 31 DOWNTO 0 );
                                    -- Adder to compute Branch Address
    Branch_Add <= PC_plus_4( 9 DOWNTO 2 ) + Sign_extend( 7 DOWNTO 0 ) ;
    Add_result   <= Branch_Add( 7 DOWNTO 0 );


PROCESS ( ALU_ctl, Ainput, Binput )
    BEGIN
                                    -- Select ALU operation
    CASE ALU_ctl IS
                                    -- ALU performs ALUresult = A_input AND B_input
        WHEN "000"  => ALU_output_mux  <= Ainput AND Binput;
                                    -- ALU performs ALUresult = A_input OR B_input
        WHEN "001"  => ALU_output_mux  <= Ainput OR Binput;
                                    -- ALU performs ALUresult = A_input + B_input
        WHEN "010"  => ALU_output_mux  <= Ainput + Binput;
                                    -- ALU performs ?
        WHEN "011"  => ALU_output_mux  <= X"00000000" ;
                                    -- ALU performs ?
        WHEN "100"  => ALU_output_mux  <= X"00000000" ;
                                    -- ALU performs ?
        WHEN "101"  => ALU_output_mux  <= X"00000000" ;
                                    -- ALU performs ALUresult = A_input - B_input
        WHEN "110"  => ALU_output_mux  <= Ainput - Binput;
                                    -- ALU performs SLT
        WHEN "111"  => ALU_output_mux  <= Ainput - Binput ;
        WHEN OTHERS => ALU_output_mux  <= X"00000000" ;
    END CASE;
END PROCESS;
END behavior;
```

14.8 The Data Memory Stage

The data memory stage of the MIPS core shown in Figure 14.8 contains the data memory. To speed synthesis and simulation, data memory is limited to 256 locations of 32-bit memory. Data memory is implemented using the Altsyncram megafunction. Memory write cycle timing is critical in any design. The Altsyncram function requires an internal address register with a clock. In this design, the falling clock edge is used to load the data memories internal address register. The rising clock edge starts the next instruction. Two M4K RAM blocks are used for data memory. Two M4K RAM blocks are also used for the 32-bit instruction memory.

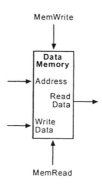

Figure 14.8 Block Diagram of MIPS Data Memory Unit.

```
-- Dmemory module (implements the data
-- memory for the MIPS computer)

LIBRARY IEEE;
USE IEEE.STD_LOGIC_1164.ALL;
USE IEEE.STD_LOGIC_ARITH.ALL;
USE IEEE.STD_LOGIC_SIGNED.ALL;
LIBRARY altera_mf;
USE altera_mf.atlera_mf_components.ALL;

ENTITY dmemory IS
    PORT(   read_data            : OUT    STD_LOGIC_VECTOR( 31 DOWNTO 0 );
            address              : IN     STD_LOGIC_VECTOR( 7 DOWNTO 0 );
            write_data           : IN     STD_LOGIC_VECTOR( 31 DOWNTO 0 );
            MemRead, Memwrite     : IN     STD_LOGIC;
            clock, reset         : IN     STD_LOGIC );
END dmemory;

ARCHITECTURE behavior OF dmemory IS
SIGNAL write_clock : STD_LOGIC;
BEGIN
```

```
    data_memory: altsyncram
    GENERIC MAP (
        operation_mode => "SINGLE_PORT",
        width_a => 32,
        widthad_a => 8,
        lpm_type => "altsyncram",
        outdata_reg_a => "UNREGISTERED",
                            -- Reads in mif file for initial data memory values
        init_file => "dmemory.mif",
        intended_device_family => "Cyclone"lpm_widthad    =>      8
    )
    PORT MAP (
        wren_a => memwrite,
        clock0 => write_clock,
        address_a => address,
        data_a => write_data,
        q_a => read_data        );
                            -- Load memory address & data register with write clock
        write_clock <= NOT clock;
END behavior;
```

MIPS data memory is initialized to the value specified in the file dmemory.mif shown in Figure 14.9. Note that the address displayed in the dmemory.mif file is a word address and not a byte address. Two values, 0x55555555 and 0xAAAAAAAA, at byte address 0 and 4 are used for memory data in the short test program. The remaining locations are all initialized to zero.

```
    -- MIPS Data Memory Initialization File
    Depth = 256;
    Width = 32;
    Content
    Begin
                    -- default value for memory
        [00..FF] : 00000000;
                        -- initial values for test program
        00 : 55555555;
        01 : AAAAAAAA;
    End;
```

Figure 14.9 MIPS Data Memory Initialization File, dmemory.mif.

14.9 Simulation of the MIPS Design

The top-level file MIPS.VHD is compiled and used for simulation of the MIPS. It uses VHDL component instantiations to connect the five submodules. The values of major busses and important control signals are output at the top level for use in simulations. A reset is required to start the simulation with PC = 0. A clock with a period of approximately 200ns is required for the simulation.

Memory is initialized only at the start of the simulation. A reset does not re-initialize memory.

The execution of a short test program can be seen in the MIPS simulation output shown in Figure 14.10. The program loads two registers from memory with the LW instructions, adds the registers with an ADD, and stores the sum with SW. Next, the program does not take a BEQ conditional branch with a false branch condition. Last, the program loops back to the start of the program at PC = 000 with another BEQ conditional branch with a true branch condition.

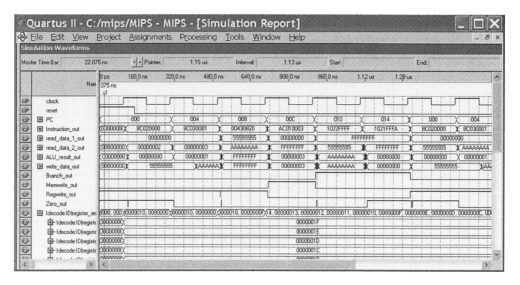

Figure 14.10 Simulation of MIPS test program.

14.10 MIPS Hardware Implementation on the UP 3 Board

A special version of the top level of the MIPS, VIDEO_MIPS.VHD, is identical to MIPS.VHD except that it also contains a VGA video output display driver. This driver displays the hexadecimal value of major busses in the MIPS processor on a monitor. The video character generation technique used is discussed in Chapter 9. It also displays the PC on the UP3's LCD displays and uses the pushbuttons for the clock and reset inputs. This top-level module should be used instead of MIPS.VHD after the design has been debugged in simulations. The final design with video output is then downloaded to the FPGA chip on the UP 3 board. The addition of the VGA video driver slows down the compile and simulation step, so it is faster to not add the video output while running initial simulations to debug a new design. The video driver uses two M4K RAM memory blocks for format and character font data.

After simulation with MIPS.VHD, recompile using VIDEO_MIPS.VHD and download the design to the UP 3 board for hardware verification. Attach a VGA monitor to the UP 3's VGA connector. Any changes or additions made to top level signal names in MIPS.VHD and other modules will need to also be cut and pasted to VIDEO_MIPS.VHD.

MIPS	COMPUTER
PC	00000008
INST	00430820
REG1	00000055
REG2	000000AA
ALU	000000FF
W.B.	000000FF
BRAN	0
ZERO	0
MEMR	0
MEMW	0
CLK	↓
RST	↓

Figure 14.11 MIPS with Video Output generated by UP 3 Board.

14.11 For Additional Information

The MIPS processor design and pipelining are described in the widely-used Patterson and Hennessy textbook, *Computer Organization and Design The Hardware/Software Interface*, Third Edition, Morgan Kaufman Publishers, 2005. The MIPS instructions are described in Chapter 2 and Appendix A of this text. The hardware design of the MIPS, used as the basis for this model, is described in Chapters 5 and 6 of the Patterson and Hennessy text.

SPIM, a free MIPS R2000 assembly language assembler and PC-based simulator developed by James Larus, is available free from http://www.cs.wisc.edu/~larus/spim.html . The reference manual for the SPIM simulator contains additional explanations of all of the MIPS instructions.

The MIPS instruction set and assembly language programming is also described in J. Waldron, *Introduction to RISC Assembly Language Programming*, Addison Wesley, 1999, and Kane and Heinrich, *MIPS RISC Architecture*, Prentice Hall, 1992.

A SMALLER VERSION OF THE MIPS PROCESSOR FOR THE UP 2 CAN BE FOUND IN THE CD-ROM \UP2 SUBDIRECTORY.

14.12 Laboratory Exercises

1. Use VHDL to synthesize the MIPS single clock cycle design in the file TOP_SPIM.VHD. After synthesis and simulation perform the following steps:

    Display and print the timing diagram from the simulation. Verify that the information on the timing diagram shows that the hardware is functioning correctly. Examine the test program in IFETCH.VHD. Look at the program counter, the instruction bus, the register file and ALU outputs, and control signals on the timing diagram and carefully follow the execution of each instruction in the test program. Label the important values for each instruction on the timing diagram and attach a short write-up explaining in detail what the timing diagram shows relative to each instruction's execution and correct operation.

    Return to the simulator and run the simulation again. Examine the ALU output in the timing diagram window. Zoom in on the ALU output during execution of the add instruction and see what happens when it changes values. Explain exactly what is happening at this point. Hint: Real hardware has timing delays.

2. Recompile the MIPS model using the VIDEO_MIPS.VHD file, which generates video output. Download the design to the UP 3 board. Attach a VGA monitor to the UP 3 board. Single step through the program using the pushbuttons.

3. Write a MIPS test program for the AND, OR, and SUB instructions and run it on the VHDL MIPS simulation. These are all R-format instructions just like the ADD instruction. Modifications to the memory initialization files, program.mif and dmemory.mif, (i.e. only if you use data from memory in the test program) will be required. Registers have been preloaded with the register number to make it easy to run short test programs.

4. Add and test the JMP instruction. The JMP or jump instruction is not PC-relative like the branch instructions. The J-format JMP instruction loads the PC with the low 26 bits of the instruction. Modifications to the existing VHDL MIPS model will be required. For a suggested change, see the hardware modifications on page 317 of *Computer Organization and Design The Hardware/Software Interface*.

5. Add and test the BNE, branch if not equal, instruction. Modifications to the existing VHDL MIPS model will be required. Hint: Follow the implementation details of the existing BEQ, branch if equal, instruction and a change to add BNE should be obvious. Both BEQ and BNE must function correctly in a simulation. Be sure to test both the branch and no branch cases.

6. Add and test the I-format ADDIU, add immediate unsigned, instruction. Modifications to the existing VHDL MIPS model will be required.

7. Add and test the R-format SLT, set if less than, instruction. As an example SLT $1, $2, $3 performs the operation, If $2<$3 Then $1 = 1 Else $1 = 0. SLT is used before BEQ or BNE to implement the other branch conditions such as less than or greater.

8. Pipeline the MIPS VHDL simulation. Test your VHDL model by running a simulation of the example program shown in Figure 6.21 using the pipeline hardware shown in Figure 6.27 in *Computer Organization and Design The Hardware/Software Interface*. To minimize changes, pipeline registers must be placed in the VHDL module that generates the input to the pipeline. As an example, all of the pipeline registers that store control signals must be placed in the control module. Synthesize and check the control module first, since it is simple to see if it works correctly when you add the pipeline flip-flops. Use the following notation which minimizes changes to create the new pipeline register signals, add a "D_" in front of the signal name to indicate it is the input to a D flip-flop used in a pipeline register. Signals that go through two D flip-flops would be "DD_" and three would be "DDD_". As an example, *instruction* would be the registered version of the signal, D_*instruction*.

Add pipeline registers to the existing modules that generate the inputs to the pipeline registers shown in the text. This will prevent adding more modules and will not require extensive changes to the MIP.VHD module. Add signal and process statements to model the pipeline modules – see the PC in the ifetch.vhd module for an example of how this can work. A few muxes may have to be moved to different modules.

The control module should contain all of the control pipeline registers – 1, 2, or 3 stages of pipeline registers for control signals. Some control signals must be reset to zero, so use a D flip-flop with a synchronous reset for these pipeline registers. This generates a flip-flop with a Clear input that will be tied to Reset. Critical pipeline registers with control signals such as regwrite or memwrite should be cleared at reset so that the pipeline starts up correctly. The MIPS instruction ADD $0, $0, $0 is all zeros and does not modify any values in registers or memory. It is used to initialize the IF/ID pipeline at reset. Pipeline registers for instruction and data memory outputs can also be added by modifying options in the Altsyncram megafunction.

The data memory clocking scheme might also change with pipelining. In Dmemory.vhd, the data memory address and data inputs are already pipelined inside the altsyncram function used for data memory (this is why it has a clock input). You will need to take this into account when you pipeline your design. High speed memory writes almost always require a clock and the design in the textbook skips over this point – since they do not have their design running on real hardware. As an example, in the Quartus software you can't even have altsyncram memory without a clock!

Currently in the original single cycle design, data memory uses NOT CLOCK as the clock input so that there is time to get both the correct ALU result loaded into the memories internal address and data pipeline registers (first half of clock cycle) and write to memory (second half of clock cycle).

Once you pipeline the model, you will probably want to have your data memory clock input use CLOCK instead of NOT CLOCK for the fastest clock cycle time. With NOT CLOCK you would be loading the ALU Result into the pipeline register in the middle of the clock cycle (not the end) – so it would slow down the clock cycle time on real hardware.

Since there is already a pipeline register in the data memory inputs, don't add another one in the address or data input paths to data memory, if you switch NOT CLOCK to CLOCK. You will still need to delay the ALU result two clocks (with two pipeline registers) for the register file write back operation.

Sections 6.2 and 6.3 of Computer Organization and Design The Hardware/Software Interface contain additional background information on pipelining.

9. Once the MIPS is pipelined as in problem 8, data hazards can occur between the five instructions present in the pipeline. As an example consider the following program:

        Sub $2,$1,$3
        Add $4,$2,$5

The subtract instruction stores a result in register 2 and the following add instruction uses register 2 as a source operand. The new value of register 2 is written into the register file by SUB $2,$1,$3 in the write-back stage after the old value of register 2 was read out by ADD $4,$2,$5 in the decode stage. This problem is fixed by adding two forwarding muxes to each ALU input in the execute stage. In addition to the existing values feeding in the two ALU inputs, the forwarding multiplexers can also select the last ALU result or the last value in the data memory stage. These muxes are controlled by comparing the rd, rt, and rs register address fields of instructions in the decode, execute, or data memory stages. Instruction rd fields will need to be added to the pipelines in the execute, data memory, and write-back stages for the forwarding compare operations. Since register 0 is always zero, do not forward register 0 values.

Add forwarding control to the pipelined model developed in problem 8. Test your VHDL model by running a simulation of the example program shown in Figure 6.29 using the hardware shown in Figures 6.32 of Computer Organization and Design The Hardware/Software Interface by Patterson and Hennessy.

Two forwarding multiplexers must also be added to the Idecode module so that a register file write and read to the same register work correctly in one clock cycle. If the register file write address equals one of the two read addresses, the register file write data value should be forwarded out the appropriate read data port instead of the normal register file read   data value. Section 6.4 of *Computer Organization and Design The Hardware/Software Interface* contains additional background information on forwarding.

10. Add LW/SW forwarding to the pipelined model. This will allow an LW to be followed by an SW that uses the same register. It is possible since the MEM/WB register contains the load instruction register write data in time for use in the MEM stage of the store. Write a test program and verify correct operation in a simulation.

11. When a branch is taken, several of the instructions that follow a branch have already been loaded into the pipeline. A process called flushing is used to prevent the execution of these instructions. Several of the pipeline registers are cleared so that these instructions do not store any values to registers or memory or cause a forwarding operation. Add branch flushing to the pipelined MIPS VHDL model as shown in Figures 6.38 of the Computer Organization and Design The Hardware/Software Interface by Patterson and Hennessy. Note that two new forwarding multiplexers at the register file outputs (not shown in the Figure, currently at ALU inputs) are needed to eliminate the new Branch data hazards that appear when the branch comparator is moved into to the decode stage. Section 6.6 *of Computer Organization and Design The Hardware/Software Interface* contains additional background information on branch hazards.

12. Use the timing analyzer to determine the maximum clock rate for the pipelined MIPS implementation, verify correct operation at this clock rate in a simulation, and compare the clock rate to the original non-pipelined MIPS implementation.

13. Redesign the pipelined MIPS VHDL model so that branch instructions have 1 delay slot as seen in Figure 6.40 (i.e. one instruction after the branch is executed even when the branch is taken). Rewrite the VHDL model of the MIPS and test the program from the problem 10 assuming 1 delay slot. Move instructions around and add nops if needed.

14. Add the overflow exception hardware suggested at the end of Chapter 6 in Figure 6.42 of *Computer Organization and Design The Hardware/Software Interface* by Patterson and Hennessy. Add an overflow circuit that produces the exception with a test program containing an ADD instruction that overflows. Display the PC and the trap address in your simulation. For test and simulation purposes make the exception address 40 instead of 40000040. Section 6.8 of *Computer Organization and Design The Hardware/Software Interface* contains additional background information on exceptions.

15. Investigate using two Altsyncram memory blocks to implement the register file in IDECODE. A single Altsyncram block can be configured to do a read and write in one clock cycle (dual port). To perform two reads, use two Altsyncrams that contain the same data (i.e. always write to both blocks).

16. Add the required instructions to the model to run the MIPS bubble sort program from Chapter 3 of *Computer Organization and Design The Hardware/Software Interface.*

After verifying correct operation with a simulation, download the design to the UP 3 board and trace execution of the program using the video output. Sort this four element array 4, 3, 5, 1.

17. Add programmed keyboard input and video output to the sort program from the previous problem using the keyboard, vga_sync, and char_rom UP3cores. Use a dedicated memory location to interface to I/O devices. Appendix A.36-38 of *Computer Organization and Design The Hardware/Software Interface* contains an explanation of MIPS memory-mapped terminal I/O.

18. The MIPS VHDL model was designed to be easy to understand. Investigate various techniques to increase the clock rate such as using two dual-port memory blocks for the register file, moving hardware to different pipeline stages to even out delays, or changing the way memory is clocked. Additional fitter effort settings may also help. Use the timing analysis tools to evaluate design changes.

19. Develop a VHDL synthesis model for another RISC processor's instruction set. Possible choices include the Nios, Microblaze, Picoblaze, PowerPC, ARM, SUN SPARC, the DEC ALPHA, and the HP PARISC. CD-ROM Appendix D of Computer Organization and Design The Hardware/Software Interface contains information on several RISC processors. Earlier hardware implementations of the commercial RISC processors designed before they became superscalar are more likely to fit on a UP 3.

CHAPTER 15

Introducing System-on-a-Programmable-Chip

A small SOPC-based aircraft autopilot system that contains an FPGA with a Nios processor core, a DSP processor, and memory is seen above. The bottom sensor board contains a GPS receiver, an A/D converter, MEMS gyros and accelerometers for all three axes, an airspeed sensor, and an altitude sensor. Photograph ©2004 courtesy of Henrik Christophersen, Georgia Institute of Technology Unmanned Aerial Research Facility.

15 Introducing System-on-a-Programmable-Chip[1]

A new technology has emerged that enables designers to utilize a large FPGA that contains both memory and logic elements along with an intellectual property (IP) processor core to implement a computer and custom hardware for system-on-a-chip (SOC) applications. This new approach has been termed system-on-a-programmable-chip (SOPC).

15.1 Processor Cores

Processor cores can be classified as either "hard" or "soft." This designation refers to the flexibility/configurability of the core. Hard cores are less configurable; however, they tend to have higher performance characteristics than soft cores.

Hard processor cores use an embedded processor core (in dedicated silicon) in addition to the FPGA's normal logic elements. Hard processor cores added to an FPGA are a hybrid approach, offering performance trade-offs that fall somewhere between a traditional ASIC and an FPGA; they are available from several manufacturers with a number of different processor flavors. For example, Altera offers an ARM processor core embedded in its APEX 20KE family of FPGAs that is marketed as an Excalibur™ device. Xilinx's Virtex-II Pro family of FPGAs include up to four PowerPC processor cores on-chip. Cypress Semiconductor also offers a variation of the SOPC system. Cypress's Programmable-System-on-a-Chip (PSoC™) is formed on an M8C processor core with configurable logic blocks designed to implement the peripheral interfaces, which include analog-to-digital converters, digital-to-analog converters, timers, counters, and UARTs.[2]

Soft cores, such as Altera's Nios II and Xilinx's MicroBlaze processors, use existing programmable logic elements from the FPGA to implement the processor logic. As seen in Table 15.1, soft-core processors can be very feature-rich and flexible, often allowing the designer to specify the memory width, the ALU functionality, number and types of peripherals, and memory address space parameters at compile time. However, such flexibility comes at a cost. Soft cores have slower clock rates and use more power than an equivalent hard processor core.

With current pricing on large FPGAs, the addition of a soft processor core costs as little as thirty-five cents based on the logic elements it requires. The remainder of the FPGA's logic elements can be used to build application-specific system hardware. Traditional system-on-a-chip devices (ASICs and custom VLSI ICs) still offer higher performance, but they also have large

---

[1] Portions reprinted, with permission, from T. S. Hall and J. O. Hamblen, "System-on-a-Programmable-Chip Development Platforms in the Classroom," *IEEE Transactions on Education*, vol. 47, no. 4, pp. 502-507, Nov. 2004. © 2004 IEEE.

[2] D. Seguine, "Just add sensor - integrating analog and digital signal conditioning in a programmable system on chip," *Proceedings of IEEE Sensors*, vol. 1, pp. 665–668, 2002.
M. Mar, B. Sullam, and E. Blom, "An architecture for a configurable mixed-signal device," *IEEE J. Solid-State Circuits*, vol. 38, pp. 565–568, Mar. 2003.

development costs and longer turnaround times.[3] For projects requiring a hardware implementation, the FPGA-based SOPC approach is easier, faster, and more economical in low to medium quantity production.

Table 15.1 Features of Commercial Soft Processor Cores for FPGAs

Feature	Nios II 5.0	MicroBlaze 4.0
Datapath	32 bits	32 bits
Pipeline Stages	1-6	3
Frequency	Up to 200 MHz[4]	Up to 200 MHz[4]
Gate Count	26,000 – 72,000	30,000 – 60,000
Register File	32 general purpose & 6 special purpose	32 general purpose & 32 special purpose
Instruction Word	32 bits	32 bits
Instruction Cache	Optional	Optional
Hardware Multiply & Divide	Optional	Optional
Hardware Floating Point	Third Party	Optional

Typically, additional software tools are provided along with each processor core to support SOPC development. A special CAD tool specific to each soft processor core is used to configure processor options, which can include register file size, hardware multiply and divide, floating point hardware, interrupts, and I/O hardware. This tool outputs an HDL synthesis model of the processor core in VHDL or Verilog. In addition to the processor, other system logic is added and the resulting design is synthesized using a standard FPGA synthesis CAD tool. The embedded application program (software) for the processor is typically written in C or C++ and compiled using a customized compiler provided with the processor core tools.

15.2 SOPC Design Flow

The traditional flow of commercial CAD tools typically follows a path from hardware description language (HDL) or schematic design entry through synthesis and place and route tools to the programming of the FPGA. FPGA manufacturers provide CAD tools such as Altera's Quartus II and Xilinx's ISE software, which step the designer through this process. As shown in Fig. 15.1, the addition of a processor core and the tools associated with it are a superset of the traditional tools. The standard synthesis, place and route, and programming

---

[3] H. Chang et al., *Surviving the SOC Revolution a Guide to Platform-Based Design*. Norwell, MA: Kluwer, 1999.

[4] This speed is not achievable on all devices for either processor core. Some FPGAs may limit the maximum frequency to as low as 50 MHz.

functionality is still needed, and in the case of both Altera and Xilinx, the same CAD tools (Quartus II or ISE) are used to implement these blocks.

Processor Core Configuration Tools

Today, a number of pre-defined processor cores are available from various sources. GPL-licensed public processor cores can be found on the web (i.e., www.opencores.org and www.leox.org), while companies such as Altera (Nios II processor), Xilinx (MicroBlaze processor), and Tensilica (Xtensa processor) provide their processors and/or development tools for a fee.

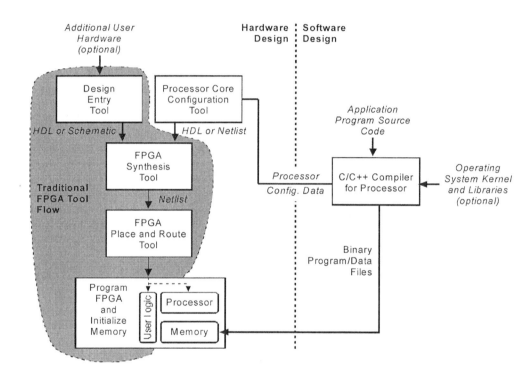

Figure 15.1 The CAD tool flow for SOPC design is comprised of the traditional design process for FPGA-based systems with the addition of the Processor Core Configuration Tool and software design tools. In this figure, the program and data memory is assumed to be on-chip for simplicity.

Processor cores provided by FPGA manufacturers are typically manually optimized for the specific FPGA family being used, and as such, are more efficiently implemented on the FPGA than a student-designed core (especially given the time and resource constraints of most class projects). The simple computer and MIPS processor cores developed earlier in this book were designed to be easy for students to understand and were not optimized for any particular FPGA. Additionally, FPGA companies provide extensive support tools to ease the customization and use of their cores, including high-level compilers targeted at the custom cores.

In the case of Altera and Xilinx, the Processor Core Configuration Tool block shown in Fig. 15.1 is realized in a user-friendly GUI interface that allows the designer to customize the processor for a particular project. The configurable parameters can include the datapath width, memory, address space, and peripherals (including arbitrarily defined general-purpose I/O, UARTs, Ethernet controllers, memory controllers, etc.). Once the processor parameters are specified in the GUI interface, the processor core is generated in the form of an HDL file (in Altera) or a netlist file (in Xilinx). This file can then be included within a traditional HDL or schematic design using the standard CAD tools. Specific pin assignments and additional user logic can be included at this point like any other FPGA design. Next, the full hardware design (processor core and any additional user logic) is compiled (synthesis, place and route, etc.), and the FPGA can be programmed with the resulting file using the standard tools. The hardware design is complete, and the FPGA logic has been determined.

High-level Compiler for Processor Core

As shown on the right side of Fig. 15.1, the next step is to write and compile the software that will be executed on the soft processor core. When the Processor Core Configuration Tool generates the HDL or netlist files, it also creates a number of library files and their associated C header files that are customized for the specific processor core generated. A C/C++ compiler targeted at this processor is also provided. The designer can then program stand alone programs to run on the processor. Optionally, the designer can compile code for an operating system targeted for the processor core. Several operating systems for the Nios II are available from third-party vendors along with the community supported open source eCos (www.niosforum.com).

15.3 Initializing Memory

Once a program/data binary file has been generated, it must be loaded into the processor's program and data memories. This loading can be done several ways depending on the memory configuration of the processor at hand.

On-chip Memory

If the application program is small and can fit into the memory blocks available on the FPGA, then the program can be initialized in the memory when the hardware configuration is programmed. This initialization is done through the standard FPGA tools, such as Altera's Quartus II software or Xilinx's ISE software. However, on-chip memory is typically very limited, and this solution is not usually an option.

Bootloader

In a prototyping environment, the application program will most likely be modified a number of times before the final program is complete. In this case, the ability to download the application code from a PC to the memory on an

FPGA board must be provided. This functionality, typically called a "bootloader" or "boot monitor," can be implemented in either software or hardware.

A software bootloader is comprised of code that is loaded into an on-chip memory and starts running on power up. This program is small enough (1-2 KB) to fit in most on-chip memories, and its primary function is to receive a program binary file over the serial port (or other interface), load it into external memory, and then start the new code executing. In this way, a new program can be stored into external memory (SRAM, SDRAM, Flash memory, etc.) by downloading it over the serial or JTAG port (or other interface) on the fly without having to reload the FPGA's hardware configuration. Xilinx provides a boot monitor for their MicroBlaze soft-core processor that includes the ability to download a program binary over the serial port (or other interface), store it in memory, and start the code executing. They also provide a more enhanced version called XMDstub that adds debugging capabilities. Altera's legacy Nios processors included a bootloader called GERMS. The Nios II processor still includes limited support for the GERMS monitor; however, a hardware bootloader is the preferred solution in Nios II.

A hardware bootloader provides functionality very similar to a software bootloader; however, it is implemented in dedicated logic within the processor core. Typically, the processor will be paused or stalled upon power up and the hardware bootloader will have direct access to memory or the memory registers in the processor's datapath. The bootloader hardware can start and stop the processor and can control the downloading of a program over the JTAG or serial interface to the desired memory locations. Altera's hardware bootloader is a part of the JTAG debug module, which resides within the Nios II processor. This logic uses the JTAG interface with the PC to receive the execution code, and it then writes the code to the appropriate memory. Finally, the bootloader hardware overwrites the processor's program counter with the start address of the code just downloaded and releases the pause bit to allow the processor to begin executing the downloaded code.

External Non-volatile Storage

The application program code can be stored on an external EEPROM, Flash memory, or other form of non-volatile memory. The application program can either be pre-programmed in the external memory module (for a production run) or a bootloader program can be used to store the application program in non-volatile storage. For low-speed applications, the code can be executed directly from the external memory. However, if high-speed functionality is required, then a designer could use three memories as shown in Fig. 15.2. In this scheme, the on-chip memory is initialized with a bootloader, which handles the movement of the application program between the memories. (On-chip memory is replaced with a hardware bootloader on some systems including the Nios II processor.)

The fast, volatile memory (i.e., SDRAM) is used to store the application program during execution. Finally, the slower, non-volatile memory (i.e., Flash or EEPROM) is used for the permanent storage of the application program. The bootloader program can be modified to initialize the system, retrieve a program from non-volatile memory, store it in the faster, volatile memory, and then start it executing from the faster memory. This scheme provides the advantages of permanent storage, fast execution, and the ability to modify the application program when needed. Of course, it comes at the expense of having additional memory.

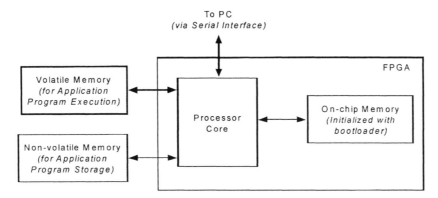

Figure 15.2 This arrangement of on-chip and external memories provides flexibility and performance to an SOPC system.

15.4 SOPC Design versus Traditional Design Modalities

The traditional design modalities are ASIC and fixed-processor design. SOPC design has advantages and disadvantages to both of these alternatives as highlighted in Table 15.2. The strengths of SOPC design are a reconfigurable, flexible nature and the short development cycle. However, the trade offs include lower maximum performance, higher unit costs in production, and relatively high power consumption.

The benefit of having a flexible hardware infrastructure can not be overestimated. In many new designs, features and specifications are modified throughout the design cycle. For example, marketing may detect a shift in demand requiring additional features (e.g., demand drops for cell phones without cameras), a protocol or specification is updated (e.g., USB 2.0 is introduced), or the customer requests an additional feature. In traditional design modalities (including ASIC and fixed-processor designs), these changes can dramatically effect the ASIC design, processor selection, and/or printed circuit board design. Since the hardware architecture is often settled upon early in the design cycle, making changes to the hardware design later in the cycle will typically result in delaying a product's release and increasing its cost.

Flexible infrastructure can also be beneficial in extending the life (and thus reducing the cost) of a product's hardware. With flexible, reconfigurable logic, often a single printed circuit board can be designed that can be used in multiple product lines and in multiple generations/versions of a single product. Using reconfigurable logic as the heart of a design, allows it to be reprogrammed to implement a wide range of systems and designs. Extending the life of a board design even one generation can result in significant savings and can largely offset the increased per-unit expense of reconfigurable devices.

Table 15.2 Comparing SOPC, ASIC, and Fixed-Processor Design Modalities

Feature	SOPC	ASIC	Fixed-Processor
S/W Flexibility	●	●	●
H/W Flexibility	●	○	○
Reconfigurability	●	○	○
Development Time/Cost	●	○	●
Peripheral Equipment Costs	●	●	○
Performance	◐	●	●
Production Cost	◐	●[5]	●
Power Efficiency	○	●	●

Legend: ● – Good; ◐ – Moderate; ○ – Poor

The SOPC approach is ideal for student projects. SOPC boards can be used and reused to support an extremely wide range of student projects at a very low cost. ASIC development times are too long and mask setup fees are too high to be considered for general student projects. A fixed-processor option will often require additional hardware and perhaps even a new printed circuit board (PCB) design for each application. Given the complexity of today's multilayer surface mount PCB designs, it is highly unlikely that students would have sufficient time and funds to develop a new printed circuit board for a design project.

15.5 An Example SOPC Design

The SOPC-based autopilot system seen in the photograph on the first page of this chapter and the sensor board that mounts below it (described earlier in Section 13.5) makes an interesting case study in SOPC design. The autopilot system continuously reads in sensor data that indicates attitude, altitude, speed, and location. It then uses this data to solve the control system motion equations for the aircraft and outputs updated signals to control the aircraft.

The flexibility of SOPC design allows the use of FPGA's logic elements to interface to a wide range of sensors without the need for additional I/O support chips that would be needed if a more traditional fixed-processor option was

---

[5] In very large quantities.

used. This results in an extremely small and low weight PCB design. An ASIC could be used instead of the FPGA, but the small production quantities needed for this system do not justify the greatly increased development time and cost needed for an ASIC.

Different types of aircraft also require markedly different I/O standards for the control outputs. Some aircraft controls use serial interfaces, while others use PWM or even parallel I/O. Here again, the flexibility of using the FPGA's logic elements to implement the I/O interface is of great benefit. By varying the logic in the interface peripherals, the same programmable processor core and PCB board can be used to support a wide range of aircraft without any hardware changes to the PCB.

The autopilot system requires intensive floating-point calculations to solve the complex control system equations for the aircraft. While it would be possible to perform floating-point calculations using a larger FPGA, the decision was made to use a fixed-processor DSP chip for the intensive floating-point calculations. By offloading the algorithmic computations to a fixed processor, the Nios II processor is primarily acting as an intelligent I/O processor for the system. This partitioning of the system between a fixed-processor DSP and soft-core processor results in higher computational performance than using just an FPGA (with floating-point hardware logic) and higher interface flexibility than using just a fixed processor in the system. However, new generations of FPGAs with DSP features such as hardware multipliers and floating-point IP cores are currently changing this set of design tradeoffs.

15.6 Hardware/Software Design Alternatives

The SOPC-based approach offers new design space alternatives. It is possible to explore design options that use software, dedicated hardware, or a mixture of both. Hardware solutions offer faster computations, but offer less flexibility and may require a larger FPGA. Implementation of solutions using software is easier to design for more complicated algorithms.

It is also possible to consider a combination of both approaches. Some processor cores allow the user to add custom instructions. If an application program requires the same calculation repeatedly in loops, adding a custom instruction using extra hardware to accelerate the inner loop code can greatly speed up the application.

15.7 For additional information

This chapter has provided a brief overview of SOPC systems and designs. More information about SOPC systems can be found from manufacturers such as Altera, Xilinx, Cypress Semiconductor, Stretch Incorporated, and Tensilica. SOPC systems are an active area of research. Publications of interest include the following:

- T. S. Hall and J. O. Hamblen, "System-on-a-Programmable-Chip Development Platforms in the Classroom," *IEEE Transactions on Education*, vol. 47, no. 4, pp. 502-507, Nov. 2004.

- C. Snyder, "FPGA processor cores get serious," in *Cahners Microprocessor Report*, http://www.MPRonline.com/, Sept. 2000.

- D. Seguine, "Just add sensor - integrating analog and digital signal conditioning in a programmable system on chip," *Proceedings of IEEE Sensors*, vol. 1, pp. 665–668, 2002.

- M. Mar, B. Sullam, and E. Blom, "An architecture for a configurable mixed-signal device," *IEEE J. Solid-State Circuits*, vol. 38, pp. 565–568, Mar. 2003.

- H. Chang and et. al., *Surviving the SOC Revolution A Guide to Platform-Based Design*. Kluwer Academic Publishers, 1999.

- J. Fisher, P. Faraboschi, and C. Young, *Embedded Computing : A VLIW Approach to Architecture, Compilers and Tools,* Morgan Kaufmann, 2004.

- A. Jerraya, H. Tenhunen, and W. Wolf, "Multiprocessor Systems-on-Chips," *IEEE Computer*, vol. 38, no. 7, pp. 36-41, July 2005.

- S. Liebson and J. Kim, "Configurable Processors: A New Era in Chip Design," *IEEE Computer*, vol. 38, no. 7, pp.51-59, July 2005.

15.8 Laboratory Exercises

1. Compare the instruction formats and the instruction set of the Nios II processor to the MIPS processor from Chapter 14. Information on the Nios II instruction set architecture is available at Altera's website (www.altera.com) in the Nios II Processor Reference Handbook.

2. A system needs a processor to run a control program, but the application also needs to compute FFTs at a somewhat high data rate. FFTs require a large number of multiply and add operations on an array in nested loops. What SOPC hardware/software design tradeoffs would you need to consider? Justify your answer.

3. List several types of products that could likely take advantage of the SOPC design approach. Explain your reasoning.

4. Compare the memory read access time of the UP 3's Flash and SRAM memory chips. Information can be found in each chip's datasheet. If the processor did not have an instruction cache, how much faster could a program read instructions from SRAM?

5. You are asked to specify the memory types and sizes for an SOPC design that will execute a program with a 60 KB length or footprint. During execution, the program requires 16 KB of data memory for the stack and heap. If the SOPC hardware mandates a single memory (for program and data memory), select the type and size of memory. Perform an online search to find a manufacturer and model number for the memory you

selected. You may have to modify your initial selection based on availability and cost of various memories. Justify your selection considering cost, specification, performance, and availability. Don't forget that you need non-volatile memory to boot the system.

6. Given the SOPC system outlined in Problem 5, select the type and size of memory needed for this system when program and data memory are separate. Justify your selection considering cost, specification, performance, and availability. Compare the single memory option from Problem 5 with the dual-memory option from this problem. Which memory configuration is preferable? Justify your answer.

7. There are a number of different non-volatile memory technologies available to SOPC designers. For a system with a 256 KB code footprint, compare the cost, reprogrammability, configuration time, access time (reading only), and longevity for PROM, EEPROM, and Flash memories.

CHAPTER 16

Tutorial III: Nios II Processor Software Development

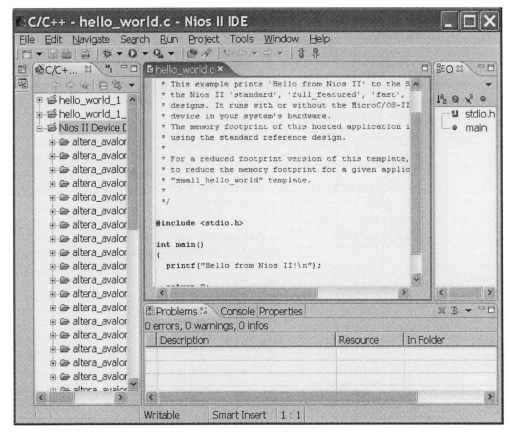

The Nios II IDE tool compiles C/C++ code for the Nios II processor and provides an integrated software development environment for Nios II systems.

16 Tutorial III: Nios II Processor Software Development

Designing systems with embeddeded processors requires both hardware and software design elements. A collection of CAD tools developed by Altera enable you to design both the hardware and software for a fully functional, customizable, soft-core processor called Nios II. This tutorial steps you through the software development for a Nios II processor executing on the UP 3 board. A Nios II processor reference design targeted for the UP 3 board is used here. To design a custom Nios II processor refer to Tutorial IV (in the following chapter), which introduces the hardware design tools for the Nios II processor.

Upon completion of this tutorial, you will be able to:

- Navigate Altera's Nios II Integrated Development Environment (IDE),

- Write a C-language software program that executes on the Nios II reference design,

- Download and execute a software program on the Nios II processor, and

- Test the peripherals and memory components of the Nios II reference design on the UP 3 board.

This tutorial will step you through writing and running two programs for the Nios II processor. First, a simple "Hello World" type of program will be written, compiled, downloaded to the UP 3 board, and run. Next, a test program that uses interrupts, pushbuttons, dipswitches, LEDs, the LCD display, SRAM, Flash memory, and SDRAM will be written that can be used to test the major peripherals on the UP 3 board.

16.1 Install the UP 3 board files

Locate the **booksoft\chap17** directory on the CD-ROM that came with the book. In this directory, there are two subdirectories called **up3_tristate_lcd** and **up3_tristate_sram**. Copy both of these subdirectories to the *quartus_install_dir***sopc_builder\components** directory on your local hard drive.

16.2 Starting a Nios II Software Project

The Nios II Integrated Development Environment (IDE) is a standalone program that works in conjunction with Quartus II. To design software in the IDE, Quartus II does not have to be installed on your system; however, you will need a valid Quartus II project with a Nios II processor in it to use the IDE. A Nios II reference design for the UP 3 board is included on the CD-ROM that came with this book. This hardware design will be used for the remainder of this tutorial. Copy the design files from **booksoft\chap16** on the CD-ROM to a working directory on your hard drive. (If you are using a UP 3 board with the 1C12 FPGA on it, then copy the files from the **booksoft\chap16\1C12** folder on the CD-ROM.) The software design files will be stored in a subdirectory of this project directory.

Open the Nios II IDE software. For the default installation, the software icon can be found under **Start** ➪ **All Programs** ➪ **Altera** ➪ **Nios II Development Kit** ➪ **Nios II IDE**.

You should be prompted to **Select a Workspace**. If the dialog box in Figure 16.1 does not appear, then select **File** ➪ **Switch Workspace…**. The workspace is a cache for information about all projects associated with a given Nios II processor design. Enter the full pathname of the Quartus II project directory you created above (the directory to which *CD-ROM*\booksoft\chap16 was copied) followed by the subdirectory **\software** as shown in Figure 16.1. Click **OK** to select the default location and continue.

Figure 16.1 Setting the Nios II IDE workspace to the Nios II reference design software directory.

To create a new project, select **File** ➪ **New** ➪ **Project…**. The **New Project** wizard will begin. On the first dialog box, select **C/C++ Application** and click **Next** to continue.

In the next dialog box, fill in the requested information as shown in Figure 16.2. The name of the project is **rpds_software**, the **SOPC Builder System** should point to the **nios32.ptf** file in your hardware project directory, and the **Project Template** should be set to **Blank Project**. If the workspace was correctly set to your project directory as detailed above, then the default location will be correct and **Use Default Location** should be checked. However, if the workspace is set to some other directory, then unselect **Use Default Location** and enter **c:*your_project_directory*\software\rpds_software** in the **Location** field. Click **Next** to continue.

In the final dialog box, select the option **Create a new system library named: rpds_software_syslib**. Click **Finish** to create and open the project. When the **New Project** dialog box disappears, click on the **Workbench** icon on the **Welcome** page in the main IDE window if it does not come up automatically.

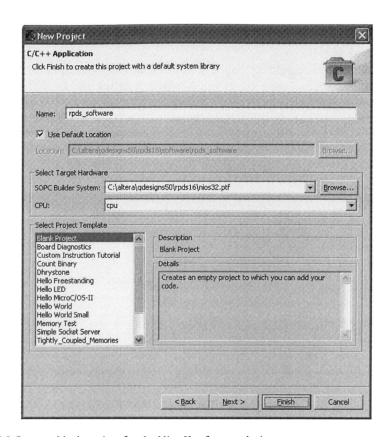

Figure 16.2 Create a blank project for the Nios II reference design.

16.3 The Nios II IDE Software

Take a few minutes to orient yourself to the Nios II IDE software. The middle of the window will display the contents of the source files when you open them. The Outline pane on the right-hand side will provide links to each of the functions that are declared in the open C source file. Clicking on a link will jump the cursor to the start of that function. On the left-hand side, a list of projects for the current workspace is shown in the **C/C++ Projects** pane.

Three projects should appear in the **C/C++ Projects** pane by default (after having created your new blank project): **Nios II Device Drivers**, **rpds_software**, and **rpds_software_syslib**.

- The **Nios II Device Drivers** library contains C and/or assembly-language source files and C header files for each of the components that can be added to a Nios II processor. This library contains all component drivers and not just the drivers for peripherals added to the Nios II reference design. Just because a device appears in this list, does not mean that the Nios II processor you are writing software for contains that peripheral.

- The **rpds_software** library is the location for your software. Since a blank project has been created, no source or header files exist in this library yet.

- The **rpds_software_syslib** library is the container for the top-level system header file (system.h) that contains the names and base addresses of peripherals in the Nios II reference design system for which you are writing software. It also contains a list of device drivers (from the Nios II Device Drivers library) for just the peripherals that are in your Nios II processor. This library will be empty until the system library is generated in the next section.

16.4 Generating the Nios II System Library

Each Nios II system is unique. It has different peripherals, different memory-mapped addresses, different interrupt settings, etc. To accommodate this flexibility, the Nios II IDE creates a system library from your Nios II hardware settings file (e.g., nios32.ptf). The system library defines the names of the peripherals in a given system and maps them to their memory addresses, and it defines several system-critical definitions that are used to make several standard C libraries compatible with your specific Nios II system.

Before the system library can be generated, several settings must be modified. Right click on **rpds_software_syslib** in the **C/C++ Projects** pane and select **Properties** from the drop-down menu. In the dialog box, select **System Library** from the list on the left to view the configuration options for the system library.

Under **System Library Contents**, select the UART device for **stdout**, **stderr**, and **stdin**. The Nios II system allows the stdout, stderr, and stdin data streams to be redirected to a UART interface using a serial cable connected to your PC. This means that the output of *printf* and other standard output functions will be displayed in a console window on your PC since there is no monitor attached to the FPGA directly at the moment. Likewise, the use of *scanf* and other standard input functions will wait for data to be transmitted from the PC to the UART. Any text that you type in the Nios II IDE's console window will be sent via the serial cable and UART to the Nios II processor.

Notice that the various segments of memory can be individually assigned to different memory devices (SRAM, SDRAM, Flash, etc.). For this tutorial, leave all of the memory segments set to **SRAM**. It is also useful to note that this dialog box contains an option to use a **Small C library** for your project. Selecting this option, removes many of the less common functions of the ANSI C standard library such as printf's floating-point number support, scanf, file seek (fseek), and more. Using a small standard library can result in a much smaller amount of memory needed for storing your software. A complete list of standard library functions affected by selecting the **Small C Library** option can be found in the *Nios II Software Developer's Handbook* available on Altera's website. For this tutorial, leave the **Small C Library** option **unchecked** as shown in Figure 16.3 and click **OK** to continue.

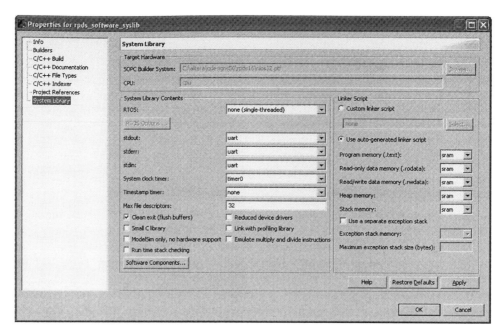

Figure 16.3 These are the system library settings that should be used for this tutorial.

To generate the system library for this Nios II system, right click on **rpds_software_syslib** in the **C/C++ Projects** pane and select **Build Project** from the drop-down menu. Once building has completed, view the files created by the build by double clicking on the **rpds_software_syslib** item in the **C/C++ Projects** pane. Under the **rpds_software_syslib** folder, several folders appear. The **Includes** folder contains links to the device drivers for peripherals in the Nios II reference design processor that you are using. The **Debug ⇨ System Description** folder contains the **system.h** header file that includes definitions for all of the peripherals in this Nios II processor.

16.5 Software Design with Nios II Peripherals

Accessing and communicating with Nios II peripherals can be accomplished in three general methods: direct register access, hardware abstraction layer (HAL) interface, and standard ANSI C library functions. Depending on the complexity of the operation and the specific device being used, a programmer will often use each of the three methods at one point or another. In this tutorial, direct register access will be used to communicate with the LEDs, dipswitches, and LCD display. The HAL interface will be used to communicate with Flash and install an interrupt handler for the pushbuttons, and standard C library conventions will be used to access the SRAM and Timer memory devices. The SRAM device driver distributed by the UP 3 board manufacturer does not currently support standard file I/O (*fread*, *fwrite*, etc.); however, support for these functions may be added in the future.

Below, each type of peripheral access is discussed. As an example, the C code necessary to provide a one second delay using each method is shown in Figures 16.4-6.

```c
#include "system.h"

#include "altera_avalon_timer_regs.h"

int main( void ) {

  IOWR_ALTERA_AVALON_TIMER_PERIODL( TIMER0_BASE,
      (48000000 & 0xFFFF) );

  IOWR_ALTERA_AVALON_TIMER_PERIODH( TIMER0_BASE,
      ((48000000>>16) & 0xFFFF) );

  IOWR_ALTERA_AVALON_TIMER_STATUS( TIMER0_BASE, 0 );

  IOWR_ALTERA_AVALON_TIMER_CONTROL( TIMER0_BASE, 0x4 );

  while( (IORD_ALTERA_AVALON_TIMER_STATUS( TIMER0_BASE ) &
          ALTERA_AVALON_TIMER_STATUS_TO_MSK) == 0 ) {}

}
```

Figure 16.4 This is the C code necessary for providing a one second delay by directly accessing the system timer's registers. The timer peripheral in this system is called *timer0*.

Direct Register Access

Each peripheral's registers can be directly accessed through read and write macros that are defined in each component's device driver header file. This type of communication is the lowest level and while it provides the most flexibility in interfacing with peripherals, it can also be the most tedious. As illustrated in Figure 16.4, interfacing with the timer device can be quite cumbersome, even to provide a relatively straight-forward function such as a one second delay. If you read the actual count to determine elapsed time, you also need to keep in mind how your code will function when the timer count wraps around and starts over.

```
#include <sys/alt_alarm.h>

int main( void ) {
  int first_val, second_val;

  second_val = 0;
  first_val  = alt_nticks();

  while( (second_val - first_val) < 1000000 ) {
    second_val = alt_nticks();
  }
}
```

Figure 16.5 This is the C code necessary for providing a one second delay by using the HAL interface functions.

HAL Interface

A layer of software called a hardware abstraction layer (HAL) has been created that resides between the user code and the peripheral hardware registers. The HAL interface contains a number of very useful functions that allow the user to communicate with peripherals at a higher functional level. For example, the HAL interface provides functions *alt_flash_open_dev*, *alt_read_flash*, *alt_write_flash*, and *alt_ flash_close_dev* for communication with Flash memory. By providing these functions, Flash memory can be accessed by opening the device and reading from it and writing to it without having to create user functions that provide the read and write functionality from low-level peripheral register accesses.

For the timer device, a function called *alt_nticks* provides convenient access to the timer. As illustrated in Figure 16.5, the HAL functions provide a more straight-forward method of creating a one second delay.

```
#include <unistd.h>

int main( void ) {
  usleep( 1000000 );
}
```

Figure 16.6 This is the C code necessary for providing a one second delay by using the standard ANSI C library functions.

Standard Library Functions

Access to most of Nios II's peripherals has been incorporated into the standard ANSI C library functions. Using standard ANSI C libraries such as *stdlib*, *stdio*, *string*, *time*, *malloc*, etc. you can manipulate strings, access memory and memory-like devices through standard file I/O functions, use the timer to add a delay using *usleep* or *wait* functions, and much more. This is the highest level of abstraction provided by the Nios II IDE. Many of these functions use the peripheral-specific HAL functions to provide a single common interface to various types of peripherals. For example, *fopen*, *fread*, *fwrite*, and *fclose* functions from the *stdio* library can be used for memory accesses on some SDRAM, Flash, or SRAM memory devices. The system library functions will use the appropriate HAL function calls for each access depending on the particular memory device being used. To create a one second delay using the timer, a single call to the standard library function *usleep* can be made as shown in Figure 16.6

```
#ifndef _RPDS_SOFTWARE_H_
#define _RPDS_SOFTWARE_H_

#include <stdio.h>
#include <unistd.h>
#include "system.h"
#include "altera_avalon_pio_regs.h"


#endif //_RPDS_SOFTWARE_H_
```

Figure 16.7 This is your first C program's main header file.

16.6 Starting Software Design – main()

Create a C header file by selecting the **rpds_software** item in the **C/C++ Projects** pane. Choose **File** ➪ **New** ➪ **Header File**. When the dialog box appears, enter **rpds_software.h** for the **Header File** and click **Finish** to continue.

Start your program's main header file by adding the *#include* and definition statements shown in Figure 16.7.

The C program that you will now write will print "Hello World" in the Nios II IDE's console window (via the UART and serial cable), and it will blink the LEDs on the UP 3 board.

Create your program's main C source file by selecting the **rpds_software** item in the **C/C++ Projects** pane. Choose **File** ➪ **New** ➪ **File**. When the dialog box appears, enter **rpds_software.c** for the **File name** and click **Finish** to continue.

Start your program by including the **rpds_software.h** header file and typing the code shown in Figure 16.8.

```
#include "rpds_software.h"

int main( void ) {
  unsigned char led_val = 1;

  /* Print message to the Nios II IDE console via UART */
  printf( "Hello World\n" );

  while(1) {
    /* Output a 4-bit value to the LEDs */
    IOWR_ALTERA_AVALON_PIO_DATA( LEDS_BASE, (led_val & 0xF) );

    if( led_val == 8 )
      led_val = 1;
    else
      led_val = led_val << 1;

    /* Wait for 0.5 seconds */
    usleep( 500000 );
  }

  return(0);
}
```

Figure 16.8 This is your first C program's main source file.

16.7 Downloading the Nios II Hardware and Software Projects

To execute your software on a Nios II processor, you must configure the FPGA with the Nios II hardware reference design and then you can download the compiled program code to the processor's program memory.

Connect the ByteBlaster cable and then connect the UP 3's serial cable from the UP 3's serial port to the PC's COM1: (default) or other COM ports. (If necessary, you can change the default COM port setting in the Nios II IDE. Select Run and use the Target connection tab's Host COM port setting)

BEFORE DOWNLOADING THE HARDWARE DESIGN AND SOFTWARE, ENSURE THAT YOU HAVE A BYTEBLASTER OR USB BLASTER CABLE AND AN RS-232 SERIAL CABLE CONNECTING YOUR PC'S COM PORT AND UP 3 BOARD.

Select **Tools** ⇨ **Quartus II Programmer...** to configure the FPGA. When the **Quartus II Programmer** appears, click on **Add File...** and select the **rpds16_time_limited.sof** file from your project directory. Click **Open** to add the selected file to the download chain. Check the **Program/Configure** box on the row with your configuration file on it, and click **Start** to begin hardware configuration.

When configuration is complete, a dialog box similar to the one in Figure 16.9 will appear. If you are using the fully licensed version of the software, this dialog box will not appear. The web/student versions of Quartus II and Nios II create time-limited program files, but they will work fine as long as your PC is attached and this dialog box remains open. Return to the **Nios II IDE** window.

Figure 16.9 Keep this dialog box open as long as the FPGA is being used.

From the **Nios II IDE** window, right click the **rpds_software** item in the **C/C++ Projects** pane and select **Build Project** from the drop-down menu. This will begin a full compilation of all libraries in your project.

To download the compiled code to the Nios II processor executing on the FPGA, right click the **rpds_software** item in the **C/C++ Projects** pane and select **Run As** ⇨ **Nios II Hardware**. The first time you select this option a run settings dialog box appears. Click the **Run** button to close this box.

16.8 Executing the Software

Once the program code has been downloaded to the Nios II processor's program memory (SRAM in this configuration), your code automatically begins executing. If you are using one of the older ByteBlaster II, ByteBlaster MV, or ByteBlaster JTAG cables, then in rare cases you may experience intermittent problems downloading your program to the Nios II processor. As a part of the normal download process, the Nios II IDE downloads your program to memory and then reads the program memory back to verify that the code in program memory is correct. If there are any problems with downloading your program, then the processor is stalled and a message that alerts you to this fact appears in the **Console** pane in the bottom right-hand side of the **Nios II IDE** window. If this happens, right click the **rpds_software** item in the **C/C++ Projects** pane and select **Run As** ⇨ **Nios II Hardware** again. If problems persist, consider upgrading to Altera's USB Blaster JTAG cable.

Once your program begins executing, the Nios II IDE's **Console** pane becomes a standard input/output terminal connected to your processor via the RS-232 UART device and cable. The text message "Hello World" should appear in the **Console** pane as soon as your program begins. Also, the four LEDs on the UP 3 board should turn on and off one at a time.

16.9 Starting Software Design for a Peripheral Test Program

Now that you have written your first program and have it successfully running on the UP 3 board, it is time to write a longer program that will test each of the major peripheral components on the UP 3 board.

A second project can be added to the current workspace, and since the same Nios II processor is being used for all projects in this workspace, the same system library can be used for them all. This option will be selected in the dialog box shown in Figure 16.10.

To create a new project, select **File** ⇨ **New** ⇨ **Project....** The **New Project** wizard will begin. On the first dialog box, select **C/C++ Application** and click **Next** to continue.

In the next dialog box, fill in the requested information. The name of the project is **rpds_up3_test**, the **SOPC Builder System** should point to the **nios32.ptf** file in your hardware project directory, and the **Project Template** should be set to **Blank Project**. If the workspace was correctly set to your project directory as detailed above, then the default location will be correct and **Use Default Location** should be checked. However, if the workspace is set to some other directory, then unselect **Use Default Location** and enter **c:*your_project_directory*\software\rpds_up3_test** in the **Location** field. Click **Next** to continue.

In the final dialog box, select the option **Select or create a system library** and choose the **rpds_software_syslib** project from the list of **Available System Library Projects for: nios32** as shown in Figure 16.10. Click **Finish** to create and open the project.

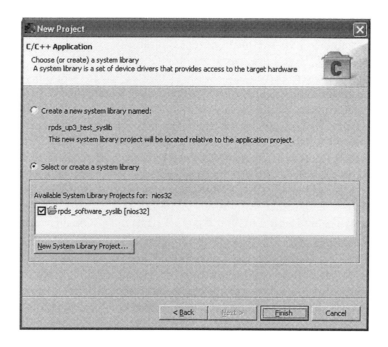

Figure 16.10 Since this project uses the same Nios II processor as your first program, the same system library can be used. Select the **rpds_software_syslib** from the list of available libraries.

Create a C header file by selecting the **rpds_up3_test** item in the **C/C++ Projects** pane. Choose **File** ⇨ **New** ⇨ **Header File**. When the dialog box appears, enter **rpds_up3_test.h** for the **Header File** and click **Finish** to continue.

Start your program's main header file by adding the *#include* and definition statements shown in Figure 16.11.

```
#ifndef _RPDS_UP3_TEST_H_
#define _RPDS_UP3_TEST_H_

#include <stdio.h>
#include <unistd.h>
#include "system.h"
#include "alt_types.h"
#include "sys/alt_irq.h"
#include "sys/alt_flash.h"
#include "altera_avalon_pio_regs.h"


#endif //_RPDS_UP3_TEST_H_
```

Figure 16.11 This is the beginning of your C program's main header file.

The C program that you will now write uses the four pushbuttons on the UP 3 board to select which device to test. When a pushbutton is pressed, it will be decoded (in an interrupt handler) and a variable will be set. The program's main thread will continuously read the function variable (at 50 ms intervals) and initiate the appropriate peripheral test. The function variable will be cleared at the end of each test routine so that buttons pressed while a peripheral is being tested will be ignored. The mapping of pushbutton to device shown in Table 16.1 will be used.

Table 16.1 Pushbutton to Device Mapping for Sample C Program

Pushbuttons (4-3-2-1)	Peripheral to Test
0001	LCD Display
0010	SRAM Memory
0100	Flash Memory
1000	SDRAM Memory

```
#include "rpds_up3_test.h"

int main( void ) {
  volatile int function = 0;
  int ret_val;

  while(1) {
    switch( function ) {
      case 0x1:           /* Test the LCD display */
        ret_val = test_lcd();
        break;
      case 0x2:           /* Test the SRAM */
        ret_val = test_sram();
        break;
      case 0x4:           /* Test the Flash memory */
        ret_val = test_flash();
        break;
      case 0x8:           /* Test the SDRAM */
        ret_val = test_sdram();
        break;
      default:            /* Do nothing */
        break;
    }
    function = 0;
    usleep( 50000 );      /* Wait 50 ms */
  }
  return(0);
}
```

Figure 16.12 This is the beginning of your C program's main source file.

Create your program's main C source file by selecting the **rpds_up3_test** item in the **C/C++ Projects** pane. Choose **File** ⇨ **New** ⇨ **File**. When the dialog box appears, enter **rpds_up3_test.c** for the **File name** and click **Finish** to continue.

Start your program by including the **rpds_up3_test.h** header file and typing the code shown in Figure 16.12.

16.10 Handling Interrupts

Inputs can be evaluated using two methods—polling and interrupts. To poll an input, your code can periodically check the value of the input device and determine if the value has changed. If a change has occurred, then the appropriate action should be taken to evaluate the input. An interrupt-driven input, however, works differently. When the value of the input changes, an interrupt signal is activated and the processor is alerted. The processor immediately performs a jump into a section of code known as the interrupt handler. This code determines which interrupt has occurred (most processors support multiple interrupt signals) and calls the appropriate interrupt service routine (a function that has been written to handle the specific interrupt signal). When the interrupt service routine has finished processing the input, the processor returns to the code it was executing before the interrupt occurred.

The program you are writing will use a combination of polling and interrupt driven inputs. The dipswitches and function variable will be polled every 50 ms. The value of the dipswitches will be displayed on the LEDs, and the value of the function variable will determine which, if any, peripheral should be tested.

The pushbuttons on the UP 3 board are represented by a 4-bit parallel I/O (PIO) peripheral called **buttons** in the Nios II reference design that you are using for this tutorial. The **buttons** PIO has been configured to generate an interrupt whenever any pushbutton is pressed and released.

To support interrupts you first must create a function that will execute when an interrupt occurs. This function is called an interrupt service routine (ISR). ISRs should generally be very short and execute quickly. Add the function **buttons_isr** as shown in Figure 16.13. The ISR function here reads the value of the PIO's edge capture register and stores it in the *function* variable. Next, it resets the edge capture register and IRQ mask register to allow the next interrupt to be captured and read properly.

```
static void buttons_isr( void* context, alt_u32 id ) {
  volatile int *function = (volatile int*) context;

  *function = IORD_ALTERA_AVALON_PIO_EDGE_CAP( BUTTONS_BASE );
  IOWR_ALTERA_AVALON_PIO_EDGE_CAP( BUTTONS_BASE, 0 );
  IOWR_ALTERA_AVALON_PIO_IRQ_MASK( BUTTONS_BASE, 0xF );
}
```

Figure 16.13 This is the interrupt service routine for the pushbuttons.

In your **main** function, you need to register your interrupt service routine and set the pushbuttons' IRQ mask register to allow interrupts to be captured. Add the two following lines before the *while* loop in your **main** function:

```
alt_irq_register(BUTTONS_IRQ, (void *) &function, buttons_isr);
IOWR_ALTERA_AVALON_PIO_IRQ_MASK( BUTTONS_BASE, 0xF );
```

16.11 Accessing Parallel I/O Peripherals

Macros are included in the **altera_avalon_pio_regs.h** file that read and write from the control and data registers in PIO components. You have already used these macros in the pushbutton's interrupt service routine to read and write the edge capture register and IRQ mask register. Now, you need to use these macros to read the values from the dipswitches and write them to the LEDs. Add the following two lines immediately above the *usleep(50000)* line in your **main** function:

```
        switches = IORD_ALTERA_AVALON_PIO_DATA( SWITCHES_BASE );
        IOWR_ALTERA_AVALON_PIO_DATA( LEDS_BASE, switches );
```

You will also need to add a declaration for the integer variable *switches* to your **main** function.

```
void lcd_init( void ) {

    /* Set Function Code Four Times -- 8-bit, 2 line, 5x7 mode */
    IOWR( LCD_BASE, LCD_WR_COMMAND_REG, 0x38 );
    usleep(4100);    /* Wait 4.1 ms */
    IOWR( LCD_BASE, LCD_WR_COMMAND_REG, 0x38 );
    usleep(100);     /* Wait 100 us */
    IOWR( LCD_BASE, LCD_WR_COMMAND_REG, 0x38 );
    usleep(5000);    /* Wait 5.0 ms */
    IOWR( LCD_BASE, LCD_WR_COMMAND_REG, 0x38 );
    usleep(100);

    /* Set Display to OFF */
    IOWR( LCD_BASE, LCD_WR_COMMAND_REG, 0x08 );
    usleep(100);

    /* Set Display to ON */
    IOWR( LCD_BASE, LCD_WR_COMMAND_REG, 0x0C );
    usleep(100);

    /* Set Entry Mode -- Cursor increment, display doesn't shift */
    IOWR( LCD_BASE, LCD_WR_COMMAND_REG, 0x06 );
    usleep(100);

    /* Set the cursor to the home position */
    IOWR( LCD_BASE, LCD_WR_COMMAND_REG, 0x02 );
    usleep(2000);

    /* Clear the display */
    IOWR( LCD_BASE, LCD_WR_COMMAND_REG, 0x01 );
    usleep(2000);
}
```

Figure 16.14 This is the LCD initialization function.

16.12 Communicating with the LCD Display

The LCD display on the UP 3 board can be treated similarly to a memory device. However, there are some additional initialization commands that must be sent to the LCD display that are not typical memory transactions. LCD initialization commands vary depending on the LCD controller chip that is on a particular LCD display. The manufacturer's datasheet will detail the proper initialization procedure their LCD displays. The initialization routine for the LCD display that ships with the UP 3 board is shown in Figure 16.14. Add this routine to your C source file. Also, add a call to this function in your **main** function preceding the line of code that calls the **test_lcd** function.

The code for **test_lcd** is shown in Figure 16.15. You will notice that this code expects several constants to be defined. Add definitions for the following constants in your **rpds_up3_test.h** header file:

- LCD_WR_COMMAND_REG = 0
- LCD_WR_DATA_REG   = 2

The **main** function in your C source file should now be complete and look similar to the code in Figure 16.16. Note that a few *printf* statements have been added to provide the user with the program's status while executing.

```
alt_u32 test_lcd( void ) {
  int i;
  char message[17] = "Counting...      ";
  char done[12] = "Done!        ";

  /* Write a simple message on the first line. */
  for( i = 0; i < 16; i++ ) {
    IOWR( LCD_BASE, LCD_WR_DATA_REG, message[i] );
    usleep(100);
  }
  /* Count along the bottom row */
  /* Set Address */
  IOWR( LCD_BASE, LCD_WR_COMMAND_REG, 0xC0 );
  usleep(1000);
  /* Display Count */
  for( i = 0; i < 10; i++ ) {
    IOWR( LCD_BASE, LCD_WR_DATA_REG, (char)(i+0x30) );
    usleep(500000);    /* Wait 0.5 sec. */
  }

  /* Write "Done!" message on first line. */
  /* Set Address */
  IOWR( LCD_BASE, LCD_WR_COMMAND_REG, 0x80 );
  usleep(1000);
  /* Write data */
  for( i = 0; i < 11; i++ ) {
    IOWR( LCD_BASE, LCD_WR_DATA_REG, done[i] );
    usleep(100);
  }
  return(0);
}
```

Figure 16.15 This is the code to test the LCD display.

```
int main( void ) {
  volatile int function = 0;
  alt_u32 switches, ret_val;

  printf("Welcome to the Nios II Test Program\n" );

  alt_irq_register(BUTTONS_IRQ, (void *) &function, buttons_isr);
  IOWR_ALTERA_AVALON_PIO_IRQ_MASK( BUTTONS_BASE, 0xF );

  while(1) {
    switch( function ) {
      case 0x1:              /* Test the LCD display */
        printf("Testing LCD Display\n" );
        lcd_init();
        ret_val = test_lcd();
        printf("...Completed.\n" );
        break;
      case 0x2:              /* Test the SRAM */
        printf("Testing SRAM\n" );
        ret_val = test_sram();
        printf("...Completed with %d Errors.\n", ret_val );
        break;
      case 0x4:              /* Test the Flash memory */
        printf("Testing Flash memory\n" );
        ret_val = test_flash();
        printf("...Completed with %d Errors.\n", ret_val );
        break;
      case 0x8:              /* Test the SDRAM */
        printf("Testing SDRAM\n" );
        ret_val = test_sdram();
        printf("...Completed with %d Errors.\n", ret_val );
        break;
      default:               /* Do nothing */
        break;
    }

    function = 0;

    switches = IORD_ALTERA_AVALON_PIO_DATA( SWITCHES_BASE );
    IOWR_ALTERA_AVALON_PIO_DATA( LEDS_BASE, switches );

    usleep( 50000 );
  }

  return(0);
}
```

Figure 16.16 This is the completed *main* function.

16.13 Testing SRAM

To test the SRAM, you will write a large number of values to memory and then read back from the same memory locations to verify that the contents of memory are what you expect. Since SRAM is currently being used for program and data memory, accessing SRAM is straight-forward. Any array that is created in a function will be stored in data memory (e.g., in SRAM). The code for **test_sram** is shown in Figure 16.17. You will notice that this code expects the constant value **SRAM_MAX_WORDS** to be defined. Add a definition for this constant to your **rpds_up3_test.h** header file and set it equal to **8000**.

This test routine assumes that there is not a data cache memory present in the Nios II system. If data cache is present, then declaring an array in a function like **test_sram** would not ensure SRAM writes, because the data cache memory could be used as a temporary buffer. Since this function is very short and the array's scope is internal to the function, it is highly likely that the array data would never be written to SRAM. To avoid these potential issues, the reference hardware design used in this tutorial does *not* include data cache.

> THERE ARE SEVERAL WAYS TO BYPASS DATA CACHE IN THE NIOS II PROCESSOR. (1) CREATE A BUFFER THAT IS LARGER THAN THE DATA CACHE TO FORCE AT LEAST SOME SRAM ACCESSES. (2) USE SPECIAL MEMORY ACCESS INSTRUCTIONS (SUCH AS LWIO AND SWIO) IN THE NIOS II INSTRUCTION SET THAT BYPASS DATA CACHE AND FORCE A MEMORY ACCESS.

```
alt_u32 test_sram( void ) {
  alt_u32 i, val;
  alt_u32 errors = 0;
  alt_u32 buffer[SRAM_MAX_WORDS];

  /* Write data to SRAM */
  for( i = 0; i < SRAM_MAX_WORDS; i++ ) {
    buffer[i] = i + 1000;
  }
  /* Check output from SRAM */
  for( i = 0; i < SRAM_MAX_WORDS; i++ ) {
    if( buffer[i] != (i+1000) )
      errors++;
  }
  return( errors );
}
```

Figure 16.17 This is the code to test the SRAM memory device.

16.14 Testing Flash Memory

Flash memory is organized into blocks of data and is accessed in a different manner than SRAM and SDRAM. The Nios II HAL interface includes memory access functions for Flash devices that conform to the Common Flash Memory Interface (CFI) standard. The functions **alt_flash_open_dev**, **alt_read_flash**, **alt_write_flash**, and **alt_flash_close_dev** provide an interface that is very similar to file I/O. These subroutines and more lower-level functions are all declared in the **sys/alt_flash.h** header file.

Flash memory write operations happen at the block level meaning that to write a block or any portion of a block (down to a single byte of data) requires the entire block of data to be erased and overwritten. When writing to a partial block of data, the user is responsible for reading the portion of the block that is not to be overwritten, storing it, and passing it with the new data as a complete block to be written. Also, keep in mind that Flash memory typically has a life expectancy of 100,000 write cycles. Because of the overhead involved in writing partial blocks of data and the finite number of write cycles for a given Flash memory device, it is best to buffer data until a full block can be written to Flash memory.

The code for **test_flash** is shown in Figure 16.18. The data to be written to flash is buffered in the in_buff array located in data memory. Once is it full, the entire buffer is sent to the alt_flash_write command which partitions it into blocks of data and writes the full blocks to Flash memory. Depending on the total length of the in_buff array the final block written may be a partial block, but at least it will only get written once. You will also notice that this code expects the constant value **FLASH_MAX_WORDS** to be defined. Add a definition for this constant to your **rpds_up3_test.h** header file and set it equal to **1000**.

UP 3 BOARDS CURRENTLY CONTAIN ONE OF TWO TYPES OF FLASH MEMORY CHIPS, TOSHIBA AND EXCELSEMI. SOME UP 3 BOARDS CONTAIN TOSHIBA FLASH MEMORY CHIPS THAT DO NOT PROGRAM AND WILL NOT PASS THE EXAMPLE FLASH MEMORY TEST. THE FLASH MEMORY CHIP, U8, IS LOCATED DIRECTLY BELOW THE LARGE SQUARE FPGA CHIP IN THE MIDDLE OF THE UP 3 BOARD. THE MANUFACTURER'S NAME IS PRINTED ON THE FLASH CHIP. THE TOSHIBA CHIPS REQUIRE NEW FLASH DRIVER SOFTWARE THAT IS BEING DEVELOPED AT THE TIME OF THIS PRINTING. IF YOUR UP 3 PROJECT REQUIRES FLASH MEMORY, VERIFY THAT YOU HAVE THE TOSHIBA CHIPS AND CHECK FOR TOSHIBA FLASH DRIVER SOFTWARE UPDATES AT THE BOOK'S WEBSITE:

HTTP://WW.ECE.GATECH.EDU/USERS/HAMBLEN/BOOK/BOOKTE.HTM

```
alt_u32 test_flash( void ) {
  alt_u32 i, errors = 0;
  alt_u32 in_buff[FLASH_MAX_WORDS], out_buff[FLASH_MAX_WORDS];
  alt_flash_fd* flash_handle;

  flash_handle = alt_flash_open_dev( FLASH_NAME );

  /* Create data buffer to write to Flash memory */
  for( i = 0; i < FLASH_MAX_WORDS; i++ ) {
    in_buff[i] = i + 1000000;
  }

  /* Write data to Flash memory */
  alt_write_flash( flash_handle, 0, in_buff, FLASH_MAX_WORDS*4 );

  /* Read data from Flash memory */
  alt_read_flash( flash_handle, 0, out_buff, FLASH_MAX_WORDS*4 );

  /* Check output from Flash memory */
  for( i = 0; i < FLASH_MAX_WORDS; i++ ) {
    if( out_buff[i] != (i+1000000) )
      errors++;
  }

  alt_flash_close_dev( flash_handle );
  return( errors );
}
```

Figure 16.18 This is the code to test the Flash memory device.

16.15 Testing SDRAM

To test the SDRAM, write a large number of values to memory and then read the same memory locations and verify that the contents of memory are the expected values. To access the SDRAM on the UP 3 board, a pointer to the SDRAM memory space can be used. Once a pointer has been initialized to an address in the SDRAM memory space, that pointer can be dereferenced like an array to store values in successive SDRAM memory locations. This method of accessing memory would use the data cache if it were present (which it is *not* in the reference example). If you are using a Nios II processor with data cache and you want to access SDRAM directly (bypassing the data cache), then use you need to use the IORD and IOWR macros as shown in the previous sections.

The code for **test_sdram** is shown in Figure 16.19. You will notice that this code expects the constant value **SDRAM_MAX_WORDS** to be defined. Add a definition for this constant to your **rpds_up3_test.h** header file and set it equal to **1000000**.

```c
alt_u32 test_sdram( void ) {
  alt_u32 i;
  alt_u32 errors = 0;
  alt_u32 *buffer = (alt_u32 *)SDRAM_BASE;

  /* Write data to SDRAM */
  for( i = 0; i < SDRAM_MAX_WORDS; i++ ) {
    buffer[i] = (i + 1000000);
  }

  /* Check output from SDRAM */
  for( i = 0; i < SDRAM_MAX_WORDS; i++ ) {
    if( buffer[i] != (i+1000000) )
      errors++;
  }
  return( errors );}
```

Figure 16.19 This is the code to test the SDRAM memory device.

Your C source and header files should now be complete. Figure 16.20 shows the final **rpds_up3_test.h** header file, and Figure 16.21 shows the final **rpds_up3_test.c** file.

```c
#ifndef _RPDS_UP3_TEST_H_
#define _RPDS_UP3_TEST_H_

#include <stdio.h>
#include <unistd.h>
#include "system.h"
#include "alt_types.h"
#include "sys/alt_irq.h"
#include "sys/alt_flash.h"
#include "altera_avalon_pio_regs.h"

/* LCD constants */
#define LCD_WR_COMMAND_REG   0
#define LCD_WR_DATA_REG      2

/* Memory constants */
#define SRAM_MAX_WORDS       8000
#define FLASH_MAX_WORDS      1000
#define SDRAM_MAX_WORDS      1000000


#endif //_RPDS_UP3_TEST_H_
```

Figure 16.20 This is the final copy of the **rpds_up3_test.h** header file.

```
#include "rpds_up3_test.h"

static void buttons_isr( void* context, alt_u32 id ) {
  volatile int *function = (volatile int*) context;

  *function = IORD_ALTERA_AVALON_PIO_EDGE_CAP( BUTTONS_BASE );
  IOWR_ALTERA_AVALON_PIO_EDGE_CAP( BUTTONS_BASE, 0 );
  IOWR_ALTERA_AVALON_PIO_IRQ_MASK( BUTTONS_BASE, 0xF );
}

void lcd_init( void ) {
  /* Set Function Code Four Times -- 8-bit, 2 line, 5x7 mode */
  IOWR( LCD_BASE, LCD_WR_COMMAND_REG, 0x38 );
  usleep(4100);    /* Wait 4.1 ms */
  IOWR( LCD_BASE, LCD_WR_COMMAND_REG, 0x38 );
  usleep(100);     /* Wait 100 us */
  IOWR( LCD_BASE, LCD_WR_COMMAND_REG, 0x38 );
  usleep(5000);    /* Wait 5.0 ms */
  IOWR( LCD_BASE, LCD_WR_COMMAND_REG, 0x38 );
  usleep(100
  /* Set Display to OFF */
  IOWR( LCD_BASE, LCD_WR_COMMAND_REG, 0x08 );
  usleep(100);
  /* Set Display to ON */
  IOWR( LCD_BASE, LCD_WR_COMMAND_REG, 0x0C );
  usleep(100);
  /* Set Entry Mode -- Cursor increment, display doesn't shift */
  IOWR( LCD_BASE, LCD_WR_COMMAND_REG, 0x06 );
  usleep(100);
  /* Set the cursor to the home position */
  IOWR( LCD_BASE, LCD_WR_COMMAND_REG, 0x02 );
  usleep(2000);
  /* Clear the display */
  IOWR( LCD_BASE, LCD_WR_COMMAND_REG, 0x01 );
  usleep(2000);
}

alt_u32 test_lcd( void ) {
  int i;
  char message[17] = "Counting...      ";
  char done[12] = "Done!       ";

  /* Write a simple message on the first line. */
  for( i = 0; i < 16; i++ ) {
    IOWR( LCD_BASE, LCD_WR_DATA_REG, message[i] );
    usleep(100);
  }
```

Figure 16.21 This is the final copy of the **rpds_up3_test.c** source file.

```
  /* Count along the bottom row */
  /* Set Address */
  IOWR( LCD_BASE, LCD_WR_COMMAND_REG, 0xC0 );
  usleep(1000);
  /* Display Count */
  for( i = 0; i < 10; i++ ) {
    IOWR( LCD_BASE, LCD_WR_DATA_REG, (char)(i+0x30) );
    usleep(500000);   /* Wait 0.5 sec. */
  }

 /* Write "Done!" message on first line. */
 /* Set Address */
 IOWR( LCD_BASE, LCD_WR_COMMAND_REG, 0x80 );
 usleep(1000);
 /* Write data */
 for( i = 0; i < 11; i++ ) {
   IOWR( LCD_BASE, LCD_WR_DATA_REG, done[i] );
   usleep(100);
 }
 return(0);
}

alt_u32 test_sram( void ) {
  alt_u32 i, val;
  alt_u32 errors = 0;
  alt_u32 buffer[SRAM_MAX_WORDS];

  /* Write data to SRAM */
  for( i = 0; i < SRAM_MAX_WORDS; i++ ) {
    buffer[i] = i + 1000;
  }
  /* Check output from SRAM */
  for( i = 0; i < SRAM_MAX_WORDS; i++ ) {
    if( buffer[i] != (i+1000) )
      errors++;
  }
  return( errors );
}

alt_u32 test_flash( void ) {
  alt_u32 i, errors = 0;
  alt_u32 in_buff[FLASH_MAX_WORDS], out_buff[FLASH_MAX_WORDS];
  alt_flash_fd* flash_handle;

  flash_handle = alt_flash_open_dev( FLASH_NAME );
```

Figure 16.21 continued

```
  /* Create data buffer to write to Flash memory */
  for( i = 0; i < FLASH_MAX_WORDS; i++ ) {
    in_buff[i] = i + 1000000;
  }

  /* Write data to Flash memory */
  alt_write_flash( flash_handle, 0, in_buff, FLASH_MAX_WORDS*4 );

  /* Read data from Flash memory */
  alt_read_flash( flash_handle, 0, out_buff, FLASH_MAX_WORDS*4 );

  /* Check output from Flash memory */
  for( i = 0; i < FLASH_MAX_WORDS; i++ ) {
    if( out_buff[i] != (i+1000000) )
      errors++;
  }

  alt_flash_close_dev( flash_handle );
  return( errors );
}

alt_u32 test_sdram( void ) {
  alt_u32 i;
  alt_u32 errors = 0;
  alt_u32 *buffer = (alt_u32 *)SDRAM_BASE;

  /* Write data to SDRAM */
  for( i = 0; i < SDRAM_MAX_WORDS; i++ ) {
    buffer[i] = i + 1000000;
  }
  /* Check output from SDRAM */
  for( i = 0; i < SDRAM_MAX_WORDS; i++ ) {
    if( buffer[i] != (i+1000000) )
      errors++;
  }
  return( errors );
}

int main( void ) {
  volatile int function = 0;
  alt_u32 switches, ret_val;

  printf( "Welcome to the Nios II Test Program\n" );
  alt_irq_register(BUTTONS_IRQ, (void *) &function, buttons_isr);
  IOWR_ALTERA_AVALON_PIO_IRQ_MASK( BUTTONS_BASE, 0xF );
```

Figure 16.21 continued

```
while(1) {
  switch( function ) {
    case 0x1:              /* Test the LCD display */
      printf("Testing LCD Display\n" );
      lcd_init();
      ret_val = test_lcd();
      printf("...Completed.\n" );
      break;
    case 0x2:              /* Test the SRAM */
      printf("Testing SRAM\n" );
      ret_val = test_sram();
      printf("...Completed with %d Errors.\n", ret_val );
      break;
    case 0x4:              /* Test the Flash memory */
      printf("Testing Flash Memory\n" );
      ret_val = test_flash();
      printf("...Completed with %d Errors.\n", ret_val );
      break;
    case 0x8:              /* Test the SDRAM */
      printf("Testing SDRAM\n" );
      ret_val = test_sdram();
      printf("...Completed with %d Errors.\n", ret_val );
      break;
    default:               /* Do nothing */
      break;
  }
  function = 0;
  switches = IORD_ALTERA_AVALON_PIO_DATA( SWITCHES_BASE );
  IOWR_ALTERA_AVALON_PIO_DATA( LEDS_BASE, switches );
  usleep( 50000 );
}
return(0);
}
```

Figure 16.21 continued

16.16 Downloading the Nios II Hardware and Software Projects

To execute your software on a Nios II processor, you must configure the FPGA with the Nios II hardware reference design and then you can download the compiled program code to the processor's program memory.

BEFORE DOWNLOADING THE HARDWARE DESIGN AND SOFTWARE, ENSURE THAT YOU HAVE A BYTEBLASTER II CABLE AND AN RS-232 SERIAL CABLE CONNECTING YOUR PC AND THE UP 3 BOARD.

Select **Tools** ⇨ **Quartus II Programmer...** to configure the FPGA. When the **Quartus II Programmer** appears, click on **Add File...**, and select the **rpds16_time_limited.sof** file from your project directory. Click **Open** to add the selected file to the download chain. Check the **Program/Configure** box on the row with your configuration file on it and click **Start** to begin hardware configuration.

When configuration is complete, a dialog box similar to the one in Figure 16.22 will appear. The student version of Quartus II and Nios II creates time-limited program files. These programs will work fine as long as your PC is attached and this dialog box remains open. Therefore, leave the window open and return to the **Nios II IDE** window.

Figure 16.22 Keep this dialog box open as long as the FPGA is being used.

From the **Nios II IDE** window, right click the **rpds_up3_test** item in the **C/C++ Projects** pane and select **Build Project** from the drop-down menu. This will begin a full compilation of all libraries in your project.

To download the compiled code to the Nios II processor executing on the FPGA, right click the **rpds_up3_test** item in the **C/C++ Projects** pane and select **Run As** ⇨ **Nios II Hardware**.

16.17 Executing the Software

Once the program code has been downloaded to the Nios II processor's program memory (SRAM in this configuration), your code should automatically start executing. If you are using one of the older ByteBlaster II, ByteBlaster MV, or ByteBlaster JTAG cables, then in rare cases you may experience intermittent problems downloading your program to the Nios II processor. As a part of the normal program download, the Nios II IDE verifies that the code in program memory is the same as downloaded program before program execution begins. If there are any problems with downloading your program then the processor is stalled and a message that alerts you to this fact appears in the **Console** pane in the bottom right-hand side of the **Nios II IDE** window. If this happens, right click the **rpds_up3_test** item in the **C/C++ Projects** pane and select **Run As** ⇨ **Nios II Hardware** again. If problems persist, consider upgrading to Altera's USB Blaster JTAG cable.

Once your program begins executing, the Nios II IDE's Console pane becomes a standard input/output terminal connected to your processor via the RS-232 UART device and cable. Press each of the four pushbuttons in turn. A different device will be tested when each button is pressed and released. Look at the text in the Console pane to verify that the proper test is being executed.

Change the dipswitches' value and verify that the appropriate LEDs light.

IMPORTANT NOTE: Peripheral ICs on the UP 3 board can come from several manufacturers. While they are generically equivalent, they may exhibit slightly different characteristics and timing. UP 3 boards currently contain one of two brands of Flash memory chips, Toshiba and ExcelSemi. UP 3 boards that contain a Toshiba Flash memory chip currently do not program and will not pass the example Flash memory test. The Flash memory chip, U8, is located directly below the large square FPGA chip in the middle of the UP 3 board, and the manufacturer's name is printed on the chip. The Toshiba chip requires a new Flash driver that is being developed at the time of this printing. If your UP 3 project requires Flash memory and you verify that you have the Toshiba chips, check for the Toshiba Flash driver, software updates, and specific information about different chips on UP 3 boards at the book's website:

http://ww.ece.gatech.edu/users/hamblen/book/bookte.htm

ALL SOURCE FILES FOR THIS UP 3 NIOS SOFTWARE REFERENCE DESIGN

CAN BE FOUND ON THE CD-ROM IN THE \CHAP16 DIRECTORY.

16.18 For additional information

This chapter has provided a brief overview of Nios II Software development. Additional information can be found at Altera's website (www.altera.com) in the *Nios II Software Developer's Handbook* and at the Nios Community forum (www.niosforum.com).

16.19 Laboratory Exercises

1. Write a C program to blink the four LEDs in a reversing shift pattern on the UP 3 board. After the last LED in each direction turns on, reverse the direction of the shift. Run and demonstrate the program on the UP 3 board. Recall that C supports shift operations ("<<" and ">>") and you will need a time delay in your code to see the LEDs blink.

2. Write a C program that displays a count of the seconds that the program has been running in the LCD display on the UP 3 board. Demonstrate the program on the UP 3 board.

3. Expand the C program in the previous problem to display the elapsed time in hours, minutes, and seconds on the LCD. Have one pushbutton reset the time to zero and another pushbutton start and stop the timer just like a stopwatch.

4. Memory test programs cannot test all possible patterns. Research the various algorithms widely used in more thorough memory test programs and write your own more advanced

memory test program for SRAM. Most memory test programs use several algorithms to check for different types of faults. Execute the test code from SDRAM.

5. Write a retro version of the 1970's classic kill the bit computer game for the UP 3. The goal in the kill the bit game is to turn off all of the four LEDs using the four pushbuttons. The game starts with an initial non-zero pattern displayed in the LEDs. The pattern constantly does a circular shift moving through the LEDs in a loop with a time delay to slow down the shifts. If you hit one of the four pushbuttons exactly when the the the same number LED is turned on, it will turn off one LED in the pattern. If you hit a pushbutton and it's LED is off another LED turns on.

Here is how the program works. Each time just before the pattern shifts, the pattern is bit-wise exclusive or'ed with one input sample from the pushbuttons to generate a new pattern. When both the pushbutton is pushed and its corresponding bit in the pattern are High, one less bit will be High in the new pattern after the exclusive or (i.e., 1 xor 1 is 0). After the shift, one less LED will be turned on since there is one less "1" in the new pattern. If your timing is off and the LED is not turned on when you hit the pushbutton, a new high bit will be generated in the pattern (i.e., 1 xor 0 is 1). When this happens, the new "1" bit in the pattern lights another LED. Note that you need a "1" when a pushbutton is pressed and a "1" to turn on an LED for the xor function to work.

Display the elapsed time in the LCD display and stop the time display when a player wins the game (turns out all LEDs). Adjust the shift time delay for reasonable game play. Blink all of the LEDs when a player wins. If you want a more challenging game, use a pattern and shift register larger than four bits and just display four bits at a time in the LEDs.

6. Port an interesting C application program to the UP 3. Execute the application from SDRAM.

CHAPTER 17

Tutorial IV: Nios II Processor Hardware Design

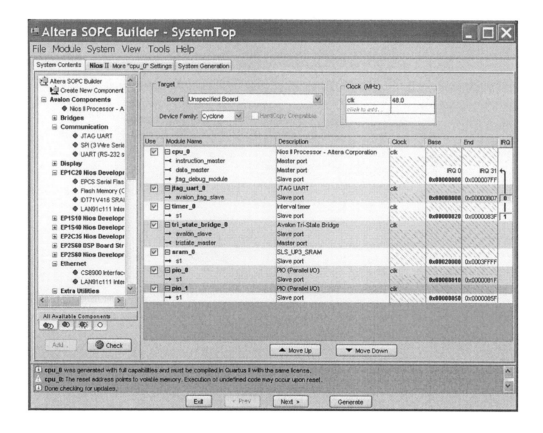

SOPC Builder is a GUI-based hardware design tool used to configure the Nios II processor core options and to design bus and I/O interfaces for the processor.

17 Tutorial IV: Nios II Processor Hardware Design

Designing systems with embeddeded processors requires both hardware and software design elements. A collection of CAD tools developed by Altera enable you to design both the hardware and software for a fully functional, customizable, soft-core processor called Nios II. This tutorial steps you through the hardware implementation of a Nios II processor for the UP 3 board, and Tutorial III (in the preceding chapter) introduces the software design tools for the Nios II processor.

Upon completion of this tutorial, you will be able to:

- Navigate Altera's SOPC Builder (Nios II processor design wizard),
- Generate a custom Nios II processor core,
- Create a PLL that supplies a clock signal for the on-board SDRAM, and
- Specify the top-level pin assignments and project settings necessary for implementing the Nios processor on the UP 3 board.

17.1 Install the UP 3 board files

Locate the **booksoft\chap17** directory on the CD-ROM that came with the book. In this directory, there are two subdirectories called **up3_tristate_lcd** and **up3_tristate_sram**. Copy both of these subdirectories to the *quartus_install_dir*\sopc_builder\components\ directory on your local hard drive.

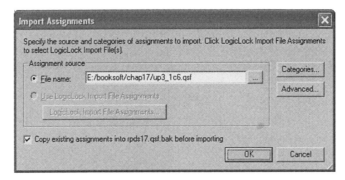

Figure 17.1 Import the default pin and project assignments for the UP 3 board.

17.2 Creating a New Project

Create a new Quartus II project as illustrated in Tutorial I (see Section 1 of Chapter 1). Use the project name **rpds17** and create a top-level Block Diagram/Schematic file named **rpds17.bdf**.

Import the pin assignments and project settings file from the CD-ROM by choosing **Assignments ⇨ Import Assignments....** Enter the full path for the **booksoft\chap17\up3_1c6.qsf** file located on the CD-ROM that came with this book as shown in Figure 17.1. (If you are using a UP 3 board with the larger 1C12 FPGA on it, then you must use the **booksoft\chap17\up3_1c12.qsf** file located on the CD-ROM.) Click on the **Advanced** button and verify that the settings match the dialog box in Figure 17.2. If different settings are used, then all of the pin and project assignments may not be made correctly, and downloading your project to the UP 3 board could damage it. When the settings are correct, click **OK** to exit the dialog box. Click **OK** in the **Import Assignments** dialog box to import the settings.

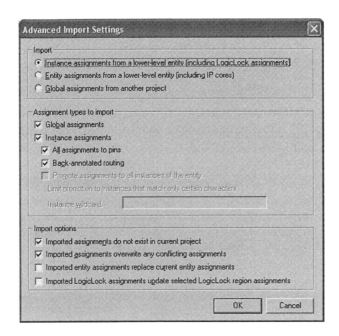

Figure 17.2 It is important that the Advanced Import Options be set as shown here.

17.3 Starting SOPC Builder

A Nios II processor is created using the SOPC Builder wizard. Within this wizard, you can specify the settings for the Nios II processor, add peripherals, and select the bus connections, I/O memory mapping, and IRQ assignments for the processor. To start the SOPC Builder, choose **Tools ⇨ SOPC Builder…**.

Figure 17.3 Specifying the name of the Nios II processor for your system.

In the **Create New System** dialog box, enter the name **nios32**, and set the **Target HDL** to **VHDL** as shown in Figure 17.3. Click **OK** to open SOPC Builder with a blank project titled nios32.

The system settings in the top part of SOPC Builder window must be set for the board and device that you are using. For the UP 3 board, the on-board clock circuit is 48.0 MHz; therefore, enter **48.0** in the **clk** field. Select **Cyclone** as **Device Family** and **Unspecified Board** in the **Board** field. When these settings have been entered, your SOPC Builder window should look similar to the screen shot in Figure 17.4.

IT IS CRITICAL THAT THE FREQUENCY SELECTED IN THE SOPC BUILDER IS THE ACTUAL CLOCK RATE USED IN YOUR HARDWARE DESIGN. IF A PLL IS USED TO GENERATE A DIFFERENT NIOS II CLOCK SIGNAL, THEN THAT CLOCK FREQUENCY MUST BE ENTERED INTO THE SOPC BUILDER BEFORE THE SYSTEM IS GENERATED. IF YOU MODIFY THE CLOCK FREQUENCY FOR THE NIOS II PROCESSOR LATER, THEN YOU MUST RE-GENERATE THE NIOS II PROCESSOR WITH THE UPDATED FREQUENCY SPECIFIED HERE.

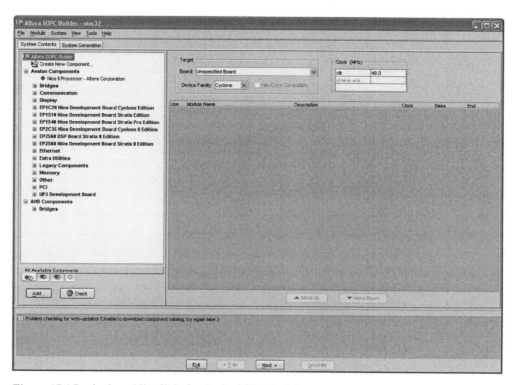

Figure 17.4 Beginning a Nios II design in the SOPC Builder.

Take a minute to familiarize yourself with the layout of the SOPC Builder window. Along the left-hand side, there is an expandable list of components organized by category that can be added to a Nios II system. Click on the "+" symbol next to the items in this list to expand the list of components for each category. Under the **Avalon Components** category, there are a number of development boards listed. Expanding these items will reveal components that are specific to these boards. If you installed the design files as discussed in Section 17.1, then the UP 3 **development board** category will appear under **Avalon Components**.

17.4 Adding a Nios II Processor

The first component that you will add to your Nios II processor design is the processor core itself. In the list of components on the left-hand side of the SOPC Builder, expand the **Avalon Components** category and select the **Nios II Processor – Altera Corporation** component. Click the **Add...** button at the bottom of the component list.

When a component is added to your system, a dialog box will appear that allows you to select options and set specific parameters for this particular implementation. For the Nios II processor, the dialog box shown in Figure 17.5 will appear. This first selection will determine the general parameters of the Nios II processor. Notice that there are three general configurations allowed that vary in size, performance, and functionality. Select the middle configuration, **Nios II/s** as shown in Figure 17.5. In the **Hardware Multiply** field, select **Logic Elements**, and click **Next** to continue.

The next dialog box allows you set the size of the instruction cache in the Nios II processor. Keep the default value (4 KB), and click **Next** to continue.

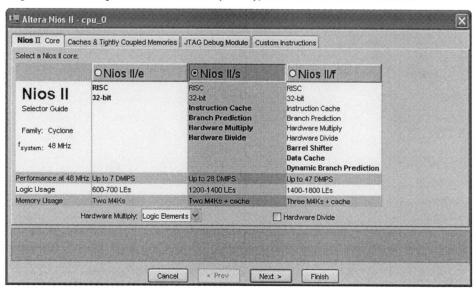

Figure 17.5 Nios II supports three different general configurations. Select Nios II/s for this tutorial.

Nios II processors can be compiled with support for one of four different debugging systems. The differences between them are shown in Figure 17.6, along with the FPGA resources required to implement each type of debugging. There is an order of magnitude difference in the number of logic elements required to implement Level 4 debugging versus Level 1 debugging. This difference is significant when compared to the overall size of the Nios II processor. The Level 4 debugging system is two to three times larger then the Nios II/s processor itself. Since the cost of FPGAs are largely based on their size, the debugging logic will typically be removed before a design enters production to minimize the number of logic elements, and thus the size of the FPGA, required for the production quantities.

The full features of Level 3 and Level 4 debugging are only available when a license from First Silicon Solutions, a third-party company, is purchased. The availability of this license within your company or school along with the complexity of your end system and the size of the FPGA available will be the primary factors in determining which debugging system should be selected for a given system. For this tutorial, select **Level 1** (the default), and click **Next** to continue.

The final option in the Nios II processor configuration is the adding of custom instructions. Nios II processors allow the addition of up to 256 custom instructions into the processor's data path. These can be used to further customize your processor for a specific task. For this tutorial, no custom instructions will be added. Click **Finish** to complete the Nios II configuration for this system.

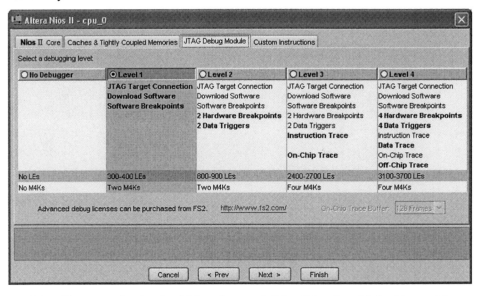

Figure 17.6 Nios II supports four levels of debugging capabilities. Select Level 1 for this tutorial.

When the SOPC Builder window reappears, the Nios II processor will appear as an added component in the main part of the window with the default module name **cpu_0**. Also, a number of error and warning messages will appear in the console at the bottom of the SOPC Builder window. These messages result from there not being any defined memory in the system yet. When memory is added in the next few sections, the messages will disappear. Right-click on the **cpu_0** name and select **Rename** from the drop-down menu. Rename the module to **cpu**.

17.5 Adding UART Peripherals

Two UART peripherals are needed for this system: a JTAG UART and an RS-232 serial UART. The ByteBlaster or USB Blaster JTAG cable that is used to configure the FPGA can also be used as a UART device after the FPGA is configured. (The JTAG cable is also used as the communication channel between the PC and the debugging logic selected for the Nios II processor.) The Nios II software integrated development environment (IDE) uses the JTAG UART as the default device for downloading your software instructions to the Nios II processor and was used for that purpose in the previous tutorial on software design.

Add the JTAG UART device by expanding **Avalon Components** ⇨ **Communication**. Select **JTAG UART** and click **Add...**. When the JTAG UART Configuration dialog box appears, click **Finish** to accept the default values for all fields and add the component. In the SOPC Builder, rename the JTAG UART module to **jtag**.

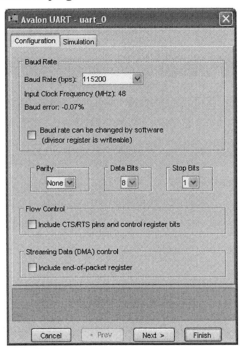

Figure 17.7 These are the settings for the RS-232 UART device to be added to the Nios II system.

If you are using a USB Blaster JTAG cable, then you can use the JTAG UART for all serial communication between the PC and the Nios II processor and can skip to the next section. However, if you are using the older ByteBlaster II, ByteBlaster MV, or ByteBlaster cables, then you need to add a second RS-232 UART for run-time serial communication. These older JTAG cables do not transmit the run-time serial data robustly. Some setups have been known to work; however, it is more reliable to use a standard serial UART and 9-pin serial cable for stdin, stdout, and stderr data.

Add the RS-232 UART peripheral by expanding **Avalon Components** ⇨ **Communication**. Select **UART (RS-232 serial port)** and click **Add...**. When the UART configuration dialog box appears, set the options as shown in Figure 17.7. Click **Finish** to add the component. In the SOPC Builder, rename the UART module to **uart**.

17.6 Adding an Interval Timer Peripheral

Most processor designs require at least one timer. This timer is used to delay the processor, coordinate transactions, timestamp events, generate time slice interrupts for an operating system scheduler, a watchdog timer, and more. The Nios II timer peripheral is flexible; it has several options and three predefined configurations. Add a full-featured interval timer to your Nios II processor by expanding **Avalon Components** ⇨ **Other**. Select **Interval Timer** and click **Add...**. When the timer configuration dialog box appears, set the options as shown in Figure 17.8. Click **Finish** to add the component. In the SOPC Builder, rename the timer module to **timer0**. The "0" is appended to the timer name here to provide a consistent naming convention for your timers if additional timers are added at a later time. It is not unusual for a processor to have two or three timers – often of different configurations for specific uses.

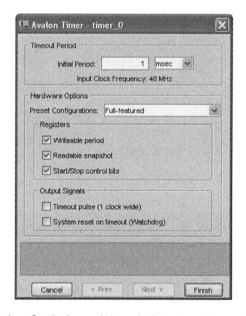

Figure 17.8 These are the settings for the interval timer device to be added to the Nios II system.

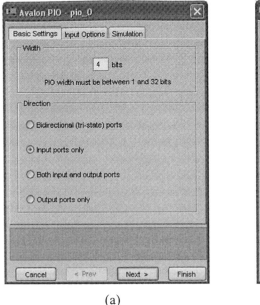
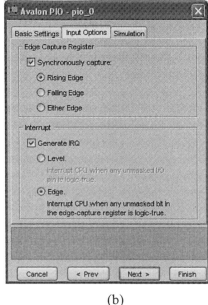

(a)   (b)

Figure 17.9 These are the settings for the pushbutton PIO device to be added to the Nios II system.

17.7 Adding Parallel I/O Components

Many processors require a certain amount of general-purpose I/O pins. These pins can be attached directly to pushbuttons, switches, LEDs, and similar I/O devices. They can also be attached to relatively simple or low bandwidth interfaces that don't have a large amount of overhead associated with data transmission. Examples of these types of interfaces include PS/2, I^2C, SPI, and parallel data interfaces.

In addition, general-purpose I/O pins can be used to pass low-bandwidth data between a custom VHDL or Verilog block and the Nios II processor. A faster method of transferring data to a VHDL block is to create a custom peripheral that can attach to the Avalon bus. Implementing a VHDL module that is compliant with the Avalon bus specification is more involved and requires more logic elements than using general-purpose I/O pins, but it does provide a faster more efficient interface.

General-purpose I/O pins are added to the Nios II processor with the PIO (Parallel I/O) component. The PIO component has a number of options for customizing general-purpose I/O interfaces. PIO interfaces can be specified as input only, output only, or bidirectional. If bidirectional is selected here, then the direction of each pin must be set in the direction register at run-time via software. Input PIO interfaces can also have various interrupt and edge capture capabilities including the capturing of either or both edges and edge or level-sensitive interrupt triggers.

For this tutorial, you will add three PIO components: one for the pushbuttons, one for the dipswitches, and one for the LEDs. First, add a PIO component for the pushbuttons to your processor design by expanding **Avalon Components** ⇨ **Other**. Select **PIO (Parallel I/O)** and click **Add…**. When the PIO configuration dialog box appears, set the **Width** of the interface to **4** bits (there are four pushbuttons) and set the **Direction** to **Input ports only** as shown in Figure 17.9(a). Click **Next** to continue. On the next configuration page, set the options as shown in Figure 17.9(b). Click **Finish** to add the component. In the SOPC Builder, rename the PIO module to **buttons**.

Using the same procedure as above, add a second PIO component for the dipswitches. The settings for the PIO devices are shown in Figure 17.10. Rename this PIO module to **switches**.

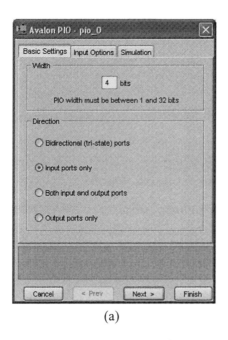

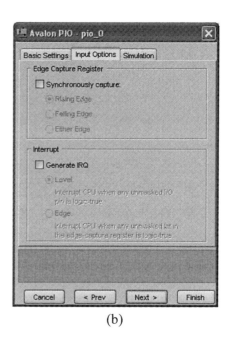

(a)   (b)

Figure 17.10 These are the settings for the dipswitch PIO device to be added to the Nios II system.

Finally, add a third PIO component for the LEDs. On the first configuration page, set **4** bits for the **Width**, and set the **Direction** to **Output ports only**. When the PIO is an output-only device, the interrupt and edge-capture options are not applicable. Rename this PIO module to **leds**.

17.8 Adding a SDRAM Memory Controller

There are three types of memory on the UP 3 board: SDRAM, SRAM, and Flash. Each type of memory requires is own unique memory controller and must be added individually. Add the SDRAM memory controller by expanding **Avalon Components** ⇨ **Memory**. Select **SDRAM Controller** and click **Add…**. The SDRAM controller must be configured for the timing requirements of the specific SDRAM brand and model being used. The configuration and timing values requested here are typically available in the datasheet for SDRAM ICs. For the SDRAM modules on the UP 3 board, set the options in the configuration dialog boxes to the values shown in Figure 17.11. Click **Finish** to add the component. In the SOPC Builder, rename the SDRAM controller module to **sdram**.

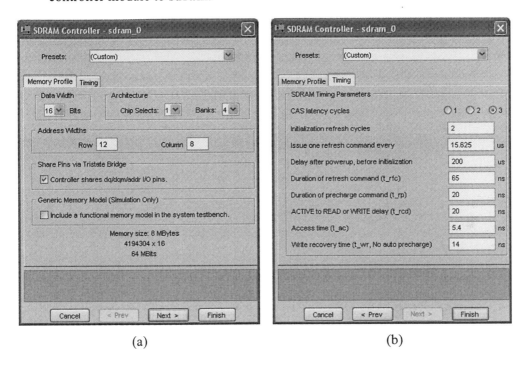

(a)   (b)

Figure 17.11 These are the SDRAM controller settings for use with the SDRAM on the UP 3 board.

17.9 Adding an External Bus

The SRAM, Flash, and LCD display devices on the UP 3 board share a common address and bidirectional data bus. Having multiple external devices like these share the same address and data bus pins can dramatically reduce the number of pins required on the FPGA, and the Nios II processor supports this type of bus sharing with its tristate bus components. To accommodate the bidirectional data bus and multiple devices on a single bus, an Avalon Tri-state Bridge component must be added. The Avalon tri-state bridge creates a peripheral (tri-state) bus to which multiple memory controllers and other external components can be attached. It also provides a seamless interface between the peripheral bus and the main system bus. A conceptual drawing of this arrangement is shown in Figure 17.12.

Add the Avalon Tri-state Bridge component by expanding **Avalon Components** ⇒ **Bridges**. Select **Avalon Tri-state Bridge** and click **Add…**. There is only one option for this component: registered or not registered. Select **Registered** and click **Finish** to add the component. In the SOPC Builder, rename the bridge module to **ext_bus**.

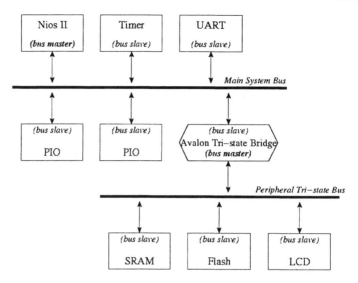

Figure 17.12 This is a conceptual drawing of the bus configuration with the Tristate Bridge connecting the main system bus and the shared peripheral bus.

17.10 Adding Components to the External Bus

Once the Avalon tri-state bridge has been added, the peripherals that are going to connect to the external peripheral bus can be added. First, add the Flash memory controller by expanding **Avalon Components ⇨ Memory**. Select **Flash Memory (Common Flash Interface)** and click **Add....** When the Flash memory configuration dialog box appears, set the options as shown in Figure 17.13.

IMPORTANT NOTE: Peripheral ICs on the UP 3 board can come from several manufacturers. While they are generically equivalent, they may exhibit slightly different characteristics and timing. UP 3 boards currently contain one of two types of flash memory chips, Toshiba and ExcelSemi. UP 3 boards that contain a Toshiba flash memory chip currently will not program and do not pass the example flash memory test. The flash memory chip, U8, is located directly below the large square FPGA chip in the middle of the UP 3 board. The manufacturer's name is printed on the flash chip. The Toshiba chips require new flash driver software that is being developed at the time of this printing. If your UP 3 project requires flash memory and you verify that you have the Toshiba chips, check for Toshiba flash driver software updates and specific information about different chips on UP 3 boards at the book's website:

http://ww.ece.gatech.edu/users/hamblen/book/bookte.htm

Add the SRAM memory controller to your Nios II processor by expanding **Avalon Components ⇨ UP3 Development Board**. Select **up3_tristate_sram** and click **Add....** When the SRAM configuration dialog box appears, leave the default values and click **Finish** to add the component. In the SOPC Builder, rename the timer module to **sram**.

Add the LCD component to your Nios II processor by expanding **Avalon Components** ⇨ **UP3 Development Board**. Select **up3_tristate_lcd** and click **Add....** This component does not have any configuration options; therefore, no dialog box will appear. The LCD component will be added to the list of peripherals in your Nios II processor. In the SOPC Builder, rename the LCD module to **lcd**.

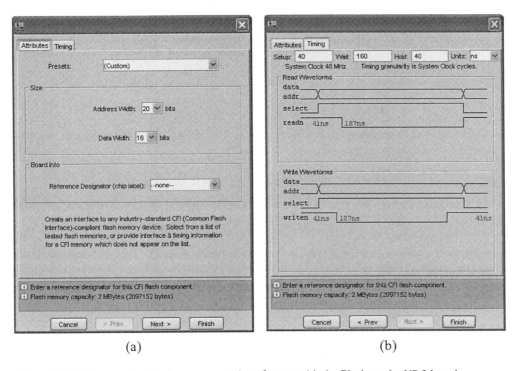

(a)   (b)

Figure 17.13 These are the Flash memory settings for use with the Flash on the UP 3 board.

17.11 Global Processor Settings

All of the necessary peripherals have been added now. The next step is to configure some global settings for your processor.

The Nios II processor uses a memory-mapped I/O scheme for accessing peripherals. Each component added to the system is assigned a unique set of memory addresses. Any device or data registers needed for a particular peripheral can be accessed by reading from or writing to its respective memory address. In general, the specific memory address assignments do not matter as long as the assigned memory address spaces do not overlap. If the Nios II system is going to be a part of a legacy system, there may be some constraints placed on the memory address assignments; however, there is nothing intrinsic within the Nios II system that restricts the settings. For this tutorial, let SOPC

Builder make the memory assignments automatically by selecting **System** ➪ **Auto-Assign Base Addresses**. Next, select **System** ➪ **Auto-Assign IRQs** to have SOPC Builder automatically assign the IRQ values to the devices that support interrupts.

To view and modify the bus connections in your processor, select **View** ➪ **Show Connections**. (If **Show Connections** is already selected, then un-select it and select it again.) This will expand the **cpu** and **ext_bus** modules in the table of peripherals and show the bus connections. The final SOPC Builder window should look like the screen shot in Figure 17.14. The three buses are displayed vertically. From left-to-right, the buses are the main system instruction, main system data, and tri-state data bus. Notice that the UARTs, timer, and PIO components are only attached to the system data bus since they don't normally interact with instruction memory. SDRAM and the Avalon Tri-state Bridge are connected to both the system instruction and system data buses, because the memory devices can store both data and instruction memory. Finally, the Flash memory, SRAM, and LCD devices are connected to the tri-state bus as expected. No modifications need to be made to the bus connections for this tutorial.

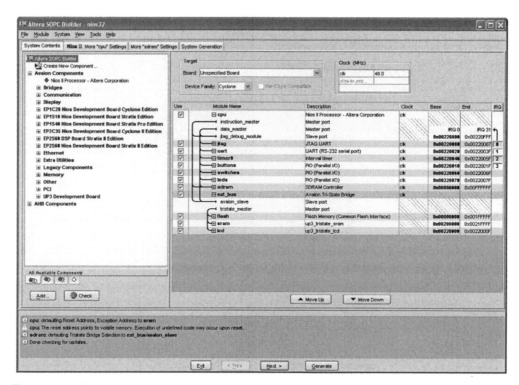

Figure 17.14 This is the completed Nios II design in SOPC Builder.

17.12 Finalizing the Nios II Processor

Click the **Next** button to modify more processor settings. The **Nios II More "cpu" settings** dialog box allows you to modify the program memory device and beginning address. For this tutorial, set the **Reset** and **Exception** addresses to **SRAM** and keep the default addresses and offsets as shown in Figure 17.15. Click **Next** to continue with the processor settings.

The **More "sdram" Settings** dialog box allows you to select the tri-state bus bridge to which to bind the SDRAM controller. There are no possible modifications for this setting in the current processor. Leave the default value and click **Next** to continue with the processor settings.

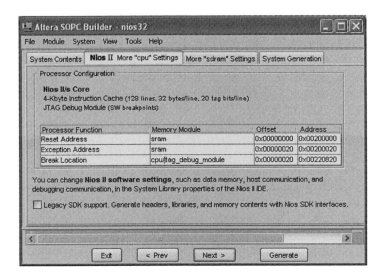

Figure 17.15 These are the processor configuration settings for the Nios II processor.

The **System Generation** dialog box is the final group of settings. In this dialog box, select the files that need to be generated. For this tutorial, you will not be simulating the processor in ModelSim or other third-party simulation tool; therefore, unselect the **Simulation. Create simulator project files** option. Verify that the option **HDL. Generate system module logic in VHDL** is selected. Click the **Generate** button to generate the design files for your Nios II processor. It will take 2-3 minutes to generate your Nios II processor. When it completes, the console should contain a message that states that your processor was generated successfully. If your system did not generate successfully, study the error log display in the console, correct the problem, and re-generate the Nios II processor. When you have successfully generated your Nios II system, click the **Exit** button to close SOPC Builder.

17.13 Add the Processor Symbol to the Top-Level Schematic

When SOPC Builder closes, return to your blank top-level schematic file, **rpds17.bdf**. Double click on a blank area of your empty top-level schematic file to add a component. In the **Libraries** pane of the **Symbol** dialog box, expand the **Project** item and select the **nios32** component. Click **OK** to add the selected component. Click in the middle of schematic file to place your Nios system.

17.14 Create a Phase-Locked Loop Component

SDRAM and the Nios II processor core operate on different clock edges. The Nios processor uses the rising edge and SDRAM the falling edge. The SDRAM would need a clock signal that is phase shifted by 180 degrees. An inverter would do this, but the phase shift also needs to be adjusted a bit to correct for the internal FPGA delays and the distance between the SDRAM and the FPGA on the UP 3 board. To create this SDRAM clock signal, a phase-locked loop (PLL) component can be implemented on the FPGA. To create a PLL, use Quartus II's MegaWizard Plug-in Manager by selecting **Tools** ⇨ **MegaWizard Plug-In Manager...**. Click **Next** on page 1 of the wizard to create a new component. On page 2, select the **Installed Plug-Ins** ⇨ **I/O** ⇨ **ALTPLL** module from the list. Enter the full path of your project directory followed by the filename **up3_pll** into the output filename field. Complete the remaining fields with the information shown in Figure 17.16. Click **Next** to continue.

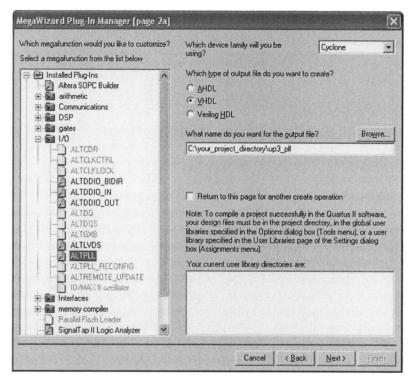

Figure 17.16 These are the initial settings for the ALTPLL module.

On page 3 of the MegaWizard manager, enter **48.00 MHz** as the **frequency of the inclock0 input**. Leave the other options set to their default values. Click

Next to continue. On page 4 of the MegaWizard manager, *un-select all checkmarks*. Click Next to continue.

On page 5 of the MegaWizard manager, enter a **Clock phase shift** of **-90 deg**. Leave the other options set to their default values. Click **Finish** to skip pages 6 and 7 and jump to page 8 of the MegaWizard manager. Click **Finish** again to complete the MegaWizard manager and create the files for the PLL component.

Double click on a blank area of the top-level schematic file. Select the **Project ⇨ up3_pll** module and add it to your top-level schematic as shown in the completed schematic in Figure 17.17(a).

IMPORTANT NOTE: Different or future versions of the Altera software may generate slightly different hardware time delays for the SDRAM clock. If you experience SDRAM errors after running memory tests on your final design or the program downloads to SDRAM do not verify, and after double checking that everything else is correct in your design, the PLL phase shift may need to be adjusted a small amount. Most designs seem to fall within about 30 degress of -90. This corresponds to a time delay adjustment of only 1 or 2 ns.

17.15 Add the UP 3 External Bus Multiplexer Component

The Flash memory, SRAM, and LCD display devices use the same FPGA pins and the same tri-state bus pins. The FPGA pins must be multiplexed to the different tri-state bus signals available on the Nios II symbol. A custom multiplexer component has been created and is located on the CD-ROM that came with this book in the **booksoft\chap17** directory. Copy the **up3_bus_mux.vhd** and **up3_bus_mux.bsf** files from the CD-ROM to your project directory.

When the **up3_bus_mux** files have been copied to your project directory, add the **up3_bus_mux** component to your top-level schematic by double clicking on a blank area of the top-level schematic. Select the **Project ⇨ up3_bus_mux** module and add it to your top-level schematic as shown in the completed schematic in Figure 17.17.

17.16 Complete the Top-Level Schematic

To complete the top-level schematic, add the input, output, and bi-directional pins (and pin names) shown in Figure 17.17(a). Also, complete the connections between the three top-level components as shown in the figure. Finally, name the **FLASH_CS_N**, **SRAM_CS_N**, and **SDRAM_CS_N** signals to make virtual connections to the output pins of the same name. If you have trouble reading the signal names in the figure, the file is available on the CD-ROM.

17.17 Design Compilation

Verify that the pin assignments discussed in Section 17.2 were made correctly by going to **Assignments ⇨ Pins**. A long list of pin numbers and names corresponding to the pin names you entered into the top-level schematic should appear. If it does not, then repeat the steps in Section 17.2 to import the pin assignments.

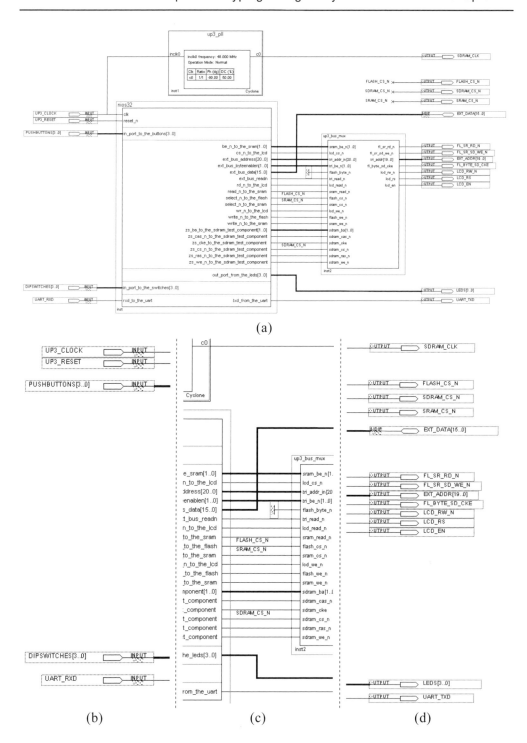

Figure 17.17 The final top-level schematic for the Nios II system is shown in (a). The figures in (b), (c), and (d) are pieces of the overall figure enlarged for better visibility of the signal names.

Verify that the global assignments discussed in Section 17.2 were made correctly by going to **Assignments ⇨ Device... ⇨ Device & Pin Options ⇨ Unused Pins**.

The **Reserve all unused pins** option should be set to **As inputs, tri-stated**. If it is not, then select this option. Click **OK** until all dialog boxes are closed.

Select **Processing ⇨ Start Compilation** to begin compiling your project.

17.18 Testing the Nios II Project

To fully test your Nios II project, you will need to write a software program to run on the Nios II processor that tests each component. To complete this task, refer to the previous chapter, which contains *Tutorial III: Nios II Processor Software Design*.

You might want to try your test program from the previous chapter first to verify that memory still works in your new design. After switching to a new workspace for the new project in Nios II IDE, you can import an existing software project into a new design project's software directory using **File ⇨ Import**. You will need to clean and rebuild the software project since the system library changes for each new hardware design.

 ALL SOURCE FILES FOR THIS UP 3 NIOS HARDWARE REFERENCE DESIGN

CAN BE FOUND ON THE CD-ROM IN THE \CHAP17 DIRECTORY.

17.19 For additional information

This chapter has provided a brief overview of Nios II hardware development. Additional information can be found at Altera's website (www.altera.com) in the *Nios II Processor Reference Handbook, Embedded Peripherals Handbook* and *Hardware Development Tutorial*. Nios II components for the UP 3 board and other reference designs can be found at the SLS website (www.slscorp.com). The Nios Community Forum (www.niosforum.com) also contains useful information and downloads for Nios II projects.

17.20 Laboratory Exercises

1. Add two 8-bit PIO to the Nios II hardware design that connect to the 5 volt I/O pins on the UP 3's J3 connector. Setup one port for input and one port for output. Connect the PIO port's I/O pins to eight input pins and eight output pins on J3. This is a handy way to interface external devices and sensors like those used in the UP 3 robot projects in Chapter 13 to the UP 3 board's Nios II processor.

2. Add a PIO port to the Nios II hardware design and use the PIO port's I/O bits to design an I²C hardware interface to the UP 3 board's real-time clock chip. Software will be needed to send I²C commands, the PIO port just provides a hardware interface to the I²C SDA and SLC bits (see Section 12.4).

3. Add a parallel port to the Nios II hardware design. Use two 8-bit ports, one for data and one for status and control bits. Connect the PIO port's I/O bits to the parallel port connector on the UP 3 board. Software will be needed to monitor and control the handshake lines (see Section 12.1) when connecting to a device like a parallel printer.

4. Add an SPI interface to the Nios II hardware design and use it to interface to an external SPI device connected to one of the UP 3 board's expansion connectors.

5. Implement one of the UP 3 robotics projects from Chapter 13 using a Nios II processor running C code. See problem 1 for robot interface suggestions.

6. Design an automatic setback HVAC thermostat using the UP 3. Interface a temperature sensor to the UP 3. Some temperature sensors are available with digital outputs that would not require a separate analog-to-digital IC. Display the current time, temperature, heat, fan, and A/C status, and the temperature settings in the LCD. Use the pushbuttons to change the temperature settings and setback times. Use the LEDs to indicate the heat, A/C, and fan control outputs from the thermostat. You can heat the temperature sensor with your finger to cycle the thermostat and cool it with ice or an aerosol spray can of dust off cleaner.

7. Interface a PS/2 keyboard or mouse to the Nios II processor using PIO ports. Write software to demonstrate the new keyboard or mouse interface. Display the output on the LCD or the UART. There are two major options to consider, use the keyboard and mouse cores from Chapter 11 or do everything in software.

8. Use the video sync core and character generation ROM from Chapter 10 to add a video text display to the Nios processor. Add a dual port memory to store a screen full of characters. Write charcters to the dual port memory from the Nios II processor using PIO ports added to the Nios II design. The video system constantly reads the characters out of the dual port memory and then uses the character generation ROM to generate the video display. Write a software driver for the video display and attach a monitor to the UP 3's VGA connector to demonstrate your design.

9. After solving the previous two problems, develop software for a video game that uses the mouse or keyboard for input and displays output on the monitor. If you need graphics for your game, consider replacing the character memory and text display with a larger memory containing only pixels used in a graphics display. Keep in mind that the internal FPGA memory is limited.

10. Add a custom instruction to the Nios II processor designed to speed up a particular application area. See the *Nios II Custom Instruction User Guide*. Demostrate the speedup obtained with the new instruction by running the application with and without the new instruction.

11. Interface the dual port video display memory used in one of the earlier problems directly to the Avalon system bus instead of using PIO ports. See the *Avalon Interface Specification Manual*.

12. Program the UP 3's serial flash device so that your Nios II hardware design loads automatically at power up. See Appendix E for instructions on programming the FPGA's serial flash configuration chip.

13. Program a complete Nios II design into both Flash memories so that the UP 3 board loads both the FPGA hardware configuration data and the software from the two Flash memories automatically at power up. See the *Nios II Flash Programmer User Guide* and study the section on how to port the Flash programmer to a new board type. A full version Altera software license is required for Flash programming of Nios II program code.

14. For a more challenging problem, port the eCos operating system to the UP 3. It is available free at www.niosforum.com. First, run a simple hello world application using the UART. For the second test, write a multithreaded application with one thread talking to the UART and a second thread blinking the LEDs.

Appendix A: Generation of Pseudo Random Binary Sequences

In many applications, random sequences of binary numbers are needed. These applications include random number generation for games, automatic test pattern generation, data encryption and decryption, data compression, and data error coding. Since a properly operating digital circuit never produces a random result, these sequences are called pseudo random. A long pseudo-random binary sequence appears to be random at first glance.

Table A.1 shows how to make an "n" bit pseudo-random binary sequence generator. Here is how it works for n = 8. Build an 8-bit shift register that shifts left one bit per clock cycle. Find the entry in Table A.1 for n = 8. This entry shows that bits 8,6,5,4 should all XORed or XNORed together to generate the next shift input used as the low bit in the shift register. Recall that the order of XOR operations does not matter. Note that the low-bit number is "1" and not "0" in this table.

A state machine that is actually a non-binary counter is produced. The counter visits all 2^n-1 non-zero states once in a pseudo-random order and then repeats. Since the counter visits every state once, a uniform distribution of numbers from 1 to 2^n-1 is generated. In addition to a shift register, only a minimal number of XOR or XNOR gates are needed for the circuit. The circuit is easy to synthesize in a HDL such as VHDL since only a few lines are required to shift and XOR the appropriate bits. Note that the next value in the random sequence is actually 2x or 2x + 1 the previous value, x. For applications that may require a more truly random appearing sequence order, use a larger random sequence generator and select a disjoint subset of the bits and shuffle the output bits.

The initial value loaded in the counter is called the seed. The seed or the random number is never zero in this circuit. If a seed of zero is ever loaded in the shift register it will stay stuck at zero. If needed, the circuit can be modified so that it generates 2^n states. For the same initial seed value, the circuit will always generate the same sequence of numbers. In applications that wait for input, a random seed can be obtained by building a counter with a fast clock and saving the value of the counter for the seed when an input device such as a pushbutton is activated.

Additional information on pseudo-random binary sequence generators can be found in *HDL Chip Design* by D.J. Smith, Doone Publications, 1996, and Xilinx Application Note 52, 1996.

Table A.1 Primitive Polynomials Modulo 2.

n	XOR from bits	n	XOR from bits	n	XOR from bits	n	XOR from bits
3	3,2	45	45,44,42,41	87	87,74	129	129,124
4	4,3	46	46,45,26,25	88	88,87,17,16	130	130,127
5	5,3	47	47,42	89	89,51	131	131,130,84,83
6	6,5	48	48,47,21,20	90	90,89,72,71	132	132,103
7	7,6	49	49,40	91	91,90,8,7	133	133,132,82,81
8	8,6,5,4	50	50,49,24,23	92	92,91,80,79	134	134,77
9	9,5	51	51,50,36,35	93	93,91	135	135,124
10	10,7	52	52,49	94	94,73	136	136,135,11,10
11	11,9	53	53,52,38,37	95	95,84	137	137,116
12	12,6,4,1	54	54,53,18,17	96	96,94,49,47	138	138,137,131,130
13	13,4,3,1	55	55,31	97	97,91	139	139,136,134,131
14	14,5,3,1	56	56,55,35,34	98	98,87	140	140,111
15	15,14	57	57,50	99	99,97,54,52	141	141,140,110,109
16	16,15,13,4	58	58,39	100	100,63	142	142,121
17	17,14	59	59,58,38,37	101	101,100,95,94	143	143,142,123,122
18	18,11	60	60,59	102	102,101,36,35	144	144,143,75,74
19	19,6,2,1	61	61,60,46,45	103	103,94	145	145,93
20	20,17	62	62,61,6,5	104	104,103,94,93	146	146,145,87,86
21	21,19	63	63,62	105	105,89	147	147,146,110,109
22	22,21	64	64,63,61,60	106	106,91	148	148,121
23	23,18	65	65,47	107	107,105,44,42	149	149,148,40,39
24	24,23,22,17	66	66,65,57,56	108	108,77	150	150,97
25	25,22	67	67,66,58,57	109	109,108,103,102	151	151,148
26	26,6,2,1	68	68,59	110	110,109,98,97	152	152,151,87,86
27	27,5,2,1	69	69,67,42,40	111	111,101	153	153,152
28	28,25	70	70,69,55,54	112	112,110,69,67	154	154,152,27,25
29	29,27	71	71,65	113	113,104	155	155,154,124,123
30	30,6,4,1	72	72,66,25,19	114	114,113,33,32	156	156,155,41,40
31	31,28	73	73,48	115	115,114,101,100	157	157,156,131,130
32	32,22,2,1	74	74,73,59,58	116	116,115,46,45	158	158,157,132,131
33	33,20	75	75,74,65,64	117	117,115,99,97	159	159,128
34	34,27,2,1	76	76,75,41,40	118	118,85	160	160,159,142,141
35	35,33	77	77,76,47,46	119	119,111	161	161,143
36	36,25	78	78,77,59,58	120	120,113,9,2	162	162,161,75,74
37	37,5,4,3,2,1	79	79,70	121	121,103	163	163,162,104,103
38	38,6,5,1	80	80,79,43,42	122	122,121,63,62	164	164,163,151,150
39	39,35	81	81,77	123	123,121	165	165,164,135,134
40	40,38,21,19	82	82,79,47,44	124	124,87	166	166,165,128,127
41	41,38	83	83,82,38,37	125	125,124,18,17	167	167,161
42	42,41,20,19	84	84,71	126	126,125,90,89	168	168,166,153,151
43	43,42,38,37	85	85,84,58,57	127	127,126		
44	44,43,18,17	86	86,85,74,73	128	128,126,101,99		

Appendix B: Quartus II Design and Data File Extensions

Project Files

Quartus II Project File (*.qpf)
Quartus II Settings File (*.qsf)
Quartus II Workspace File (*.qws)
Quartus II Default Settings File (*.qdf)

Design Files

Altera Design File (*.adf)
Block Design File (*.bdf)
EDIF Input File (*.edf)
Graphic Design File (*.gdf)
OrCAD Schematic File (*.sch)
State Machine File (*.smf)
Text Design File (*.tdf)
Verilog Design File (*.v)
VHDL Design File (*.vhd)
Waveform Design File (*.wdf)
Xilinx Netlist File (*.xnf)

Ancillary Data Files

Assignment and Configuration File (*.acf)
Assignment and Configuration Output (*.aco)
Block Symbol File (*.bsf)
Command File (*.cmd)
EDIF Command File (*.edc)
Fit File (*.fit)
Intel Hexadecimal Format File (*.hex)
History File (*.hst)
Include File (*.inc)
JTAG chain file (*.jcf)
Library Mapping File (*.lmf)
Log File (*.log)
Memory Initialization File (*.mif)
Memory Initialization Output File (*.mio)
Message Text File (*.mtf)

Programmer Log File (*.plf)
Report File (*.rpt)
Simulator Channel File (*.scf)
Standard Delay Format (*.sdf)
Standard Delay Format Output File (*.sdo)
Symbol File (*.sym)
Table File (*.tbl)
Tabular Text File (*.ttf)
Text Design Export File (*.tdx)
Text Design Output File (*.tdo)
Timing Analyzer Output File (*.tao)
Vector File (*.vec)
Verilog Quartus Mapping File (*.vqm)
VHDL Memory Model Output File (*.vmo)

Non-Editable Ancillary File Types

Compiler Netlist File (*.cnf)
Hierarchy Interconnect File (*.hif)
JEDEC file (*.jed)
Node Database File (*.ndb)
Programmer Object File (*.pof)
Raw Binary File (*.rbf)
Serial Bitstream File (*.sbf)
Simulator Initialization File (*.sif)
Simulator Netlist File (*.snf)
SRAM Object File (*.sof)

Appendix C: UP 3 Pin Assignments

Table C.1 UP 3 board's Cyclone FPGA I/O pin assignments.

Pin Name	FPGA Pin#	Pin I/O Type	Function of Pin
PARALLEL_D[1]	1	Bidirectional	Parallel Port Connector Pin 3
PARALLEL_S[3]	2	Input	Parallel Port Connector Pin 15
PARALLEL_D[3]	3	Bidirectional	Parallel Port Connector Pin 5
PARALLEL_C[3]	4	Bidirectional	Parallel Port Connector Pin 3
PARALLEL_D[2]	5	Output	Parallel Port Connector Pin 17
PARALLEL_D[0]	6	Bidirectional	Parallel Port Connector Pin 2
PARALLEL_C[1]	7	Output	Parallel Port Connector Pin 14
PARALLEL_C[0]	8	Output	Parallel Port Connector Pin 1
	9		Not available for I/O pin use
	10		Not available for I/O pin use
SDRAM_CLK	11	Output	SDRAM Clock
PS2_CLK	12	Bidirectional	PS2 Connector
PS2_DATA	13	Bidirectional	PS2 Connector
USB_VMO	14		USB Phy Chip
USB_VPO	15		USB Phy Chip
USB_OE	16		USB Phy Chip
USB_RCV	17		USB Phy Chip
USB_VP	18		USB Phy Chip
USB_VM	19		USB Phy Chip
I2C_SCL	20	Bidirectional	I2C Bus EEPROM
I2C_SDA	21	Bidirectional	I2C Bus EEPROM
	22		Not available for I/O pin use
RESET	23	Input	Power on or SW8 pushbutton reset (Reset = 0)
	24		CS0# EPCS1
	25		DATA EPCS1
	26		Not available for I/O pin use
	27		Not available for I/O pin use
	28		
USB_CLK	29	Input	USB 48MHz Clock
	30		Not available for I/O pin use
	31		Not available for I/O pin use
	32		CE#
	33		Not available for I/O pin use
	34		Not available for I/O pin use
	35		Not available for I/O pin use
	36		DCLK EPCS1
	37		ASDO EPCS1
USER_CLOCK	38	Input	External Clock from J2 Pin 2
SERIAL_DCD	39	Input	MAX 3243 to Serial Port Connector Pin 1
	40		
SERIAL_DSR	41	Input	MAX 3243 to Serial Port Connector Pin 6

Pin Name	FPGA Pin#	Pin I/O Type	Function of Pin
SERIAL_RX	42	Input	MAX 3243 to Serial Port Connector Pin 2
SERIAL_CTS	43	Input	MAX 3243 to Serial Port Connector Pin 8
SERIAL_RI	44	Output	MAX 3243 to Serial Port Connector Pin 9
SERIAL_RTS	45	Output	MAX 3243 to Serial Port Connector Pin 7
SERIAL_DTR	46	Output	MAX 3243 to Serial Port Connector Pin 4
SERIAL_TX	47	Output	MAX 3243 to Serial Port Connector Pin 3
PBSWITCH_4	48	Input	Pushbutton SW4 (non-debounced, 0 = button hit)
PBSWITCH_5	49	Input	Pushbutton SW5 (non-debounced, 0 = hit)
LCD_E	50	Output	LCD Enable line
	51		Not available for I/O pin use
	52		Not available for I/O pin use
LED_D6	53	Output	LED D3 (1 = LED ON, 0= LED OFF)
LED_D5	54	Output	LED D4 (1 = LED ON, 0= LED OFF)
LED_D4	55	Output	LED D5 (1 = LED ON, 0= LED OFF)
LED_D3	56	Output	LED D6 (1 = LED ON, 0= LED OFF)
PBSWITCH_6	57	Input	Pushbutton SW6 (non-debounced, 0 = hit)
DIPSWITCH_1	58	Input	DIP Switch SW3 #1 (ON = 1, OFF = 0)
DIPSWITCH_2	59	Input	DIP Switch SW3 #2 (ON = 1, OFF = 0)
DIPSWITCH_3	60	Input	DIP Switch SW3 #3 (ON = 1, OFF = 0)
DIPSWITCH_4	61	Input	DIP Switch SW3 #4 (ON = 1, OFF = 0)
PBSWITCH_7	62	Input	Pushbutton SW7 (non-debounced, 0 = hit)
MEM_AD[7]	63	Output	Memory Address Bus
MEM_AD[8]	64	Output	Memory Address Bus
MEM_AD[9]	65	Output	Memory Address Bus
MEM_AD[10]	66	Output	Memory Address Bus
MEM_AD[11]	67	Output	Memory Address Bus
MEM_AD[12]	68	Output	Memory Address Bus
	69		Not available for I/O pin use
	70		Not available for I/O pin use
	71		Not available for I/O pin use
	72		Not available for I/O pin use
LCD_RW	73	Output	LCD R/W control line
MEM_AD[13]	74	Output	Memory Address Bus
MEM_AD[14]	75	Output	Memory Address Bus
MEM_AD[15]	76	Output	Memory Address Bus
MEM_AD[16]	77	Output	Memory Address Bus, SRAM LB, SDRAM LDQM
MEM_AD[19]	78		Memory Address Bus
MEM_WE	79	Output	Memory Write Enable
FLASH_RY/BY	80(126)*	Output	FLASH
MEM_AD[18]	81(125)*	Output	Memory Address Bus FLASH only
MEM_AD[17]	82	Output	Memory Address Bus, SRAM UB, SDRAM UDQM
MEM_AD[6]	83	Output	Memory Address Bus
MEM_AD[5]	84	Output	Memory Address Bus
MEM_AD[4]	85	Output	Memory Address Bus
MEM_AD[3]	86	Output	Memory Address Bus

Pin Name	FPGA Pin#	Pin I/O Type	Function of Pin
MEM_AD[2]	87	Output	Memory Address Bus
MEM_AD[1]	88	Output	Memory Address Bus
	89		Not available for I/O pin use
	90		Not available for I/O pin use
	91		Not available for I/O pin use
	92		Not available for I/O pin use
MEM_AD[0]	93	Output	Memory Address Bus
MEM_DQ[0]	94	Bidirectional	Memory & LCD Data Bus
MEM_DQ[8]	95	Bidirectional	Memory Data Bus
MEM_DQ[1]	96(133)*	Bidirectional	Memory & LCD Data Bus
MEM_DQ[9]	97(132)*	Bidirectional	Memory Data Bus
MEM_DQ[2]	98	Bidirectional	Memory & LCD Data Bus
MEM_DQ[10]	99	Bidirectional	Memory Data Bus
MEM_DQ[3]	100	Bidirectional	Memory & LCD Data Bus
MEM_DQ[11]	101	Bidirectional	Memory Data Bus
MEM_DQ[4]	102(128)*	Bidirectional	Memory & LCD Data Bus
MEM_DQ[12]	103(127)*	Bidirectional	Memory Data Bus
MEM_DQ[5]	104	Bidirectional	Memory & LCD Data Bus
MEM_DQ[13]	105	Bidirectional	Memory Data Bus
MEM_DQ[6]	106	Bidirectional	Memory & LCD Data Bus
MEM_DQ[14]	107	Bidirectional	Memory Data Bus
LCD_RS	108	Output	LCD Register Select Line
	109		Not available for I/O pin use
	110		Not available for I/O pin use
	111		Not available for I/O pin use
	112		Not available for I/O pin use
MEM_DQ[7]	113	Bidirectional	Memory & LCD Data Bus
MEM_DQ[15]	114	Bidirectional	Memory Data Bus
SDRAM_CKE	115	Output	Also FLASH Byte
SRAM_CE	116	Output	
	117		
SRAM_OE	118	Output	Also FLASH OE
SDRAM_CE	119	Output	SDRAM Clock Enable
I2C_SDA	120	Bidirectional	I2C RT Clock Chip
I2C_SDL	121	Bidirectional	I2C RT Clock Chip
VGA_GREEN	122	Output	VGA Connector Green Video Signal
PCI_STOP	123	Output	STOP PCI Clock Outputs
	124		
	125(NA)*		J1 Pin 38
	126(NA)*		J1 Pin 37
	127(NA)*		J1 Pin 36
	128(NA)*		J1 Pin 35
	129		Not available for I/O pin use
	130		Not available for I/O pin use
PROTO_CLKOUT	131		J4 Pin 13
	132(NA)*		J1 Pin 34
	133(NA)*		J1 Pin 33
	134		J1 Pin 32

Pin Name	FPGA Pin#	Pin I/O Type	Function of Pin
	135		J1 Pin 31
	136		J1 Pin 30
	137		J1 Pin 29
	138		
	139		J1 Pin 4
	140		J1 Pin 5
	141		J1 Pin 6
	142		Not available for I/O pin use
	143		J1 Pin 7
PROTO_CLKIN	144		J4 Pin 11
	145		CONF_DONE LED D15
	146		Not available for I/O pin use
	147		TCLK JTAG
	148		TMS JTAG
	149		TD0 JTAG
	150		Not available for I/O pin use
	151		Not available for I/O pin use
PCI_CLKE	152	Input	PCI Clock 33MHz
CPU_CLOCK	153	Input	CPU Clock 100 or 66MHz
	154		Not available for I/O pin use
	155		TDI JTAG
	156		J1 Pin 8
	157		Not available for I/O pin use
	158		J1 Pin 9
	159		J1 Pin 10
	160		J1 Pin 11
	161		J1 Pin 12
	162		J1 Pin 15
	163		J1 Pin 16
	164		J1 Pin 17
	165		J1 Pin 18
	166		J1 Pin 19
	167		J1 Pin 20
	168(NA)*		J1 Pin 21
	169(NA)*		J1 Pin 22
VGA_BLUE	170	Output	VGA Connector Blue Video Signal
	171		Not available for I/O pin use
	172		Not available for I/O pin use
IDE_DASO#	173		IDE Disk Connector J3
IDE_CS0#	174		IDE Disk Connector J3
	175(NA)*		J1 Pin 25
IDE_A[1]	176	Output	IDE Disk Connector J3
	177(NA)*		J1 Pin 26
	178		
	179		
	180(NA)*		J1 Pin 23
	181(NA)*		J1 Pin 24

Pin Name	FPGA Pin#	Pin I/O Type	Function of Pin
	182		J2 Pin 12
	183		J2 Pin 10
	184		J2 Pin 8
	185		J2 Pin 6
	186		J2 Pin 4
	187		J2 Pin 13
	188		J2 Pin 11
	189		Not available for I/O pin use
	190		Not available for I/O pin use
	191		Not available for I/O pin use
	192		Not available for I/O pin use
	193		J2 Pin 9
	194		J2 Pin 7
	195		J2 Pin 5
	196		
	197		
	198(181)*		J2 Pin 14
IDE_DMARQ	199(180)*	Input	IDE Disk Connector J3
IDE_WE#	200	Output	IDE Disk Connector J3
IDE_OE#	201	Output	IDE Disk Connector J3
IDE_IORDY	202	Output	IDE Disk Connector J3
IDE_DMACK	203	Output	IDE Disk Connector J3
IDE_INTRQ	204(177)*	Input	IDE Disk Connector J3
IDE_A[0]	205(175)*	Output	IDE Disk Connector J3
IDE_D[4]	206	Bidirectional	IDE Disk Connector J3
IDE_D[3]	207	Bidirectional	IDE Disk Connector J3
IDE_D[2]	208	Bidirectional	IDE Disk Connector J3
	209		Not available for I/O pin use
	210		Not available for I/O pin use
	211		Not available for I/O pin use
	212		Not available for I/O pin use
	213		
	214		
IDE_D[5]	215	Bidirectional	IDE Disk Connector J3
IDE_D[6]	216	Bidirectional	IDE Disk Connector J3
IDE_D[7]	217	Bidirectional	IDE Disk Connector J3
IDE_D[10]	218	Bidirectional	IDE Disk Connector J3
IDE_D[9]	219	Bidirectional	IDE Disk Connector J3
IDE_D[8]	220(169)*	Bidirectional	IDE Disk Connector J3
IDE_D[11]	221(168)*	Bidirectional	IDE Disk Connector J3
IDE_D[12]	222	Bidirectional	IDE Disk Connector J3
IDE_D[13]	223	Bidirectional	IDE Disk Connector J3
IDE_D[14]	224	Bidirectional	IDE Disk Connector J3
IDE_D[15]	225	Bidirectional	IDE Disk Connector J3
VGA_VSYNC	226	Output	VGA Connector Vertical Sync Signal
VGA_HSYNC	227	Output	VGA Connector Horizontal Sync Signal
VGA_RED	228	Output	VGA Connector Red Video Signal
	229		Not available for I/O pin use

Pin Name	FPGA Pin#	Pin I/O Type	Function of Pin
	230		Not available for I/O pin use
	231		Not available for I/O pin use
	232		Not available for I/O pin use
PARALLEL_S[4]	233	Input	Parallel Port Connector Pin 13
PARALLEL_S[5]	234	Input	Parallel Port Connector Pin 12
PARALLEL_S[7]	235	Input	Parallel Port Connector Pin 11
PARALLEL_S[6]	236	Input	Parallel Port Connector Pin 10
PARALLEL_D[6]	237	Bidirectional	Parallel Port Connector Pin 8
PARALLEL_D[5]	238	Bidirectional	Parallel Port Connector Pin 7
PARALLEL_D[7]	239	Bidirectional	Parallel Port Connector Pin 9
PARALLEL_D[4]	240	Bidirectional	Parallel Port Connector Pin 6

* **NOTE:** In the Table, differences in Pin assignments on the larger UP3 1C12 board are noted in "(..)"s. NA indicates not available. The larger Cyclcone 1C12 has twice the logic, but 12 fewer I/O pins than the 1C6 since it needs more power and ground pins.

If you switch a design from a 1C6 board to a 1C12 board, you will need to change the device type, fix the few pin assignments that are different, and recompile. Typically, only designs using the LCD or external memory devices will be affected by the pin changes.

When connecting external hardware to the UP 3's header pins, note that many of the pins on the J3 header provide 5V logic levels by going through a voltage level conversion chip. Most other pins connected to the J1 & J2 headers have 3V logic levels coming directly from FPGA pins. Check the UP 3 schematic for more details.

Appendix D: ASCII Character Code

Table D.1 ASCII Charcter to Hexadecimal Conversion Table.

	0	1	2	3	4	5	6	7	8	9	A	B	C	D	E	F
0	NUL	SOH	STX	ETX	EOT	ENQ	ACK	BEL	BS	HT	LF	VT	FF	CR	SO	SI
1	DLE	DC1	DC2	DC3	DC4	NAK	SYN	ETB	CAN	EM	SUB	ESC	FS	GS	RS	US
2	SP	!	"	#	$	%	&	'	(	)	*	+	,	-	.	/
3	0	1	2	3	4	5	6	7	8	9	:	;	<	=	>	?
4	@	A	B	C	D	E	F	G	H	I	J	K	L	M	N	O
5	P	Q	R	S	T	U	V	W	X	Y	Z	[	\	]	^	_
6	`	a	b	c	d	e	f	g	h	i	j	k	l	m	n	o
7	p	q	r	s	t	u	v	w	x	y	z	{			}	~	DEL

The American Standard Code for Information Interchange (ASCII) is a standard seven-bit code for computer equipment adopted in 1968. In Table D.1, locate "A". It is in row 4 in column 1 and the hexadecimal value for "A" is therefore 41. The UP 3's LCD and most RS-232C serial devices and printers use the ASCII character code. The eighth bit may be used for parity. Codes below 0x20 are called control codes. Control codes perform operations other than printing a character. Several of the most common control codes are described below:

NUL (null) – all zeros, sometimes used for end of strings.

BEL (bell) - Causes a beep in terminals and terminal emulation programs.

BS (backspace) - Moves the cursor move backwards (left) one space.

HT (horizontal tab) - Moves the cursor right to the next tab stop. The spacing of tab stops is dependent on the output device, but is often 8 or 10.

LF (line feed) - Moves the cursor down to a new line, but not to the left.

VT (vertical tab)

FF (form feed) - Advances paper to the top of the next page on printers.

CR (carriage return) - Moves the cursor all the way to the left, but does

not advance to the next line. For a new line, both CR and LF are used.

ESC (escape) – Sometimes used to terminate program commands

SP (space) prints a blank space and cursor moves to the next character position.

Note that the decimal digit characters "0" to "9" range in value from 0x30 to 0x39. The code is setup so that "A"<"B"<"C"... so that numeric values can be used for alphabetical ordering. A single bit in the code changes the case of a character (i.e. see "A" and "a"). Extended ASCII codes use an eight bit code to display another 128 special graphics characters. There are several different standards for these new graphics characters, so check the device manual for details. The first 128 characters are the same as the 7-bit original ASCII code standard.

Appendix E: Programming the UP 3's Flash Memory

In deployed systems, FPGAs are normally automatically programmed at power on using a small serial non-volatile memory chip. The serial configuration device used on the UP 3 1C6 is the EPCS1 1-Mbit flash memory or on the UP 3 1C12 the larger EPCS4 4-Mbit flash memory. Once a design has been tested, you may want to demonstrate it without having to download the design using the ByteBlaster and a PC each time you power on the board. One common example would be a UP 3 robot project. If you program the UP 3's flash configuration memory, the FPGA will load the design each time the robot is turned on.

Steps to program the UP 3's flash memory configuration device:

Power down the UP 3 board and unplug the ByteBlaster II cable connector from the UP 3. Plug the ByteBlaster II cable into the second connector immediately to the left and nearer the edge of the board than the JTAG connector normally used for programming. This connector must be used for programming the serial configuration memory. Turn the UP 3 board's power back on.

Select **Tools** ⇨ **Programmer** and then change the Mode pull down box setting to **Active Serial Programming**. Click **yes** on the dialog box that appears. Select **Add File** and add *yourdesign.pof* (not *.sof as normally used) to the file list. Select *yourdesign.pof* and click **Open** so that *yourdesign.pof* appears as the programming file. Now **click on the file's Program/Configure** check box. An EPCS1 is used on the UP 3 1C6 and an EPCS4 is used on the larger UP 3 1C12. If you see the correct configuration device in the file's device entry column, **click start** to program the flash memory device.

If the correct configuration device does not appear in the *.pof file's device entry or you see a can't recognize silicon ID error when you try to program, you will need to change the project's configuration device type. Use **Assignments** ⇨ **Device** and **click on the Device and Pin Options button**. Next, **Click on the Configuration tab** and **select the correct device in the configuration device pull down box**. Also make sure that the **Generate compressed bitstreams box is checked. Click OK twice** to exit and **recompile** the project. When you **reopen the Programmer window** and the *yourdesign.pof* file you should now see the correct device type. **Click start** and the device should now program without errors.

After successfully downloading the FPGA's flash configuration memory, the UP 3 board will now start running the design every time it is turned on. Turn the UP 3 board off and then on to verify that the design is running.

Glossary

Assignment & Configuration File (.acf): An ASCII file (with the extension .acf) used by the older MAX+PLUS tool to store information about probe, pin, location, chip, clique, logic option, timing, connected pin and device assignments, as well as configuration settings for the Comp iler, Simulator, and Timing Analyzer for an entire project. The ACF stores information entered with menu commands in all MAX+PLUS II applications. You can also edit an ACF manually in a Text Editor window. This same information is now found in the *.q* files in Quartus II.

Active-high (Active-low) node: A node that is activated when it is assigned a value one (zero) or Vcc (Gnd).

AHDL: Acronym for Altera Hardware Description Language. Design entry language that supports Boolean equation, state machine, conditional, and decode logic. It also provides access to all Altera and user-defined macrofunctions.

Ancillary file: A file that is associated with a Quartus II project, but is not a design file in the project hierarchy tree.

Antifuse: Any of the programmable interconnect technologies forming electrical connection between two circuit points rather than making open connections.

Architecture: Describes the behavior, RTL or dataflow, and/ or structure of a VHDL entity. An architecture is created with an architecture body. A single entity can have more than one architecture. In some VHDL tools, configuration declarations are used to specify which architectures to use for each entity.

Array: A collection of one or more elements of the same type that are accessed using one or more indices depending on dimension of array. Array data types are declared with an array range and array element type.

ASIC: Acronym for Application-Specific Integrated Circuit. A circuit whose final photographic mask process is user design dependent.

ASM: Acronym for Algorithmic State Machine Chart. A flow-chart based method used to represent a state diagram.

Assert: A statement that checks whether a specified condition is true. If the condition is not true, a report is generated during simulation.

Assignment: In VHDL, assignment refers to the transfer of a value to a symbolic name or group, usually through a Boolean equation. The value on the right side of an assignment statement is assigned to the symbolic name or group on the left.

Asynchronous input: An input signal that is not synchronized to the device Clock.

Attribute: A special identifier used to return or specify information about a named entity. Predefined attributes are prefixed with the ' character.

Back annotation: Process of incorporating time delay values into a design netlist reflecting the interconnect capacitance obtained from a completed design. Also, in Altera's case, the process of copying device and resource assignments made by the Compiler into the Assignment and Configuration File for a project. This process preserves the current fit in future compilations.

Block: A feature that allows partitioning of the design description within an architecture.

Buried node: A combinatorial or registered signal that does not drive an output pin.

Cell: A logic function. It may be a gate, a flip-flop, or some other structure. Usually, a cell is small compared to other circuit building blocks.

Cell library: The collective name for a set of logic functions defined by the manufacturer of an FPGA or ASIC. Simulation and synthesis tools use cell libraries when simulating and synthesizing a model.

CLB: Acronym for Configurable Logic Block. This element is the basic building block of the Xilinx FPGA product family.

Clock: A signal that triggers registers. In a flip-flop or state machine, the clock is an edge-sensitive signal. In edge-triggered flip-flops, the output of the flip-flop can change only on the clock edge.

Clock enable: The level-sensitive signal on a flip-flop with E suffix, e.g., DFFE. When the Clock enable is low, clock transitions on the clock input of the flip-flop are ignored.

Compiler Netlist File (.cnf): A binary file (with the extension .cnf) that contains the data from a design file. The CNF is created by the Compiler Netlist Extractor module of the MAX+PLUS II Compiler.

Component: Specifies the port of a primitive or macrofunction in VHDL. A component consists of the name of the primitive or macrofunction, and a list of its inputs and outputs. Components are specified in the Component declaration.

Component instantiation: A concurrent statement that references a declared component and creates one unique instance of that component.

Configuration EPROM: A serial EPROM designed to configure (program) a FPGA.

Concurrent statements: HDL statements that are executed in parallel.

Configuration: It maps instances of VHDL components to design entities and describes how design entities are combined to form a complete design. Configuration declarations are used to specify which architectures to use for each entity.

Configuration scheme: The method used to load configuration (programming) data into an FPGA.

Constant: An object that has a constant value and cannot be changed.

Control unit: The hardware of a machine that controls the data path.

Cyclone: The FPGA family used on the UP 3 boards.

CPLD: Acronym for complex programmable logic device. CPLDs include an array of functionally complete or universal logic cells with an interconnection network.

Data Path: The hardware path that provides data processing and transfer of information in a machine, as opposed to the controller.

Design entity: A file that contains description of the logic for a project and is compiled by the Compiler.

Design library: Stores VHDL units that have already been compiled. These units can be referenced in VHDL designs.

Design unit: A section of VHDL description that can be compiled separately. Each design unit must have a unique name within the project.

Dual-purpose pins: Pins used to configure an FPGA device that can be used as I/O pins after initialization.

Dynamic reconfigurability: Capability of an FPGA to change its function "on -the-fly"

Embedded Array Block (EAB): A physically grouped set of 8 embedded cells that implement memory (RAM or ROM) or combinatorial logic in a Cyclone 10K device. A single EAB can implement a memory block of 256 x 8, 512 x 4, 1,024 x 2, or 2,048 x 1 bits.

EPLD: Acronym for EPROM programmable logic devices. This is a PLD that uses EPROM cells to internally configure the logic function. Also, erasable programmable logic device.

Event: The change of value of a signal. Usually refers to simulation.

Event scheduling: The process of scheduling of signal values to occur at some simulated time.

Excitation function: A Boolean function that specifies logic that directs state transitions in a state machine.

Exit condition: An expression that specifies a condition under which a loop should be terminated.

FLEX 10K and FLEX 10KA: An Altera device family based on Flexible Logic Element MatriX architecture. This SRAM-based family offers high-performance, register-intensive, high-gate-count devices with embedded arrays. The Cyclone 10K device family includes the EPF10K100, EPF10K70, EPF10K50, EPF10K40, EPF10K30, EPF10K20, and EPF10K10 devices. The FPGA used on the UP 2 board.

Fan-out: The number of output signals that can be driven by the output of a logic cell.

Fast Track interconnect: Dedicated connection paths that span the entire width and height of a Cyclone device. These connection paths allow the signals to travel between all LABs in device.

Field name: An identifier that provides access to one element of a record data type.

File type: A data type used to represent an arbitrary-length sequence of values of a given type.

For loop: A VHDL loop construct in which an iteration scheme is a for statement.

Finite state machine: The model of a sequential circuit that cycles through a predefined sequence of states.

Fitting: Process of making a design fit into a specific FPGA architecture. Fitting involves technology mapping, placement, optimization, and partitioning among other operations.

Flash: A non-volatile memory technology that also can be programmed in-circuit.

Flip-flop: An edge-sensitive memory device (cell) that stores a single bit of data.

Floorplan: Physical arrangement of functions within a connection framework of signal routing channels.

FPGA: Acronym for field programmable gate array. A regular array of logic cells that is either functionally complete or universal with an interconnection network of signal routing channels.

FPLD: Acronym for field programmable logic device. An integrated circuit used for implementing digital hardware that allows the end user to configure the chip to realize different designs. Configuring such a device is done using either a special programming unit or by doing it " in system". FLPDs include both CPLDs and FPGAs.

Functional simulation: A simulation mode that simulates the logical performance of a project without timing information.

Functional test vector: The input stimulus used during simulation to verity a VHDL model operates functionally as intended.

Functionally complete: Property of some Boolean logic functions permitting them to make any logic function by using only that function. The properties include making the AND function with an invert or the OR function with an invert or the OR function with an invert.

Fuse: A metallic interconnect point that can be electrically changed from short circuit to an open circuit by applying electrical current.

Gate: An electronic structure, built from transistors that performs a basic logic function.

Gate array: Array of transistors interconnected to form gates. The gates in turn are configured to form larger functions.

Gated clock: A clock configuration in which the output of an AND or OR gate drives a clock.

Generic: A parameter passed to a VHDL entity, component or block that describes additional, instance-specific information about that entity, component or block.

Glitch or spike: A narrow output pulse that occurs when a logic level changes two or more times over a short period.

Global signal: A signal from a dedicated input pin that does not pass through the logic array before performing its specified function. Clock, Preset, Clear, and Output Enable signals can be global signals.

GND: A Low-level input voltage. It is the default inactive node value.

Graphic Design File (.gdf): A schematic design file (with the extension .gdf) created with the MAX+PLUS II Graphic Editor.

HDL: Acronym for Hardware Description Language. A special language used to describe digital hardware.

Hexadecimal: The base 16 number system (radix). Hexadecimal digits are 0 through 9 and A through F.

Hierarchy: The structure of a design description, expressed as a tree of related components.

Identifier: A sequence of characters that uniquely identify a named entity in a design description.

Index: A scalar value that specifies an element or range of elements within an array.

Input vectors: Time-ordered binary numbers representing input values sequences to a simulation program.

I/O cell register: A register on the periphery of a Cyclone 8000 device or a fast input-type logic cell that is associated with an I/O pin.

IP core: An intellectual property (IP) core is a previously developed synthesizable hardware design that provides a widely used function. Commercially licensed IP cores include functions such as microprocessors, microcontrollers, bus interfaces, multimedia and DSP operations, and communications controllers.

LAB: Acronym for Logic Array Block. The LAB is the basic building block of the Altera MAX family. Each LAB contains at least one macrocell, an I/O block, and an expander product term array.

Latch: A level-sensitive clocked memory device (cell) that stores a single bit of data. A High to low transition on the Latch Enable signal fixes the contents of the latch at the value of the data input until the next Low-to-High transition on Latch Enable.

Latch enable: A level-sensitive signal that controls a latch. When it is High, the input flows through the output; when it is Low, the output holds its last value.

Library: In VHDL a library statement is used to store analyzed design units.

Literal: A value that can be applied to an object to some type.

Logic Synthesizer: The Compiler module that uses several algorithms to minimize gate count, remove redundant logic, and utilize the device architecture as efficiently as possible. Processing can be customized with logic options and logic synthesis style assignments. This module also applies logic synthesis techniques to help implement timing requirements for a project.

Least Significant Bit (LSB): The bit of a binary number that contributes the smallest quantity to the value of that number, i.e., the last member in a bus or group name. For example, the LSB for a bus or group named a[31..0] is a[0] (or a0).

Logic Cell (LC): The generic term for a basic building block of an Altera device. In MAX devices, a logic cell (also called a macrocell) consists of two parts: combinatorial logic and a configurable register. The combinatorial logic allows a wide variety of logic functions. In Cyclone and FLEX devices, a logic cell (also called a logic element) consists of a look-up table (LUT) and a programmable register to support sequential functions.

Logic element: A basic building block of an Altera Cyclone device. It consists of a look-up table i.e., a function generator that quickly computes any function of four variables, and a programmable flip-flop to support sequential functions.

LPM: Acronym for library of Parameterized Modules. Denotes Altera's library of design units that contain one or more changeable parts, and parameters that are used to customize a design unit as the application requires.

Macro: When used with FPGAs, a logic cell configuration that can be repeated as needed. It can be a Hard or a Soft macro. Hard macros force predefined place and route rules between logic cells.

Macrocell: In FPGAs, a portion of the FPGA that is smallest indivisible building block. In MAX devices it consists of two parts: combinatorial logic and a configurable register.

MAX: Acronym for Multiple Array MatriX, which is an Altera product family. It is usually considered to be a CPLD.

MAX+PLUS II: Acronym for multiple array matrix programmable logic user system II. An older set of computer aided design (CAD) tools that allow design and implementation of custom logic circuits with Altera's MAX and Flex FPGA devices.

Memory Initialization File (.mif): An ASCII file (with the extension .mif) used by Quartus II to specify the initial content of a memory block (RAM or ROM), i.e., the initial data values for each memory address. This file is used during project compilation and/or simulation.

Mealy state machine: A type of state machine in which the outputs are a function of the inputs and the current state.

Microblaze: A soft core RISC processor supported on Xilinx FPGAs.

Moore state machine: A state machine in which the present state depends only on its previous input and previous state, and the present output depends only on the present state. In general Moore states machines have more states than a Mealy machine.

Most Significant Bit (MSB): The bit of a binary number that contributes the greatest quantity to the value of that number, and the first member in a bus or group name. For example, the MSB for a bus named a[31..0] is a[31].

Mode: A direction of signal (either in, out, inout or buffer) used as subprogram parameter or port.

Model: A representation that behaves similarly to the operation of some digital circuit.

MPLD: Acronym for Mask Programmed Logic Device.

Netlist: A text file that describes a logic design. Minimal requirements are identification of function elements, inputs, outputs, and connections.

Netist synthesis: Process of deriving a netlist from an abstract representation, usually from a hardware description language.

Nios: A soft core RISC processor supported on Altera FPGAs.

NRE: Acronym for Non-Recurring Engineering expense. It reefers to one-time charge covering the use of design facilities, masks and overhead for test development of ASICs.

Object: A named entity of a specific type that can be assigned a value. Object in VHDL include signals, constants, variables and files.

Octal: The base 8 number system (radix). Octal digits are 0 though 7.

One Hot Encoding: A design technique used more with FPGAs than CPLDs. Only one flip-flop output is active at any time. One flip-flop per state is used. State outputs do not need to be decoded and they are hazard free.

Package: A collection of commonly used VHDL constructs that can be shared by more than one design unit.

PAL: Acronym for programmable array logic. A relatively small FPLD containing a programmable AND plane followed by a fixed-OR plane.

Parameter: An object or literal passed into a subprogram via that subprogram's parameter list.

Partitioning: Setting boundaries between subsections of a system or between multiple FPGA devices.

Physical types: A data type used to represent measurements.

Pin Number: A number used to assign an input or output signal in a design file, which corresponds to the pin number on an actual device.

PLA: (programmable logic array) a relatively small FPLD that contains two levels of programmable logic-an AND plane and an OR plane.

PLL: (phase locked loop) a device that can be used to multiply and divide clock signals and adjust the phase delay.

Placement: Physical assignment of a logical function to a specific location within an FPGA. Once the logic function is placed, its interconnection is made by routing.

PLD: Acronym for programmable logic device. This class of devices is comprised of PALs, PLAs, FPGAs, and CPLDs.

Port: A symbolic name that represents an input or output of a primitive or of a macrofunction design file.

Primitive: One of the basic functional blocks used to design circuits with Quartus II software. Primitives include buffers, flip-flops, latch, logical operators, ports, etc.

Process: A basic VHDL concurrent statement represented by a collection of sequential statements that are executed whenever there is an event on any signal that appears in the process sensitivity list, or whenever an event occurs that satisfies condition of a wait statement within the process.

Product Term: Two or more factors in a Boolean expression combined with an AND operator constitute a product term, where "product" means "logic product."

Programmable switch: A user programmable switch that can connect a logic element or input/output element to an interconnect wire or one interconnect wire to another.

Project: A project consists of all files that are associated with a particular design, including all subdesign files and ancillary files created by the user or by Quartus II software. The project name is the same as the name of the top-level design file without an extension.

Propagation delay: The time required for any signal transition to travel between pins and/or nodes in a device.

Radix: A number base. Group logic level and numerical values are entered and displayed in binary, decimal, hexadecimal, or octal radix in Quartus II.

Reset: An active-high input signal that asynchronously resets the output of a register to a logic Low (0) or a state machine to its initial state, regardless of other inputs.

Range: A subset of the possible values of a scalar type.

Register: A memory device that contains several latches or flip-flops that are clocked from the same clock signal.

Resource: A resource is a portion of a device that performs a specific, user-defined task (e.g., pins, logic cells, interconnection network).

Retargetting: A process of translating a design from one FPGA or other technology to another. Retargetting involves technology-mapping optimization.

Reset: An active-high input signal that asynchronously resets the output of a register to a logic Low (0) or a state machine to its initial state, regardless of other inputs.

Ripple Clock: A clocking scheme in which the Q output of one flip-flop drives the Clock input to another flip-flop. Ripple clocks can cause timing problems in complex designs.

RTL: Acronym for Register Transfer Level. The model of circuit described in VHDL that infers memory devices to store results of processing or data transfers. Sometimes it is referred to as a dataflow-style model.

Scalar: A data type that has a distinct order of its values, allowing two objects or literals of that type to be compared using relational operators.

Semicustom: General category of integrated circuits that can be configured directly by the user of an IC. It includes gate arrays and FPGA devices.

Signal: In VHDL a data object that has a current value and scheduled future values at simulation times. In RTL models signals denote direct hardware connections.

Simulation: Process of modeling a logical design and its stimuli in which the simulator calculates output signal values.

Slew rate: Time rate of change of voltage. Some FPGAs permit a fast or slow slew rate to be programmed for an output pin.

Slice: A one-dimensional, contiguous array created as a result of constraining a larger one-dimensional array.

SOPC: Acronym for System On-a Programmable Chip. SOPC systems contain a hard or soft core processor in the FPGA in addition to other user logic.

Speed performance: The maximum speed of a circuit implemented in an FPGA. It is set by the longest delay through any for combinational circuits, and by maximum clock frequency at which the circuit operates properly for sequential circuits.

State transition diagram: A graphical representation of the operation of a finite state machine using directed graphs.

State: A state is implemented in a device as a pattern of 1's and 0's (bits) that are the outputs of multiple flip-flops (collectively called a state machine state register).

Structural-type architecture: The level at which VHDL describes a circuit as an arrangement of interconnected components.

Subprogram: A function or procedure. It can be declared globally or locally.

Sum-of-products: A Boolean expression is said to be in sum-of-products form if it consists of product terms combined with the OR operator.

Synthesis: The process of converting the model of a design described in VHDL from one level of abstraction to another, lower and more detailed level that can be implemented in hardware.

Test bench: A VHDL model used to verify the correct behavior of another VHDL model, commonly known as the unit under test.

Tri-state Buffer: A buffer with an input, output, and controlling Output Enable signal. If the Output Enable input is High, the output signal equals the input. If the Output Enable input is Low, the output signal is in a state of high impedance. Tri-state outputs can be tied together but only one should ever be enabled at any given time.

Timing Simulation: A simulation that includes the actual device delay times.

Two's Complement: A system of representing binary numbers in which the negative of a number is equal to its logic inverse plus 1. In VHDL, you must declare a two's complement binary number with a signed data type or use the signed library.

Type: A declared name and its corresponding set of declared values representing the possible values the type.

Type declaration: A VHDL declaration statement that creates a new data type. A type declaration must include a type name and a description of the entire set of possible values for that type.

Universal logic cell: A logic cell capable of forming any combinational logic function of the number of inputs to the cell. RAM, ROM and multiplexers have been used to form universal logic cells. Sometimes they are also called look-up tables or function generators.

Usable gates: Term used to denote the fact that not all gates on an FPGA may be accessible and used for application purposes.

Variable: In VHDL, a data object that has only current value that can be changed in variable assignment statement.

Verilog: An HDL with features similar to VHDL with a syntax reminiscent of C.

VCC: A high-level input voltage represented as a High (1) logic level in binary group values.

VHDL: Acronym for VHSIC (Very High Speed Integrated Circuits) Hardware Description Language. VHDL is used to describe function, interconnect and modeling.

VITAL: Acronym for VHDL Initiative Toward ASIC Libraries. An industry-standard format for VHDL simulation libraries.

Index

A

Altera Cyclone Architecture, 50
  Embedded memory blocks, 50
  Input output elements (IOEs), 52
  logic array block (LAB), 52
  Logic elements, 50
  PLLs, 50
Altera FLEX 10K70 CPLD, 48
Altera MAX 7000S Architecture, 48
ALTSYNCRAM, 104, 123
antifuse, 58
application specific integrated circuits (ASICs), 44
arithmetic logic unit (ALU), 99, 120
ASCII, 355

C

case statement, 91, 114
cathode ray tube (CRT), 168
clock edge, 94, 117
clocking in VHDL, 94, 117
color in video display, 168
complex programmable logic devices (CPLDs), 44
component, 106
computer aided design (CAD) tools, 55, 56
concurrent assignment statement, 91
conv_integer, 89
conv_std_logic_vector, 89

D

digital oscilloscope, 67
dithering, 183

E

electric train
  direction, 132
  example controller, 134
  I/O summary, 134
  sensors, 133
  simulation, 140
  switches, 133
  track power, 132
  video output, 142
electrically erasable programmable read only memory
  (EEPROM), 48
EPCS1, 357
EPCS4, 357

F

field programmable gate arrays (FPGAs), 44
field programmable logic devices (FPLDs)
  applications, 57
floating point hardware, 101, 122
for loop, 268

G

gate arrays, 44
global clock buffer lines, 47, 52

H

hardware emulator, 57
H-bridge, 245

I

I²C Bus Interface, 211
if statement, 93, 116

K

keyboard. *See* PS/2 keyboard

L

logic element (LE), 47
look-up table (LUT), 50
LPM_DIV, 100, 121
LPM_MULT, 100, 121
LPM_RAM_DQ, 104, 123
LPM_ROM, 177

M

macrocell, 49
metastability, 47
MicroBlaze, 282
MIPS, 256
  control, 263
  decode, 268
  dmemory, 272
  execute, 270
  execution on UP 1, 274
  hardware implementation, 257
  ifetch, 265
  instruction formats, 256

instructions, 257
pipelined implementation, 258
simulation, 273
top_spim, 260
VHDL synthesis model, 259
mouse. *See* PS/2 mouse
multiply and divide hardware, 100, 121

N

Nios, 282
Nios Hardware, 324
Nios II IDE Software, 296
Nios II Processor, 327
    Flash, 334
    Interval Timer, 330
    JTAG UART, 329
    LCD, 335
    Parallel I/O, 331
    pin assignments, 324
    SDRAM, 332
    SDRAM PLL, 338
    SRAM, 334
    UART, 329
    UP 3 External Bus Multiplexer, 339
Nios II Software
    Flash, 312
    Handling Interrupts, 306
    LCD Display, 308
    Parallel I/O, 307
    Peripherals, 298
    SRAM, 311
    timer, 299
Nios II System Library, 297
Nios Software, 294

O

open collector, 227
open drain, 227

P

Parallel I/O Interface, 206
pin grid array package (PGA), 48
pixels, 168
plastic J-lead chip carrier (PLCC), 48
plastic quad flat pack (PQFP), 48
port map, 107
process, 90
process sensitivity list, 91, 94, 117
processor fetch, decode and execute cycle, 150
programmable array logic (PALs), 45
programmable interconnect array (PIA), 49
programmable logic, 44
programmable logic arrays (PLAs), 45

programmable logic devices, (PLDs), 45
PS/2 keyboard, 188
    communications protocol, 190
    connections, 188
    make and break codes, 189
    scan codes, 189
    VHDL example, 195
PS/2 mouse, 198
    commands and messages, 198
    data packet format, 199
    data packet processing, 201
    example design, 202
    initialization, 200

Q

Quartus II
    assigning a device, 8
    assigning pins, 10
    buses, 65
    compilation, 13
    connecting signal lines, 11
    entering pin names, 11
    errors and warnings, 13
    file extensions, 347
    floorplan editor, 28
    graphic editor, 7
    hierarchy, 63
    Quartus settings file (*.qsf), 12, 23, 26
    report file (*.rpt), 13
    schematic capture. *See* graphic editor
    simulation, 14
    simulation test vectors or stimulus, 14
    simulaton vector file (*.vec), 140
    symbol editor, 30
    symbol entry, 9
    timing analysis, 27, 66
    tutorial, 2, 62
    waveform editor file (*.scf), 14

R

radio-controlled (R/C) car, 242
reconfigurable computing, 57
reduced instruction set computer (RISC), 256
refresh. *See* VGA video display refresh
robot, 216
    assembly, 233
    battery, 237
    battery charger, 238
    communication, 223
    electronic compass, 229
    expansion header, 241
    GPS and DGPS receivers, 231
    gyros and accelerometers, 229
    infrared poximity detector, 221
    IR ranging, 225

line tracker sensor, 221
magnetic compass, 228, 232
modifying servos, 217
parts list, 237
sensors, 220
servo drive motors, 216
solid state cameras, 232
sonar, 225
thermal image sensors, 231
VHDL servo driver, 218
wheel encoder, 224
RS-232C Serial I/O Interface, 207
run length encoding (RLE), 182

S

seven segment decoder, 91, 114
shift operation in VHDL, 100
SOPC, 282
SOPC Builder, 325
SPI Bus Interface, 209
SR latch, 68
standard cells, 44

T

testbench, 107
to_stdlogicvector, 89
train. *See* electric train
tri-state, 49, 94, 116, 227, 366

U

UART, 207
unit under test (UUT), 107
UP 2, 3
    attaching power, 18
    downloading, 18
UP 3, 5, 36
    attaching power, 15
    Cyclone device, 38
    downloading, 17
    FPGA I/O pins, 37, 38, 40, 240
    jumper setup, 36
    LEDs, 6
    longer cable, 41
    other devices, 38
    Pin Assignments, 349
    power supplies, 41
    Programming Flash, 357
    pushbutton contact bounce, 68
    pushbuttons, 5
UP3 computer, 148
    fetch, decode, and execute, 150
    instructions, 149
    VHDL model, 157

UP3-bot. *See* robot
UP3core, 74
UP3core library, 74
    char_ROM, 82
    clk_div, 69, 79
    debounce, 68, 77
    dec_7seg, 76
    installation, 62
    keyboard, 83, 195
    mouse, 84, 200
    onepulse, 78
    tutorial, 63
    vga_sync, 80, 172

V

Verilog
    always statement, 114
    compilation, 26
    continuous assignment statement, 25
    data types, 112
    errors and warnings, 27
    hierarchy in models, 125
    inferred latches, 118
    operators, 113
    reg type, 112
    shift operations, 113
    structural model, 125
    synthesis of a counter, 118
    synthesis of a multiplexer, 115
    synthesis of a state machine, 119
    synthesis of a tri-state output, 116
    synthesis of an adder, 120
    synthesis of an ALU, 120
    synthesis of an incrementer, 118
    synthesis of an subtractor, 120
    synthesis of digital hardware, 112
    synthesis of flip-flops and registers, 117
    synthesis of gate networks, 114
    synthesis of memory, 122
    synthesis of multiply and divide hardware, 121
    synthesis of seven segment decoder, 114
    tutorial, 24
    wire statement, 114
    wire types, 112
VGA video display, 168
    bouncing ball example, 183
    character based, 176
    character font ROM, 178, 185
    color mixing using dithering, 183
    data compression, 182
    generation using an FPGA, 171
    graphics display, 181
    horizontal sync, 168
    pin assignments, 174
    refresh, 168
    RGB signals, 168

using a final output register, 174
vertical sync, 168
video examples, 175
VHDL
    Architecture body, 22
    compilation, 23
    conversion of data types, 90
    data types, 88
    editor syntax coloring, 21, 25
    Entity, 21, 24
    errors and warnings, 24
    hierarchy in models, 105
    inferred latches, 96
    libraries, 88, 96
    operators, 89
    shift operations, 89
    standard logic (STD_LOGIC), 88
    structural model, 105
    synthesis of a counter, 96
    synthesis of a multiplexer, 93
    synthesis of a state machine, 97
    synthesis of a tri-state output, 94
    synthesis of an adder, 99
    synthesis of an ALU, 99
    synthesis of an incrementer, 96

    synthesis of an subtractor, 99
    synthesis of digital hardware, 90
    synthesis of flip-flops and registers, 94
    synthesis of gate networks, 90
    synthesis of memory, 101
    synthesis of multiply and divide hardware, 100
    synthesis of seven segment decoder, 91
    train state machine, 138
    tutorial, 20
    using templates for entry, 21, 24
video display. See VGA video display

W

wait statement, 95
wired-AND, 227
with statement, 93

X

Xilinx 4000 Architecture, 53
    configurable logic block (CLB), 53
    Input output blocks (IOBs), 55
Xilinx Virtex, 58

About the Accompanying CD-ROM

Rapid Prototyping of Digital Systems, Quartus II Edition, includes a CD-ROM that contains Altera's QUARTUS II 5.0 SP1 Web Edition, Nios II IDE, SOPC Builder, and source code for all of the text's example VHDL, Verilog, Nios II SOPC reference designs, and Nios C/C++ example programs.

QUARTUS® II Software

The free Quartus® II 5.0 SP1 Web Edition software includes everything you need to design for Altera's low-cost FPGA and CPLD families. Features include:

- Schematic- and text-based design entry
- Integrated VHDL and Verilog HDL logic synthesis and simulation
- SOPC Builder system generation software for the Nios II Processor
- C/C++ Compiler and debugger for Nios II Processor systems
- Place-and-route, verification, and FPGA programming functions
- Timing Optimization Advisor
- Resource Optimization Advisor

Installing the QUARTUS® II Software

Insert the textbook's CD-ROM in your CD-ROM drive. Browse the file, **index.html**, on the CD-ROM using a web browser for complete step by step instructions. Click on the link to the book's website at the end of the index.html file to check for new software updates and any errata.

Source Code for Design Examples from the Book

Browse the file, **index.html,** on the CD-ROM using a web browser for complete step by step instructions. Design examples from the book are located in the **booksoft.zip** file, in subdirectories \chap$x$, where x is the chapter number. To use the design files, unzip them to the hard disk drive to your project directory or a subdirectory. In addition to *.bdf, *.vhd, *.v, and *.mif design files, be sure to copy any *.qpf, *.qsf, or *.qws files for each Quartus project. If you want to download a demo file, be sure to copy the *.sof device programming file. For Nios projects copy the entire project directory including subdirectories. The UP3core library files are in \chap5. In each \chap$x$ directory, the subdirectory \UP2 contains files already setup for **UP 2** board users. For users with the larger **1C12 UP 3**, use the subdirectory \1C12.

Users switching existing projects from the original UP 3 1C6 board to the larger **UP 3 1C12** board will need to change the device type to an **EP1C12Q240C8** and change a few pin assignments on each CD-ROM design and recompile when moving a project from a 1C6 board to a larger 1C12 board – See Sections 1.1, 2.3 and Appendix C.